S.-W. Breckle M. Veste W. Wucherer (Eds.): Sustainable Land Use in Deserts

Springer
*Berlin
Heidelberg
New York
Barcelona
Hong Kong
London
Milan
Paris
Singapore
Tokyo*

Siegmar-W. Breckle · Maik Veste · Walter Wucherer (Eds.)

Sustainable Land Use in Deserts

With 122 Figures, 16 in Colour

 Springer

Professor Dr. Siegmar-W. Breckle
Maik Veste
Dr. Walter Wucherer
Department of Ecology
University of Bielefeld
Universitätsstr. 25
33615 Bielefeld
Germany

Publication in part funded by UNESCO, UNCCD and BMBF

ISBN 3-540-67762-3 Springer-Verlag Berlin Heidelberg New York

Library of Congress Cataloging-in-Publication Data

Sustainable land use in deserts / Siegmar-W. Breckle, Maik Veste, Walter Wucherer (eds.).
  p. cm. Includes bibliographical references.
  ISBN 3540677623
1. Desertification – Congresses. 2. Desertifaction – Control – Congresses. 3. Desert reclamation – Congresses. 4. Desert ecology – Congresses. I. Breckle, Siegmar-W. II. Veste, Maik, 1963– III. Wucherer, Walter, 1957–
GB611.S87 2000
333.73′615–dc21   00-049707

This work is subject to copyright. All rights are reserved, whether the whole or part of the material is concerned, specifically the rights of translation, reprinting, reuse of illustrations, recitation, broadcasting, reproduction on microfilm or in any other way, and storage in data banks. Duplication of this publication or parts thereof is permitted only under the provisions of the German Copyright Law of September 9, 1965, in its current version, and permission for use must always be obtained from Springer-Verlag. Violations are liable for prosecution under the German Copyright Law.

Springer-Verlag Berlin Heidelberg New York
a member of BertelsmannSpringer Science+Business Media GmbH

© Springer-Verlag Berlin · Heidelberg 2001
The use of general descriptive names, registered names, trademarks, etc. in this publication does not imply, even in the absence of a specific statement, that such names are exempt from the relevant protective laws and regulations and therefore free for general use.

Cover design: design & production, Heidelberg
Typesetting: Camera ready by editors
SPIN 10752641   31/3130/xz   Printed on acid-free paper   5 4 3 2 1 0

Preface of the United Nations Educational, Scientific and Cultural Organization (UNESCO)

Deserts and desertification are the most prominent examples of the pending water crisis which threatens to become one of the major obstacles of sustainable development in the new century. The water crisis is exacerbated by the widening gap between the most advanced nations and the most destitute parts of the world, to which many of the desert regions belong. The pending crisis of global dimensions can be averted only through solidarity between nations, and the international scientific community is best suited to lead the way in how to make it efficient and creative, from exchange of information to real partnership in striving towards common goals.

UNESCO, as the science arm of the United Nations system, has the task to promote and catalyze scientific cooperation where and when needed. This is why arid zone research has been for a long time a permanent preoccupation of UNESCO's science programmes, MAB (Man and Biosphere) and IHP (International Hydrological Programme). Among others, UNESCO supported the cooperative efforts for the study and rehabilitation of the desertified regions of the Aral Sea basin, which constitute a considerable part of the present volume, along with relevant information from other desertified areas of the world.

The reports presented in this volume eloquently speak about the similarities and differences between various desert regions of the world, and describe the variety of inter-disciplinary measures by which the desertification processes might be mitigated. In the complex matter of land use in the vulnerable environment of the arid realm, the best way to promote scientific knowledge is to exchange experiences through cooperation between scientists and researchers from different countries, within and outside of the arid realm. The contributions presented in this volume illustrate the advantages of combining advanced know-how with local knowledge and expertise, resulting from genuine partnership between research institutions in different countries of the world. This partnership could not have been established without the generous aid of the German institutions for international cooperation.

M. El Tayeb
Director
Division for Policy Analysis & Operations
UNESCO Science Sector

Preface of the Secretariat of the United Nations Convention to Combat Desertification (UNCCD)

Over the past two decades, the problem of desertification and land-degradation in dryland regions continued to worsen. Desertification is increasingly a global problem and influence the ecological, economical and social situation of peoples in countries affected by ecological degradation and poverty. Desertification is primarily a problem of sustainable development. More than 250 million people in over 110 countries are directly affected by desertification, and about one billion are at risks. The challenge takes a geo-strategic dimension as the environmental scarcity factors linked to desertification and drought may led to forced migrations and conflicts.

Desertification was also major concern for the 1992 United Nations Conference on Environment and Development (UNCED), held in Rio de Janeiro. The international community intiated on this conference the Convention to Combat Desertification in Those Countries Experiencing Serious Drought and/or Desertification, Particulary in Africa. The Convention promotes a new and co-ordinated approach to manage dryland ecosystems and to prevent desertification in the future. In December 1992, the General Assembly agreed by adopting the UN Resolution 47/188. It entered into force on 26 December 1996, 90 days after the 50th ratification was received. The Convention is now international binding law for more than 160 countries.

In the framework of UNCCD science and technology as well as education are important tools in the fight against desertification. Success in combating desertification will require an improved understanding of its causes and impacts , and the development of new technologies but also the integration of traditional knowledge with new technologies. The Convention established a Committee on Science and Technology (CST) to advise the parties on scientific and technological issues relevant to desertification and drought. Clearly the international cooperation in scientific research, observation and information collection and exchange must be strenghtened. In this respect thematic programme networks are being established in various regions to foster decentralized co-operation and partnership

The international seminar "Ecological Problems of Sutainable Land-Use" was an important step forward to promote the international scientific cooperation and information exchange. This meeting also linked the German and international scientific communities as well as the Federal Ministry for Education and Research (BMBF) with the newly established UNCCD Secretariat in Bonn. We heartily welcome such developments. In the future continued and intensified scientific efforts are needed for success to combat desertification.

Hama Arba Diallo
Executive Secretary
UNCCD, Bonn

Preface of the Project-Management BEO of the German Federal Ministry for Education and Research

Desertification is a major environmental, social and economic problem to many countries in all parts of the world. Research on sustainable land use and resources management as well as the better understanding of natural processes in deserts can contribute to develop strategies to combat desertification and for improving peoples live.

Germanys Federal Ministry for Education and Research (BMBF) supports several international research projects on ecological problems and the development of scientific concepts for a sustainable use of natural resources in arid regions.

One of these projects, running from 1992 till 2000, is dealing with the ecological crisis caused by the desiccation of the Aral Sea in Central Asia. Within the project financed from a special German program for scientists of the Commonwealth of Independent States more than 130 researchers from Russia, Kazakhstan and Uzbekistan dealt in 40 sub-projects with major aspects of the wide spectra of environmental problems caused by unsustainable land use practices in the Aral Sea Basin. Research was focussed mainly to the water resources, the terrestrial ecosystems, peoples health, agriculture and vegetation in the delta areas of the two rivers feeding the Aral Sea – the Amudarya and the Syrdarya.

Developed environmental research institutions in the countries affected by the Aral Sea crisis are a prerequisite to overcome the situation in Central Asia. This was a main reason to establish the BMBF-support project which is coordinated by UNESCO.

Another example of BMBF-activities focussed on ecosystems in arid and semiarid regions is the German-Israeli cooperation in the „Nizzana sand dune" projects in the Negev Desert. Scientists from Israel and Germany working together to achieve information about the natural ecological processes on different scales and for desert dune stabilisation. The results are important in sense of stabilizing desert ecosystems and to prevent desertification also at other places in the world suffering from drought.

An example for multilateral cooperation within the Middle East is a project, where scientists from Egypt, Israel, Jordan and Germany working together on the ecological optimization of water use of crops under irrigation.

In all these projects, the international cooperation in science and technology is a key to achieve the goals of sustainable development.

It was considered by BMBF as especially important to make scientists of the former Soviet Union dealing with the Aral Sea crisis familiar with the latest knowledge and experiences of experts in other countries in the arid zones of the world.

Therefore the BMBF-seminar "Ecological Problems of Sustainable Land Use in Deserts" in cooperation with UNESCO, the University of Bielefeld and the new established UN Secretariat for the Convention to Combat Desertification (UNCCD) in Bonn was organized as a platform for exchange and linking scientific ideas and to promote the development of new concepts regarding the environmental situation in arid and semiarid regions.

Joachim Kutscher
BMBF-Project-Management BEO
National Research Centre Jülich
Jülich, Germany
June 2000

Contents

Introduction   **1**

Deserts, Land Use and Desertification   3
S.-W. Breckle, M. Veste, W. Wucherer

River Diversion, Irrigation, Salinization, Desertification
– an Inevitable Succession   14
D. Keyser

Part I: The Aral Sea Crisis   **25**

The Aral Sea Crisis Region   27
S.-W. Breckle, W. Wucherer, O. Agachanjanz, B. Geldyev

Flora of the Dry Seafloor of the Aral Sea   38
W. Wucherer, S.-W. Breckle, L. Dimeyeva

Vegetation Dynamics on the Dry Seafloor of the Aral Sea   52
W. Wucherer, S.-W. Breckle

Methods of Conservation and Restoration of
Vegetation Cover on the Aral Sea Coast   69
L. A. Dimeyeva

Vegetation Dynamics on the Syrdarya Delta and Modern Land Use   74
N. P. Ogar

Ecological Basis for Botanical Diversity Conservation
within the Amudarya and Syrdarya River Deltas   84
N.M. Novikova

The Tugai Forests of Floodplain of the Amudarya River:
Ecology, Dynamics and Their Conservation   95
S. Y. Treshkin

Soil Crusts in the Amudarya River Delta: Properties and Formation　103
A. Singer, A. Banin, L. Poberezsky, M. Gilenko

Irrigation and Land Degradation in the Aral Sea Basin　115
N. Orlovsky, M. Glantz, L. Orlovsky

Ecology of the Aqueous Medium in the Area Surrounding the Aral Sea　126
G. A. Lyubatinskaya

Potable Water: Research into Seasonal Changes and
Conditions in the Aral Sea Region　132
S. Sokolov

Part II: Salt Stress　**137**

Halophytes on the Dry Sea Floor of the Aral Sea　139
S.-W. Breckle, W. Wucherer, A. Scheffer

Halophytes: Structure and Adaptation　147
A. A. Butnik, U. N. Japakova, G. F. Begbaeva

Environmental State and an Analysis of Phytogenetic Resources
of Halophytic Plants for Rehabilitation and Livestock Feeding
in Arid and Sandy Deserts of Uzbekistan　154
C. Toderich, R. L. Goldshtein, W.B. Aparin, K. Idzikowska, G. S. Rashidova

Salinity: A Major Enemy of Sustainable Agriculture　166
Y. Waisel

Part III: Impact of Grazing　**175**

Remarkable Differences in Desertification Processes in the Northern and
Southern Richtersveld (Northern Namaqualand, Republic of South Africa)　177
N. Jürgens

Colour Plates　189

How Grazing Turns Rare Seedling Recruitment Events to
Non-Events in Arid Environments　197
S. J. Milton, T. Wiegand

Vegetation Degradation in Northeastern Jordan　208
O. Sharkas

Impact of Grazing on the Vegetation of South Sinai, Egypt 218
A. R. A. Moustafa

Arid Rangeland Management Supported by Dynamic Spatially
Explicit Simulation Models 229
F. Jeltsch, T. Stephan, T. Wiegand, G. E. Weber

Part IV: Desertification Processes and Monitoring 241

Remote Sensing of Surface Properties.
The Key to Land Degradation and Desertification Assessments 243
J. Hill

Evaluation of Potential Land Use Sites in Dry Areas of Burkina Faso
with the Help of Remote Sensing 255
M. Kappas

Degradation of the Vegetation in the Central Kyzylkum Desert (Uzbekistan) 265
L. Kapustina

Modern Geomorphological Processes on the Kazakhstanian Coast
of the Caspian Sea and Problems of Desertification 269
F. Akijanova

Anthropogenic Transformation of Desert Ecosystems in Mongolia 275
E. I. Rachkovskaya

Assessment of the Modern State of Sand–Desert Vegetation in Kazakhstan 281
G. K. Bizhanova

Part V: Reclamation 287

Water-Harvesting Efficiency in Arid and Semiarid Areas 289
A. Yair

The Effects of Landscape Structure on Primary Productivity
in Source–Sink Systems 303
K. Nadrowski, G. Jetschke

Sedimentary Environments in the Desiccated Aral Sea Floor:
Vegetation Recovery and Prospects for Reclamation 310
A. Yair

Seeding Experiments on the Dry Aral Sea Floor for Phytomelioration 318
G. T. Meirman, L. Dimeyeva, K. Dzhamantykov, W. Wucherer, S.-W. Breckle

Rehabilitation of Areas of Irrigation Now Derelicted Because
of Strong Salinization in Ecologically Critical Zones of Priaralia 323
A. Rau

Desert Soil Recultivation and Monitoring of (Phyto-) Toxicity:
Pilot Project in Three Phases Lasting for 4 Years 329
G.K. Hartmann, J. U. Kügler, P. Belouschek, L. Weissflog, K. H. Weiler,
H. Ch. Heydecke, G. Reisinger, G. S. Golitsyn, I. Granberg, N. P. Elansky,
E. B. Gabunshina, V. V. Alekseev, E. Putz, G. Pfister, A. Steiner

Contributions to a Sustainable Management of the Indigenous Vegetation in
the Foreland of Cele Oasis – A Project Report from the Taklamakan Desert 343
M. Runge, S. Arndt, H. Bruelheide, A. Foetzki, D. Gries, J. Huang,
M. Popp, F. Thomas, G. Wang, X. Zhang

The Control of Drift Sand on the Southern Fringe of the Taklamakan Desert
– an Example from the Cele-Oasis 350
X. Zhang, X. Li, H. Zhang

The Role of Biological Soil Crusts on Desert Sand Dunes in the
Northwestern Negev, Israel 357
M. Veste, T. Littmann, S.-W. Breckle, A. Yair

Restoration of Disturbed Areas in the Mediterranean
– a Case Study in a Limestone Quarry 368
C. Werner, A. S. Clemente, P. M. Correia, P. Lino, C. Máguas,
A.I. Correia, O. Correia

Indigenous Agroforestry for Sustainable Development
of the Area around Lake Nasser, Egypt 377
I. Springuel

Ziziphus - a Multipurpose Fruit Tree for Arid Regions 388
S. K. Arndt, S. C. Clifford, M. Popp

Root Morphology of Wheat Genotypes Grown in Residual Moisture 400
G. Manske, N. Tadesse, M. van Ginkel, M. Reynolds, P. L. G. Vlek

Field Studies in Solar Photocatalysis for Detoxification of
Organic Chemicals in Waters and Effluents 405
L. Muszkat, L. Feigelson, L. Bir

Part VI: National Programs   413

Activity of the Consulting Centre to Combat Desertification in Turkmenistan   415
C. Muradov

Desertification in China and Its Control   418
S. Baoping, F. Tianzong

Environmental Problems of the Southern Region of Kazakhstan   427
K. N. Karibayeva

National Strategy and Action Plan to Combat Desertification in Kazakstan   441
I. O. Baitulin

Part VII: Social and Economic Aspects   449

Economic-Demographic Strategies and Desertification:
Interactions in Low-Income Countries   451
B. Knerr

Final Remarks   465

List of Contributors

Agachanjanz, O.
University of Minsk, Department of Physical Geography
Gamarnikstr. 9-1-31, 220090 Minsk, Belarus

Akijanova, F.
Laboratory of Geomorphology, Institute of Geography, Kazakh National Academy of Sciences
99 Pushkin Street, Almaty 480100, Republic of Kazakhstan (FSU)
f_akijanova@mail.ru

Alekseev, V.V.
Moscow University, Dept. Geography
119899 Moscow, Russia

Aparin, W.B.
Experimental Institute of Biology, Adam Mickiewicz University
ul. Grunwaldska, 6, 60-780 Poznan, Poland
adam@main.amu.edu.pl

Arndt, S.K.
Institute of Plant Physiology, University of Vienna
A-1090 Vienna, Austria
sarndt@pflaphy.pph.univie.ac.at

Baitulin, I.O.
Institute of Botany and Phytointroduction
44, Timiryasev Str., 480090 Almaty, Kazakhstan

Banin, A.
Seagram Center for Soil and Water Sciences, Faculty of Agricultural, Food and Environmental Quality Sciences, Hebrew University of Jerusalem
P.O. Box 12, Rehovot 76100, Israel

Baoping, S.
College of Soil and Water Conservation, Beijing Forestry University
100083, Beijing, China
zhtning@beilin.bjfu.edu.cn

Begbaeva, G.F.
Botanical Institute and Botanical Garden, Uzbek Academy of Sciences
32, F.Khodzhaev Str., 700143, Tashkent, Uzbekistan
post@botany.org.uz

Belouschek, P.
Eng. bureau Kügler
Im Teelbruch 61, D-45219 Essen, Germany

Bizhanova, G.K.
Institute of Botany and Phytointroduction
44 Timiryazeva Str., 480090 Almaty, Kazakhstan
envirc@nursat.kz

Breckle, S.-W.
Department of Ecology, Faculty of Biology, University of Bielefeld
P.O. Box 10 01 31, 33501 Bielefeld, Germany
sbreckle@biologie.uni-bielefeld.de

Bruelheide, H.
Universität Göttingen, Albrecht-von-Haller-Institut für Pflanzenwissenschaften,
Abt. Ökologie und Ökosystemforschung
Untere Karspüle 2, 37073 Göttingen, Germany

Butnik, A.A.
Botanical Institute and Botanical Garden, Uzbek Academy of Sciences
32, F.Khodzhaev Str., 700143, Tashkent, Uzbekistan
post@botany.org.uz

Clemente, A.S.
Departamento de Biologia Vegetal, Faculdade Ciências
Campo Grande, C2, 4 Piso, 1700 Lisboa, Portugal

Clifford, S.C.
Horticulture Research International
Wellesbourne, Warwick, CV35 9Ef, UK

Correia, A.I.
Departamento de Biologia Vegetal, Faculdade Ciências
Campo Grande, C2, 4 Piso, 1700 Lisboa, Portugal

Correia, O.
Departamento de Biologia Vegetal, Faculdade Ciências, Campo Grande
C2, 4 Piso, 1700 Lisboa, Portugal

Correia, P.M.
Departamento de Biologia Vegetal, Faculdade Ciências
Campo Grande, C2, 4 Piso, 1700 Lisboa, Portugal

Dimeyeva, L.A.
Institute of Botany and Phytointroduction
44 Timiryazeva Str., 480090 Almaty, Kazakhstan

Dzhamantykov, K.
Priaral Research Institute of Agroecology and Agriculture
25 Abai St., 460007, Kyzylorda, Kazakstan

Elansky, N.P.
Institute of Atmospheric Physics
3 Pyzhevsky, 109017 Moscow, Russia

Foetzki, A.
Universität Göttingen, Albrecht-von-Haller-Institut für Pflanzenwissenschaften,
Abt. Ökologie und Ökosystemforschung
Untere Karspüle 2, 37073 Göttingen, Germany

Gabunshina, E.B.
Kalmykian Arid Centre
Ilishkina Str. 8, 358000 Elista, Kalmykia, Russia

Geldyev, B.
University of Almaty, Department of Physical Geography
71, Al-Farabi av., 480121 Almaty, Kazakhstan
geldyev@hotmail.com

Gilenko, M.
Waterproject
Victor Malaysov Str. 3, Tashkent 70000, Uzbekistan

Ginkel, M. van
International Maize and Wheat Improvement Center (CIMMYT)
Ado. Postal 6-641, 06600 Mexico, DF Mexico

Glantz, M.
Environmental and Societal Impacts Group, National Center for Atmospheric
Research
Box 3000, Boulder, Colorado USA 80307
glantz@ucar.edu

Goldshtein, R.I.
State Geological Enterprise Kyzyltepageologiya
7a, Navoi street, 700017, Tashkent, Uzbekistan
geoeco@globalnet.uz

Golitsyn, G.S.
Institute of Atmospheric Physics
3 Pyzhevsky, 109017 Moscow, Russia

Granberg, I.
Institute of Atmospheric Physics
3 Pyzhevsky, 109017 Moscow, Russia

Gries, D.
Universität Göttingen, Albrecht-von-Haller-Institut für Pflanzenwissenschaften,
Abt. Ökologie und Ökosystemforschung
Untere Karspüle 2, 37073 Göttingen, Germany

Hartmann, G.K.
Max-Planck-Institut für Aeronomie
Max-Planck Str. 2, 37191 Katlenburg-Lindau, Germany
ghartmann@linmpi.mpg.de

Heydecke, H.C.
Math.-Web. Eldingen
Knackendöffelstr. 36, 29351 Eldingen, Germany

Hill, J.
Remote Sensing Department, University of Trier
54286 Trier, Germany
hill@fews05.uni-trier.de

Huang, J.
Xinjiang Academy of Social Sciences, Economic Institute
16, South Beijing Road 830011 Urumqi, PR China

Idzikowska, K.
Experimental Institute of Biology, Adam Mickiewicz University
ul. Grunwaldska, 6, 60-780 Poznan, Poland
adam@main.amu.edu.pl

Japakova, U.N.
Botanical Institute and Botanical Garden, Uzbek Academy of Sciences
32, F.Khodzhaev Str., 700143, Tashkent, Uzbekistan
post@botany.org.uz

Jeltsch, F.
UFZ-Centre for Environmental Research Leipzig-Halle, Department of Ecological
Modelling
P.O. Box 2, D-04301 Leipzig, Germany
flori@oesa.ufz.de

Jetschke, G.
Theoretical Ecology Group, Institute for Ecology
07743 Jena, Germany
gottfried.jetschke@rz.uni-jena.de

Jürgens, N.
Institute of Botany, University of Köln
Gyrrhofstr. 15, Köln, Germany
njuerg@biolan.uni-koeln.de

Kappas, M.
Department of Physic. Geography, University of Mannheim
68131 Mannheim, Germany
kappas@rumms.uni-mannheim.de

Kapustina, L.
Botanical Institute and Botanical Garden
32 Khodzaev Str., Tashkent, Uzbekistan
kapl@map.silk.org

Karibayeva, K.N.
National Environmental Center, Institute of Ecology and Sustainable
Development of the Republic of Kazakhstan
85, Dostyk Av., 480100 Almaty
kkaribayeva@neapsd.kz

Keyser, D.
Zoologisches Institut und Museum
Martin-Luther-King-Platz 3, 20146 Hamburg, Germany
keyser@zoologie.uni-hamburg.de

Knerr, B.
University of Kassel, Faculty 11, Dept. of Development Economics and
Agricultural Policy
Steinstrasse 19, 37213 Witzenhausen, Germany
knerr@wiz.uni-kassel.de

Kügler, J.U.
Eng. bureau Kügler
Im Teelbruch 61, 45219 Essen, Germany

Lino, P.
Departamento de Biologia Vegetal, Faculdade Ciências
Campo Grande, C2, 4 Piso, 1700 Lisboa, Portugal

Littmann, T.
Martin–Luther University Halle-Wittenberg, Department of Geography
Domstrasse 5, 06108 Halle (Saale), Germany
Littmann@geographie.uni-halle.de

Lyubatinskaya, G.A.
Vodnik Scientific Production Company
Mira, 24, Almaty 483335 Talgar reg., Kainar, Baiserke, Kazakhstan

Máguas, C.
Departamento de Biologia Vegetal, Faculdade Ciências
Campo Grande, C2, 4 Piso, 1700 Lisboa, Portugal

Manske, G.
Center for Development Research (ZEF), University of Bonn
Walter-Flex-Str. 3, 53113 Bonn, Germany
gmanske@uni-bonn.de

Meirman, G.T.
Priaral Research Institute of Agroecology and Agriculture
25 Abai St., 460007, Kyzylorda, Kazakstan

Milton, S.J.
Nature Conservation Department, University of Stellenbosch
Private Bag X01 Matieland 7602, South Africa
lycium@mweb.co.za

Moustafa, A.-R.A.
Botany Department, Faculty of Science, Suez Canal University
Ismailia, Egypt
raoufmoustafa@hotmail.com

Muradov, C.
National Institute of Deserts, Flora and Fauna, Ministry of Environmental
Protection of Turkmenistan
15, Bitarap Turkmenistan St., Ashkhabad, Turkmenistan 744000
babaev@desert.ashgabad.su

Nadrowski, K.
Theoretical Ecology Group, Institute for Ecology
07743 Jena, Germany
karin.nadrowski@rz.uni-jena.de

Novikova, N.M.
Water Problems Institute, Russian Academy of Sciences
3, Gubkina Str., 117735, Moscow, Russia
novikova@novikova.msk.ru

Ogar, N.P.
Institute of Botany and Phytointroduction
44 Timiryazeva Str., Almaty 480070, Kazakhstan
envirc@nursat.kz

Orlovsky, N. and L.
J. Blaustein Institute for Desert Research, Ben Gurion University of the Negev,
Sede Boker Campus
84990, Israel
orlovsky@bgumail.bgu.ac.il

Pfister, G.
IGAM University of Graz
Halbärthgasse 1, A- 8010 Graz, Austria

Poberejzsky, L.
Waterproject
Victor Malaysov Str. 3, Tashkent 70000, Uzbekistan

Popp, M.
Institute of Plant Physiology, University of Vienna
A-1090 Vienna, Austria

Putz, E.
IGAM University of Graz
Halbärthgasse 1, A- 8010 Graz, Austria

Rachkovskaya, E.I.
Kunaeva Str. 119 - 1, Almaty 480100, Kazakhstan
envirc@nursat.kz

Rashidova, G.S.
Agency of Transfer Technology, State Committee of Science and Technology
Suleimanova, 29 Street, 700017 Tashkent
Uzbekistan att@ transfer.silk.org

Reisinger, G.
Eng. bureau Reisinger
Lehenstr. 3, D- 89257 Illertissen, Germany

Reynolds, M.
International Maize and Wheat Improvement Center (CIMMYT)
Ado. Postal 6-641, 06600 Mexico, DF Mexico

Runge, M.
Universität Göttingen, Albrecht-von-Haller-Institut für Pflanzenwissenschaften, Abt. Ökologie und Ökosystemforschung
Untere Karspüle 2, 37073 Göttingen, Germany
mrunge@gwdg.de

Scheffer, A.
Department of Ecology, Faculty of Biology, University of Bielefeld
P.O. Box 10 01 31, 33501 Bielefeld, Germany
anja.scheffer@biologie.uni-bielefeld.de

Sharkas, O.
Dept. of Geography, Birzeit Unversity
Birzeit P.O. Box 14, West Bank, Palestine
osharkas@birzeit.edu

Singer, A.
The Seagram Center for Soil and Water Sciences, Faculty of Agricultural, Food and Environmental Quality Sciences, The Hebrew University of Jerusalem
P.O. Box 12, Rehovot 76100, Israel
singer@agri.huji.ac.il

Sokolov, S.
Institute for Geography, Academy of Sciences
99 Pushkin Str., 480100 Almaty, Kazakhstan

Springuel, I.
UESD in Aswan, South Valley University, Faculty of Science
81528 Aswan, Egypt
irina44@yahoo.com

Steiner, A.
IGAM University of Graz
Halbärthgasse 1, A- 8010 Graz, Austria

Stephan, T.
UFZ-Centre for Environmental Research Leipzig-Halle, Department of Ecological Modelling
P.O. Box 2, D-04301 Leipzig, Germany

Tadesse, N.
Plant Science Dept., South Dakota State University
Brookings, South Dakota 57007 USA

Thomas, F.
Universität Göttingen, Albrecht-von-Haller-Institut für Pflanzenwissenschaften,
Abt. Ökologie und Ökosystemforschung
Untere Karspüle 2, 37073 Göttingen, Germany

Tianzong, F.
College of Soil Water Conservation, Beijing Forestry University
100083, Beijing, China

Toderich, C.N.
Department of Desert Research, Complex Research Institute of Regional
Problems, Samarkand Division of Uzbek's Academy of Science
Timur Malik, 3 street, 703000, Samarkand, Uzbekistan
muhidin@samuni.silk.org

Treshkin, S.Y.
Institute of Bioecology
Berdakha pr., 41, Nukus, 742000 Uzbekistan
Sergei@bioeco.nukus.silk.org

Veste, M.
Department of Ecology, Faculty of Biology, University of Bielefeld
P.O. Box 10 01 31, 33501 Bielefeld, Germany
maik.veste@biologie.uni-bielefeld.de

Vlek, P.L.G.
Center for Development Research (ZEF), University of Bonn
Walter-Flex-Str. 3, 53113 Bonn, Germany
p.vlek@uni-bonn.de

Waisel, Y.
Department of Plant Sciences, Tel Aviv University
Tel Aviv 69978, Israel
Waisel@post.tau.ac.il

Wang, G.
Lanzhou University, State Key Laboratory of Arid Agroecology
298 Tian Shui Lu, 730000 Lanzhou, PR China

Weber, G.E.
UFZ-Centre for Environmental Research Leipzig-Halle, Department of Ecological Modelling
P.O. Box 2, D-04301 Leipzig, Germany

Weiler, K.H.
Fachhochschule Emden
Constantia Pl. 4, D-26723 Emden, Germany

Weissflog, L.
UFZ- Umweltforschungszentrum Leipzig–Halle GmbH
Permoserstr. 15, 04318 Leipzig, Germany

Werner, C.
Universität Bielefeld, Lehrstuhl für exp. Ökologie und Ökosystembiologie
Universitätsstr. 25, 33615 Bielefeld, Germany
christiane.werner@biologie.uni-bielefeld.de

Wiegand, T.
UFZ-Centre for Environmental Research Leipzig-Halle, Department of Ecological Modelling
P.O. Box 2, 04301 Leipzig, Germany
towi@oesa.ufz.de

Wucherer, W.
Department of Ecology, Faculty of Biology, University of Bielefeld
P.O. Box 10 01 31, 33501 Bielefeld, Germany
walter.wucherer@biologie.uni-bielefeld.de

Yair, Y.
The Hebrew University of Jerusalem, Department of Geography
Mount Scopus, Jerusalem 91905, Israel
Aaron@vms.huji.ac.il

Zhang, H.
Xinjiang Institute of Ecology and Geography, Academia Sinica
No 40-3 Beijing South Road, Urumqi, Xinjiang 830011, P.R.China

Zhang, Ximing
Xinjiang Institute of Ecology and Geography, Academia Sinica, No 40-3 Beijing South Road, Urumqi, Xinjiang 830011, P.R.China
Lixm@public.wl.xj.cn

Introduction

Deserts, Land Use and Desertification

Siegmar–W. Breckle, Maik Veste and Walter Wucherer

Keywords. Aral Sea crisis, aridity, desert ecology, human impact, salinity, waste water

Abstract. Deserts are arid areas on the globe where plant growth is scarce. The lack of water during longer periods is due to climatic conditions. This is known from all parts of the climatic temperature zones of the globe. Deserts and their adjacent semidesertic regions, such as the Sahara, the Negev and Sinai, the Namib, the Atacama and Altiplano, Central Australia, the Mohave in Southwestern USA, the Kyzylkum and Aralkum, and the Kawir in Iran and the Afghan deserts often exhibit severe changes in their environmental design according to human impact. Fluvial and aeolian soil erosion, enhanced salinity by waterlogging, pollution by pesticides and other toxics, thus, the loss of productive areas, are some of the severe effects of inadequate use by man. Desertification often takes place by this human influence in ecotopic areas where a shift to desertic conditions by a slight change in environmental factors may cause a severe additional degradation.

Introduction

In deserts distinct climatic factors prevail. In this short introduction a review of the main factors is given and examples will be mentioned as the basic knowledge for evaluation of desertification processes. Relevant or specific literature is cited.

Deserts are arid areas on the globe where plant growth is scarce (Breckle et al. 1994, Breckle 2000a). The lack of water during longer periods of the year is mainly due to climatic conditions. This is known from many parts of the climatic zones of the globe to a smaller or greater extent (Wallace and Romney 1972; Walter 1974; Walter and Breckle 1983, 1984, 1985, 1986a, b, 1991, 1999; Agachanjanz 1986; Archibold 1995). Various examples of warm or hot deserts, excluding the arctic and antarctic polar cold deserts, have similar features; some characters differ, especially the floristic realm, depending on the vegetation history.

The Abiotic Factors in Deserts

Climatic conditions of aridity lead to high radiation and high evaporation. Lack of water by climate can be the main reason under all temperature regimes (see Table 1). Arid regions are defined as areas where potential evapotranspiration (ET) is higher than precipitation (P) (Fig. 1).

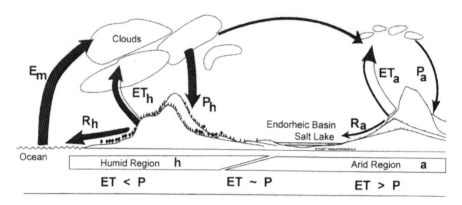

Fig. 1. Global and regional water cycling in arid and humid areas. E Evaporation; ET Evapotranspiration; P precipitation; R run-off; a arid; h humid; m marine

Aridity is not only controlled by temperature and rainfall, but also by the soil conditions, the transformation of rainfall into runoff and the nature of the rainfall (Yair and Berkowicz 1989). For a distinct area the hydrological water balance can be expressed by the formula given in Fig. 2.

In each landscape the various terms of the water balance formula can differ and are characteristic for the region (Walter and Breckle 1991). This is very basic, but often forgotten.

According to rainfall distribution or temperature regime, various desert types can be distinguished.

Equation of water balance:

$N = \Delta W + V_E + V_T + V_A + V_G$

Water input: N = precipitation (rain, snow, dewfall)
Water output: V = water losses
    E = evaporation   T = transpiration
    A = surface runoff   G = below ground losses, seepage
    ΔW = water storage within the system, buffer term (+,-)

Fig. 2. Water budget of plant cover, or of a distinct part of land surface

On a smaller scale, the geomorphology with the main processes of weathering, erosion and accumulation of substrate material leads to a typical landscape pattern, where also desert types can be distinguished. According to particle size, those desert types have been named according to Arabic or Persian names (Table 1).

Table 1. Abiotic parameters in deserts and various desert types

Climate	Desert types
High radiation	
Lack of water (aridity, high evaporation)	
Distinct rainfall distribution, fog →	Summer rain desert
	Winter rain desert
	Fog desert
	Extreme desert (without rain)
Temperature regime →	Hot desert
	Temperate desert
	Cold desert
	Polar desert (arctic, antarctic)
Wind	
Geomorphology	
Predominance of physical weathering	
Erosion and accumulation, causing particle size sequences: catena →	Hamada rocks, blocs
	reg — Stones and sand
	serir — Pebbles
	erg — Sand
	takyr — Clay
	sebkha salt
Sand movement (deflation, accumulation: dunes)	
Episodic rivers, wadis (oeds, rivieres)	
Soils	
Slow soil formation, low organic matter	
Main capillary water movement upwards: formation of crusts	
Inorganic crusts: rock varnish, limestone, gypsum, salt crusts	

One of the striking features in deserts is the salt factor (Waisel 1972; Chapman 1974; Breckle 1982, 1989, 1990, 1992, 1995, 2000b; Shainberg and Shalhevet 1984; Ungar 1991).Salinization in endorrheic basins is a natural factor. Salt in small quantities is transported by rainwater and thus accumulates in arid basins. Evolution of halophytes has taken place in several arid regions, the Chenopodiaceae have one of their evolutionary centers in the Caspian and Aral Sea area. The other striking geomorphological factor is the accumulation of sand, the formation of sand dune areas, according to the specific wind regime in the respective area (Freitag 1986; Mandaville 1986).

The Biotic Factors in Deserts

Flora and fauna exhibit strong adaptations to desert conditions, mainly the avoidance of stresses, so that many examples of tolerance to drought or heat are known. The biotic processes which play an important role in desert organisms are shown in Table 2. The adaptation to salt, heat, drought etc. has led to many specific desert organisms and an often typical flora and fauna for each desert.

Table 2. Biotic processes as desert characteristics

Flora and vegetation, fauna	Special features
• Adaptation to drought (CAM, C$_4$, succulence; life forms) and to high radiation	
• High root/shoot ratio	
• Contracted vegetation	
• Night activity (mainly in animals)	
• Adaptation to salinity (halophyte types)	
• Biodiverity in most taxa low, but very specific according to desert type	
• Convergent evolutionary processes (in every floristic kingdom)	High degree of endemism
• Important role of cryptogams	Formation of biotic crusts
Human interference	
• Oasis economy	
• Nomadism	
• Hunting	
• Disappearance of wild animals	
• Overgrazing	
• Change of albedo	Climate change
• Disappearance of woody species	Degradation
• Desertification	• Soil deterioration, erosion, salinization, aggressive neophytes, deposition of pesticides, etc.

Examples of Deserts

A short list with the relevant bibliography will be given here. The Department of Ecology has done preliminary or larger casestudies mainly to identify typical halophytic flora and their ecology, ecophysiology of typical desert species, transects of climatic gradients etc. in the following deserts:

Deserts, Land Use and Desertification

Sahara (Egypt, Wadi Allaqi; South–Tunisia) [Batanouny 1983; Batanouny and Ghabbour 1996; Barakat and Hegazy 1997; Breckle and Veste 1995; Evenari 1985; Knapp 1973; Spooner and Mann 1982; Walter and Breckle 1984, 1986a]

Area wise the biggest desert on earth, with huge areas almost rainless today, with contracted vegetation and traces of the effects of water everywhere, sand seas and huge hamada bloc types and with black rock varnish. Beduins are an example of humans adapted to a very specific desert life style, The Nile is an example of a tropical river crossing the desert and feeding Egypt and the huge city of Cairo. SouthTunisia is an example with sophisticated irrigation systems (Chenini) with three-storey cultivation (Nefta), but also with strong sand storms and large sebkhas (Shott el Djerid) and an interesting long gradient to arid mediterranean landscape pattern.

Negev and Sinai [Zohary 1973, 1982, 1983; Evenari et al. 1982; Walter and Breckle 1984, 1986a; Danin 1986; Yair and Berkowicz 1989; Blume and Berkowicz 1995; Breckle and Veste 1995; Veste and Breckle 1995, 1996, 2000]

The Negev, the eastern part of the Sinai, with a small-scale diverse mosaic of landscape, exhibiting all types of desert geomorphology, deep rooting shrubs and shaded therophytes, salt and limestone crusts, but also biotic crusts stabilizing sanddunes, antique runoff farming techniques and their restoration in new experiments (Avdat, Sede Boqer)

Namib and *Karoo* (Namibia; South Africa) [Knapp 1973; Walter and Breckle 1984, 1986a; Jürgens 1986, 1991; von Willert et al. 1992; Herppich et al. 1996; Jürgens et al. 1997; Cowling et al. 1997; Dean & Milton 1999]

The Namib, a very diverse, very old desert with a great part of sandy desert with huge dunes, salt pans, inselbergs (Spitzkoppe) and their typical rock varnish, lichen beds, lacking higher plants, succulents, the vast stretching "nara", the *Acanthosycios horrida* (Cucurbitaceae) and the relict gymnosperm, the unique *Welwitschia mirabilis*, fog basking beetles (tenebrionids), with a small population of desert elephants and giraffes in the wadis (riviere) and old rock paintings in the Brandberg and Twyfelfontain, demonstrating long-lasting human influence. The Succulent Karoo, a winter-rainfall desert is characterized by a high biodiversity and a large number of succulent species, in particular in the families of the Mesembryanthemaceae and the Crassulaceae.

Atacama (Chile); *Altiplano* (Bolivia, Chile) [Chong–Diaz 1984; Walter and Breckle 1984, 1986a, 1991]

The Atacama, one of the most extreme deserts, small, but very mountainous, up to the Andes, with soda lakes at 4200 m (Lago Minuchi), with many active volcanoes, causing destructive effects, with lava bombs and lava streams, geysers, with many trace elements in the soils and rocks (borax, Li, As), with cacti cushions and *Azorella*, with lichens on columnar cacti (fog desert) and nitrate-rich brine in salt crusts (Salar de Atacama, Pampa de Tamarugal).

Australia (Central Australia) [Stocker 1982; Beadle 1981; Walter and Breckle 1984, 1986a]

The sandy semidesert, the blue bush country (*Maireana sedifolia*), Gilgai plains, Olga mountains, Ayers Rock and Mt. Connor (inselbergs) indicate a moderate desert and a mosaic of dry savanna types with summer rains in the north (billabong formations) and winter rains in the south and southwest. The red soils, the Red Heart of the country, again indicate an old origin. Many very specific plant species (*Acacia, Eucalyptus*) of the Australian floristic kingdom and the typical fauna indicate an old separate development of the area.

Mohave (USA) [Moore et al. 1972; Wallace and Romney 1972; Breckle 1976; Wood 1980; Tiedemann et al. 1983; West 1983; Osmond et al. 1990; Walter and Breckle 1991, 1999]

A temperate desert with cold winters, with a mosaic of mountain deserts and semidesert, shrub steppes, with a "deep hole", the Death Valley, and very high summer temperatures, with specific succulents (cacti, *Yucca*) or C_4–plants (*Tidestroemia, Atriplex hymen-elytra*), with the Great Salt Lake and a great variety of halophytic communities

Kyzylkum (Uzbekistan); *Aralkum* (Kazakhstan) [Walter 1974; West 1983; Walter and Breckle 1986b, 1989; Breckle et al. 1998; Bruk et al. 1998]

Very continental with very hot summers and icy winters, with salty dust storms in spring (biskunaq), with all desert types, one of the evolutionary centres of Chenopodiaceae, where *Haloxylon aphyllum* can become a small tree, and where *Tamarix* also has its highest diversity. The Bactrian camel feeds on many geophytes and *Salsola* species. The dry sea floor of the retreating Aral Sea has become the biggest succession experiment that mankind is currently undertaking. The present area of the Aral Sea is less than half the original one, the Great and Small Aral Seas are now separate. Some fishing villages from 1960 are now located almost 100 km east of the current coast line, thus anchors and boats are rotting in the desert far from water, where huge terraces of *Zostera* indicate many old coast lines. The old sea floor is covered by *Cardium edule* shells in between many halophytes, where now the dam between the Small and the Great Aral Seas has broken a second time (April 20, 1999), and where the use of water of the Amudarya and the Syrdarya should be modernized. The Aral Sea crisis will be one of the focus points in this symposium.

Kawir (Iran); *Afghanistan* (north and south) [Breckle et al. 1969, 1975; Walter 1974; Breckle 1981, 1982, 1983, 1986, 1989; Spooner and Mann 1982; West 1983; Freitag 1986; Walter and Breckle 1986b, 1989]

Huge catenas from the mountains to the flat Kawir areas, with immense amounts of accumulated salts, with wild camels (dromedars), with rivers and salt flats, clay pans, devoid any vegetation, but with some populations of the wild ass (*Hemionus*) and again interesting endemics in the Iranian and Afghan deserts (*Halarchon vesiculosum*).

Many results are included in synthetic surveys and textbooks (Walter and Breckle 1983, 1984, 1985, 1986a, b, 1989, 1991, 1995, 1999); other examples of specific or interesting papers are already mentioned above.

Conclusions

The basic knowledge of ecological factors and ecosystem processes is a necessary precondition, if means and techniques for preventing environmental damages are to be developed and applied in a proper, sustainable way. Humans and scientists know many facts (West 1983; Evenari 1985; Burrows 1990; Archibold 1995; Walter and Breckle 1999) and have a good knowledge of important functions and processes in ecosystems by the great advancement of science. We can now see many things from the distance by fascinating satellite remote sensing techniques. We need, however, more efforts to train the young generation better (UNESCO), especially to train them in basics in ecology and biology and practical applications. This might lead to an increased consciousness that each one of the natural resources is limited. We have a good knowledge of the long-lasting (historical) (Waisel 1986) and the ever-increasing human impacts in the past few decades, most severe in deserts (Batanouny 1983), we must use this knowledge to find better means for a sustainable use and maximizing the use of renewable resources and minimizing human impacts and desertification processes, especially in cases of irreversable damage. The Aral Sea crisis is a very recent example, through the overgrazing of many huge areas and the salinization by inadequate irrigation, monitoring of desertification processes. Means of reclamation and combating desertification, even a renewable desert agriculture by national and international efforts and programs should lead to better social and economical conditions in the affected countries. All these aspects are tackled in this seminar.

Mankind is spending a lot of money in the wrong way, and we do not use it properly and early enough to combat environmental impacts (UNCCD) or desertification problems (McKell et al. 1972; Spooner and Mann 1982; Mooney and Godron 1983; Shainberg and Shalhevet 1984; Breckle 1983, 1999; Breckle and Veste 1995; Bruk et al. 1998). However, we also have to keep in mind that desertification processes are not only taking place in desert areas or on the fringe of deserts, but also in many other semiarid or semihumid regions.

Humans must discuss problems and find new solutions. The use of renewable energy is of focal interest, yet in the desert areas, where solar energy is plentiful. And even more, a very rational use of water with an optimal waste water use and recycling is necessary. Better cooperation and new joint projects, irrespective of how small the scale of this work and cooperation for mutual understanding might be. This fascinating planet, between Mars and Venus, surrounded by deserts, is so unique that we must take care of it. This international seminar is one step.

Acknowledgements. The research in various deserts and on the ecology of halophytes was partly funded by the DFG, the Schimper–Foundation and the BMBF. For technical assistance in many ways we have to thank Uta Breckle, Irmingard Meier and Anja Scheffer. Additionally, we have to thank many colleagues in their respective countries for their hospitality and for their help in many ways.

References

Agachanjanz OE (1986) Botanical geography of the USSR. Bischeischaja Schkola, Minsk (in Russian)

Archibold OW (1995) Ecology of world vegetation. Chapmann & Hall, London

Barakat HN, Hegazy AK (eds) (1997) Reviews in ecology, desert conservation and development. Metropole, Cairo

Batanouny KH (1983) Human impact on desert vegetation. In: Holzner W, Werger MJA, Ikusina I (eds) Man's impact on vegetation. Junk, The Hague, pp 139–149

Batanouny KH, Ghabbour SI (eds) (1996) Arid lands biodiversity in North Africa. IUCN, Cairo

Beadle NCW (1981) The vegetation of Australia. In: Walter H, Breckle S–W (eds): Vegetationsmonographien der einzelnen Großräume, Band IV. Fischer, Stuttgart

Blume H–P, Berkowicz SM (1995) Arid ecosystems. Advances in Geoecology 28:1–229

Breckle S–W (1976) Zur Ökologie und den Mineralstoffverhältnissen absalzender und nicht absalzender Xerohalophyten (unter besonderer Berücksichtigung von Untersuchungen an *Atriplex confertifolia* und *Ceratoides lanata* in Utah/USA). Diss.-Bot. 35:169

Breckle S–W (1981) The time-scale of salt–accumulation in a desertic hydrotope in North –Central Iran. Tübinger Atlas des Vorderen Orients, Reihe A8:51–60, Reichert, Wiesbaden

Breckle S–W (1982) The significance of salinity. In: Spooner B, Mann HS (eds): Desertification and development: dryland ecology in social perspective. Academic Press, London: pp 277–292

Breckle S–W (1983) Temperate deserts and semideserts of Afghanistan and Iran. In: West NE (ed) Temperate deserts and semideserts. Ecosystems of the world, vol.5:271–319. Elsevier/Amsterdam

Breckle S–W (1986) Studies on halophytes from Iran and Afghanistan. II. Ecology of halophytes along saltgradients. Proc R Bot Soc Edinb 89B:203–215

Breckle S–W (1989) Role of salinity and alkalinity in the pollution of developed and developing countries. Int Symp on the effect of pollutants to plants in developed and developing countries in Izmir, Turkey 22.–28.8.1988. In: Öztürk, MA (ed), pp 389–409

Breckle S–W (1990) Salinity tolerance of different halophyte types. In: Bassam NEl, Dambroth M, Loughman BC (eds) Genetic aspects of plant mineral nutrition. Plant Soil 148:167–175

Breckle S–W (1992) Salinitystress and saltrecretion in plants. Bielefelder Ökologische Beiträge (BÖB), Band 6, pp 39–52, Universität Bielefeld

Breckle S–W (1995) How do plants cope with salinity? In: Khan MA, Ungar IA (eds): Biology of salt-tolerant plants (Proc Int Symp) Dept of Botany, Univ of Karachi, Pakistan, pp 199–221

Breckle S–W (1999) Halophytic and gypsophytic vegetation of the EbroBasin at Los Monegros. Bol Soc Entomol, Aragonesa 24:101–104

Breckle S–W (2000a) Biodiversität von Wüsten und Halbwüsten. Hannover, Ber Reinh –Tüxen–Ges, 12:135–152 (in press)

Breckle S–W (2000b) Salinity, halophytes and salt affected natural ecosystems. In: Läuchli A, Lüttge U (eds): Salinity: environment – plants – molecules. Kluwer, Dordrecht (in press)

Breckle S–W, Veste M (eds) (1995) BMBF–Workshop: Sustainable land use in the arid Near East. ZiF/Bielefeld 28.–30. Nov 1994, Bielefelder Ökologische Beiträge (BÖB), Band 9, 77, Universität Bielefeld

Breckle S–W, Frey W, Hedge IC (1969, 1975) Botanical literature of Afghanistan. Notes R Bot Gard Edinb 29:357–371; 33:503–521

Breckle S–W, Dickman CR, Houérou HNL, Whitford, WG (1994) The diversity of deserts, Asian deserts. In: Seely M (ed): Deserts – the illustrated library of the earth. Weldon Owen, Sydney, pp 40–47, 50–54

Breckle S–W, Agachanjanz OE, Wucherer W (1998) Der Aralsee: Geoökologische Probleme. Naturwiss Rundsch 51:347–355

Bruk S, Keyser D, Kutscher J, Moustafaev V (eds) (1998) Ecological research and monitoring of the Aral Sea deltas, a basis for restoration. UNESCO Aral Sea Project, Paris

Burrows CJ (1990) Processes of vegetation change. Hyman, London

Chapman VJ (1974) Salt marshes and salt deserts of the world. Cramer, Lehre

Chong–Diaz G (1984) Die Salare in Nordchile – Geologie, Struktur und Geochemie. Geotekt Forsch (Stuttgart) 67:1–146

Cowling RM, Ricchardson DM, Pierce SM (eds) (1997) The vegetation of Southern Africa. Cambridge University Press, Cambridge, pp 131–166

Danin A (1986) Flora and vegetation of the Sinai. Proc R Soc Edinb 89B:159–168

Dean RW, Milton S (eds) The Karoo, ecological patterns and processes. Cambridge University Press, Cambridge

Evenari M, Shanan L, Tadmor N (1982) The Negev, the challenge of a desert. Harvard University Press, Cambridge

Evenari M (1985) Hot deserts and arid shrubland. In: Goodall DW (ed) Ecosystems of the World, vol 12, Elsevier, Amsterdam

Freitag H (1986) Notes on the distribution, climate and flora of the sand deserts of Iran and Afghanistan. Proc R Soc Edinb 89B:135–146

Herppich WB, Midgley G, von Willert DJ, Veste M (1996) CAM variations in the leafsucculent *Delosperma tradescantinoides* (Mesembryanthemaceae), native to southern Africa. Physiol Plant 98:485–492

Jürgens N (1986) Untersuchungen zur Ökologie sukkulenter Pflanzen des südlichen Afrika. Mitteilungen des Instituts für Allgemeine Botanik Hamburg 21:139–365

Jürgens N (1991) A new approach to the Namib region, I: Phytogeographic subdivision. Vegetatio 97:21–38

Jürgens N, Burke A, Seely MK, Jacobsen KM (1997) The Namib desert. In: Cowling RM, Richardson D (eds) Vegetation of Southern Africa. Cambridge University Press, Cambridge, pp 189–214

Knapp R (1973) Die Vegetation von Afrika. In: Walter H (ed) Vegetations–Monographien der einzelnen Großräume, Bd III. Fischer, Stuttgart

Mandaville JP (1986) Plant life in the Rubal–Khali (the Empty Quarter), south-central Arabia. Proc R Soc Edinb 89B:147–157

McKell CM, Blaisdell JP, Goodin JR (1972) Wildland shrubs – their biology and utilization. UNDP, Ogden, Utah

Mooney HA, Godron M (1983) Disturbance and ecosystems. Ecological Studies 44. Springer, Berlin Heidelberg New York

Moore RT, Breckle S–W, Caldwell MM (1972) Mineral ion composition and osmotic relations of *Atriplex confertifolia* and *Eurotia lanata*. Oecologia 11:67–78

Osmond CB, Pitelka LF, Hidy GM (1990) Plant biology of the basin and range. Ecological Studies 80. Springer, Berlin Heidelberg New York

Shainberg I, Shalhevet J (1984) Soil salinity under irrigation. Ecological Studies 51. Springer, Berlin Heidelberg New York

Spooner B, Mann HS (eds) (1982) Desertification and development: Dryland ecology in social perspective. Academic Press, London

Stocker O (1928) Der Wasserhaushalt ägyptischer Wüsten- und Salzpflanzen. Bot Abh 13. Fischer, Stuttgart

Tiedemann AR, Mc Arthur ED, Stutz HC et al. (eds) (1983) Proceed Symp on the biology of *Atriplex* and related chenopods. Provo, Utah

Ungar IA (1991) Ecophysiology of vascular halophytes. CRC Press, Boca Raton

Veste M, Breckle S–W (1995) Xerohalophytes in a sandy desert ecosystem. In: Khan MA, Ungar IA (eds): Biology of salt-tolerant plants. Proc Int Symp, Dept of Botany, Univ of Karachi, Pakistan, pp 161–165

Veste M, Breckle S–W (1996) Gaswechsel und Wasserpotential von *Thymelaea hirsuta* in verschiedenen Habitaten der Negev–Wüste. Verh Ges Ökol 25:97–103

Veste M, Breckle S–W (2000) Negev – pflanzenökologische und ökosystemare Analysen. Geogr Rdschau 9:24-29

Waisel Y (1972) Biology of halophytes. Academic Press, New York

Waisel Y (1986) Interactions among plants, man and climate: historical evidence from Israel. Proc R Soc Edinb 89B:255–264

Wallace A, Romney EM (1972) Radioecology and ecophysiology of desert plants at the Nevada test site. University of California, Riverside

Walter H (1974) Die Vegetation Osteuropas, Nord- und Zentralasiens. In: Walter H (ed) Vegetationsmonographien der einzelnen Großräume, Bd VII. Fischer, Stuttgart

Walter H, Breckle S–W (1983, 1991) Ökologie der Erde, Bd 1. Ökologische Grundlagen in globaler Sicht. (2. Aufl 1991) Fischer, Stuttgart

Walter H, Breckle S–W (1984, 1991) Ökologie der Erde, Bd 2. Spezielle Ökologie der Tropischen und Subtropischen Zonen. (2. Aufl 1991) Fischer, Stuttgart

Walter H, Breckle S–W (1985) Ecological systems of the geobiosphere, vol 1. Ecological principles in global perspective. Springer, Berlin Heidelberg New York

Walter H, Breckle S–W (1986) Ecological systems of the geobiosphere, vol 2. Ecology of the tropical and subtropical zonobiomes. Springer, Berlin Heidelberg New York

Walter H, Breckle S–W (1986, 1994) Ökologie der Erde, Bd 3. Spezielle Ökologie der Gemäßigten und Arktischen Zonen Eurasiens. Fischer, Stuttgart; (2.Aufl: Breckle S–W, Agachanjanz OE 1994, Fischer, Stuttgart)

Walter H, Breckle S–W (1989) Ecological systems of the geobiosphere, vol 3. Ecology of the moderate and arctic zonobiomes in Northern Eurasia. Springer, Berlin Heidelberg New York

Walter H, Breckle S–W (1991) Ökologie der Erde, Bd 4. Spezielle Ökologie der Gemäßigten und Arktischen Zonen außerhalb Eurasiens. Fischer, Stuttgart

Walter H, Breckle S–W (1999) Vegetation und Klimazonen. 7. Aufl Ulmer, Stuttgart, 544pp

West N (1983) Temperate deserts and semi-deserts. In: Goodall DW (ed) Ecosystems of the world, vol 5, Elsevier, Amsterdam

von Willert DJ, Eller BM, Werger MJA, Brinckmann E, Ihlenfeldt HD (1992) Life strategies of succulents in deserts, with special reference to the Namib desert. Cambridge Studies in Ecology, Cambridge University Press, Cambridge

Wood SL (ed) (1980) Soil-plant-animal relationships bearing on re-vegetation and land reclamation in Nevada deserts. Great Basin Nat Mem 4

Yair A, Berkowicz A (1989) Climatic and non-climatic controls of aridity: the case of the northern Negev of Israel. In: Yair A, Berkowicz S (eds) Arid and semi-arid environments – geomorphological and pedelogical aspects. Catena Suppl 14:145–158

Zohary M (1973) Geobotanical foundations of the Middle East. Fischer, Stuttgart

Zohary M (1982) Vegetation of Israel and adjacent areas. Beih Tübinger Atlas Vorderer Orient Heft A7, Reichert, Wiesbaden

Zohary M (1983) Man and vegetation in the Middle East. Geobotany 5. Junk, The Hague

River Diversion, Irrigation, Salinization, Desertification – an Inevitable Succession

Dietmar Keyser

Keywords. Irrigation problems, deserts, management problems, arid land use

Abstract. Landuse in arid regions is mainly applied for agriculture; recreational exploitation is not very often found; technical usage of these regions now has started. Three different methods of agriculture are used: animal husbandry, precipitation-fed agriculture and irrigation agriculture.

Irrigation systems are some of the most comprehensive undertakings of mankind, not only involving technical and scientific measures, but alsoinflicting social and cultural changes. Economic alterations for the population and the whole society are likely. Management questions are one of the essential points in success or disaster. A historical look at some examples of irrigation culture gives an idea of the difficulties and hazards with which irrigation societies have to cope. Differences in recently installed systems and their performance are evaluated and judged against their proposed aims. Shortcomings are identified in several fields, i.e. in the technical approach, in planned economic possibilities, in social changes and influence on cultural questions, in water management tasks or in the recognition of environmental hazards.

Some recommendations are offered to moderate the negative effects. Financing of the irrigation operating system must be guaranteed. Workers and management must be well-trained and motivated. Rotation patterns of crops must be strictly enforced. The market for the goods produced must repay the input. New technologies have to be evaluated and installed. In the case of financial problems, reduction of areas of cultivation as well as of animal relying on the fodder, is essential. Ingenious ideas of new products, i.e. making use of local weeds or agricultural use of saline lakes, must not be rejected but strongly supported. A larger diversity of crops will help to reduce fertilizer and pestizide impact as well as the dependence on a certain market.

Introduction

The world has an overall surface of 510.1 million km² of which 149.3 million km² are terrestrial surface. Of this about 20 million km² are deserts in the tropical, subtropical and boreal zones. Climatological as well as geological factors are natural reasons for these arid landscapes.

River Diversion, Irrigation, Salinization, Desertification – an Inevitable Succession 15

Deserts are thought of as landscapes in arid and semiarid regions. About 5% of the Earth's surface receives such small amount of rain that they are called real deserts. In historic times, the desertification of semiarid and arid lands has progressed with increasing speed. The reasons are not all known yet, but to a great extent these changes can be traced back to human activity. In the actual situation of the world with its 6 billion human population, even these marginal lands need to be used to provide part of the food essential for mankind.

These areas could be used in at least three ways:

- Agricultural (crop growing, animal husbandry)
- Technical (mining, use of solar energy, use of wind energy, industrial product complexes, rocket launching sites, test sites of various kinds
- Recreational and health facilities (tourism, medical uses like sanatoria and hospitals)

All of these possibilities require attention, but this chapter will cover only the agricultural part.

Agricultural Use of Land

Farming in semiarid and arid lands can be grouped into three basic methods:

- Precipitation-fed agriculture
- Nomadic livestock husbandry
- Irrigated agriculture

Normally, societies living in these areas use all three ways of subsistence, often one beside the other, as integrated possibilities. Nowadays, the dominating form is irrigation, with its great influence on the environmental issues of these regions.

Ancient Societies

The basis of several ancient societies was irrigation agriculture, and high cultural standards were developed under such conditions. The knowledge of science, jurisdiction, art, philosophy, astronomy, economy increased tremendously in these societies. It is worthwhile to just briefly look at them in order to view some of the advantages and shortcomings in these either long-forgotten or still prosperous communities.

The Egyptian culture was based on the yearly high flood of the Nile river, which permitted a highly diversified society to develop on the basis of a stable oversupply of food. This enabled trading of goods, either direct trading of crops or manufactured goods by people not directly involved in food production.

The Sumers living in the lower region of the Euphrates and Tigris rivers originally also developed their irrigation agriculture from natural flooding, but they soon realised that they could use natural channels from the slightly more elevated Euphrates to the lower-running Tigris for the water supply of their fields. This led eventually to building of dams and additional channels. This civilization also thrived over several hundred years.

The third method of irrigation was used by the people of Sheba in South Arabia. Their irrigation system and their wealth were based on the distribution of the sometimes heavy rains by a variety of dams in the wadies to terraces with agricultural crops. Their main commercial products were special plants of their region, the incense and balsa, which they successfully traded in North and East Africa and the Near East. On their trading routes they used their camels and, due to this they maintained a high living standard and high income by relying on their local specialities. However, when the Roman Empire enabled safe ways of transport by ship in the Red Sea and also when the emerging Christian religion used less incense, the Kingdom of Saba deteriorated quickly and lost not only its wealth but also the irrigation facilities. This society was never able to reestablish the irrigation system again. Obviously, the main economic basis had been lost and the people saw no sense in keeping up all the installations necessary for the well-working system which they had used for hundreds of years.

All of these societies have in common that they were very sophisticated communities. There was a central power, implementing and running all irrigation structures by means of a highly bureaucratic elite. These persons helt great responsibility and also high social status. A finely coordinated taxation and a specialized, very well-organised legal system were present. Moreover, a market economy evolved. Goods were exchanged and traded, which was the beginning of money exchange. Culture and religion were based on items or incident which were connected with important seasonal agricultural dates. All these disciplines were an intrinsic part of each irrigation culture. If only one part was missing, most likely the whole system broke down.

A good example is Timurlane, when he conquered the town Kunja–Urgentsch in Central Asia by building a dam and diverting the Amudarya (Aladin 1998). All the highly sophisticated irrigation systems dried up, as well as a part of the Aral Sea too. Ever since, this area has never attained its former performance.

The fact is that these and also the present irrigation societies are economically efficient but very fragile systems. They depend on a variety of different factors behaving like wheels in a machine, and if one wheel is broken, the whole gear system does not work.

To realize this could help us to understand many of the difficulties evolving in our newly installed irrigation schemes or in our positive attempts to help to improve old and so-called inefficient systems.

Industrial Farming

While in ancient and also in mediaeval times irrigation systems were mainly used for the subsistence of the local population, in the 19th century a new factor was added. With the rise of industrial processes in Europe and later in North America, the necessity of a huge supply of agricultural raw material became evident.

The British companies were among the first who installed in their colonies mainly in India, large irrigation systems to grow cotton for their industrial calico production (Islam 1997). Here, production of raw material and final product was divided completely. People working in these schemes were simply hired as labourers, only the impersonal Company was the owner of the land. Certainly a problem of acceptance.

This industrial landuse held a rank different, for instance, from the local silk production in China, were mostly local manufacturers were included in the original irrigation society.

Due to the well-chosen sites, the skilled local labourers and the strict operational management, these newly constructed schemes were very productive. These good examples encouraged followers. Especially together with the rapid technological development irrigation schemes were installed wherever possible. In planning and installing these new schemes, the main interest was laid on technical performance; the production tool human labour was not considered.

Finding, storage and distribution of water was the most important task, as certain economic success would follow. Only during the past 20 years did a change in attitude take place. It was realised that even very well-planned systems failed, for no known reason. Now scientists had to find the causes. They identified several areas of shortcoming in the different schemes, which led to a faster and more extended desertification process within areas used for centuries for low but stable production.

Shortcomings

Planning stage

1. Overestimation of water resources
- Not taking into account tail-end users, who are also relying on the same resources (Ghezae 1998)
- Just plain oversizing the irrigated area
- Overestimation of recharge of used aquifers, no planned artificial recharge of the aquifers (Brown 1990)
- Ignoring the hydro-geological system
- Not considering the changes in the regime which are depending on the original groundwater system and drainage of the irrigated lands (Sydykov et al. 1998)

- Misinterpretation of soil characteristics and the potential of air- or rain-borne erosion
- No consideration of climatic specialities
- Unpredictable heavy rain or long draughts. Changing of the temperature gradient by the installed system itself

Agricultural Shortcomings

Crops (Ghezae 1998)

- No crop rotation: due to economic pressure no crop rotation, which means no restoration of the soil
- Economically wrong mixture of crops: crops have to be sold, so they have to be marketable. There must be the possibility of mixing the crops so that the watering time could be prolonged, as some of the plant need water earlier, some later
- No watering schedules; farmers are left alone with their irrigation needs, which leads to excessive watering

Fields

- Excessive watering
- Waterlogging
- No or not enough leaching
- Salinization
- Inadequate drainage

Economic Shortcomings (White 1978; Islam 1997; Ghezae 1998)

- Operational management has to be very well trained and the information has to be transmitted through all channels
- Responsibility sharing is often not done, so that in the worst case the central government has to decide when, what, at which field has to be planted
- Prime cost and market oversight missing. It is not known how much it costs to produce 1 ton of crop. No information is given about the prices on the market

- Production balance sheet: if there is one at all, it is not known to the people in charge
- Stable financing facilities; it cannot be that the money spent for the scheme is only given during the installation time and is then suddenly withdrawn, leaving the scheme without the necessary compensation by their own income
- Adaptation to new ways of technology and management; there is only little understanding that constant change is necessary to cope with the requirements of the world market today
- Taxation; The procedures and the fairness of prices for services are not always acceptable, e.g. if a tail-end farmer does not get any water and has to pay for it. On the other hand, the prices must be in accordance with the income of the people

Operational Shortcomings

Manpower (Ghezae 1998)

- Labourers are not motivated, due to lack of participation in decision-making processes
- Labourers are not educated and trained, although they have often been hired from rain-fed agricultural places and no elder people can train them
- Scientits are often missing, so information for action at the right time, e.g. of pestcontrol, is not given

Techniques

- Machinery wrong or not adjusted to the special task in the fields, e.g. equipment either to heavy or to large
- Drainage not deep enough or too far apart. Even specialists from the FAO realised that their recommendation for the maximal depth of drainage channels had to be changed either to be deeper (up to 3 m) or to minimise the distance between them
- Bad maintenance of channels and drainage. Often the channels are not cleaned or cleaned in such a way that lining is destroyed and seepage is promoted
- Problems of distribution of water. Tail-end users are getting too late, too little or even no water at the correct time

- Storage and transport problems of harvested crops. The deterioration of the harvest may be up to 30% or even higher if relevant facilities are not present

Social and Health Problems

- Drinking water supply: often not even foreseen
- Wastewater treatment: is often missing and has a major effect on the health
- Danger of water-borne diseases: extreme danger for the people, which has already led to resigning from several schemes (White 1978; Asimov et al. 1998)
- Health care and education: is normally below standard, like all activities concerning the people in the schemes
- Subsistence farming: is only marginally foreseen in cash-crop irrigation systems, has the effect that people do not feel at home and do not care
- Change of life style: year-round irrigation has a great influence on farmers coming from rain-fed lands. They have to work in a different way at different times
- Disruption of sociocultural ties: all the learned and important facts have changed or are either worth nothing or altered. Care has to be taken in substitution (Ghezae 1998)
- Overpopulation: a well-kept scheme will attract people from outside. It will enable the farmers to have more children and thus all the achieved improvements in income will be distributed among more people so that in the long run, it might even come to a worsening of conditions

Irrigation is obviously hampered by a variety of problems. It is noteworthy that the majority of these are not in the field of technical shortcomings but mainly in the management and human field. Obviously, the industrial method of raising cashcrops seems to conflict with the normal way of human life, as likewise the industrial production of goods in the Western countries.

We have to realise that we cannot continue to blame only technical structures for problems arising in irrigated regions. The man-made destruction is penetrating deeper into the environmental setup than we have so far realized.

Ecological Problems

Some of the most striking environmental and ecological problems caused by certain shortcomings of irrigation and river diversion are compiled here. Irrigation-borne ecological problems

Water

1. Waterlogging

2. Salinization
3. Solonchaks and takyrs
4. Lowering of groundwater level (in adjacent deserts and aquifers)
5. Erosion by changes in vegetation cover
6. Climatic changes, due to higher evaporation

Soil

1. Deterioration
2. Exposure of bare soil and old dunes
3. Air-borne erosion and threat of new dunes
4. Sand and salt dust

Biological

1. Introduction of imported species (aliens)
2. Increasing biodiversity
3. Water-borne diseases
4. Monoculture
5. Reduction of flora and fauna in deserts (Novikova et al. 1998)

Anthropogenic

1. Overgrazing in areas with livestock
2. Collection of natural plants and wood
3. Destruction by transport vehicles
4. Pesticide enrichment (Bogdasarov et al. 1998)

Certainly there are additional problems, but this list states at least some of the most important ones which are common throughout the irrigation systems. All these problems will be discussed in the present volume as well as the causes, the monitoring and strategies against it. Many of the research activities fall short, for they are fighting the consequences and not the causes of the problems.

Also the proposals to return to the former way of life with its well-adjusted production and ecosystems sounds attractive, but is not a feasible alternative. The world population is growing tremendously and will double by the year 2150, when a stable population is expected. At this point there is no alternative to irrigation, so care has to be taken that it can be run in a sustainable and also environmentally healthy manner.

Additional ways must be found to minimise the inevitable alteration of the influenced desert landscapes. To fight excessive water use, waterlogging and salinization in the production areas must have highest priority, as these are the starting points of deterioration; but even these are already consequences of mismanagement, so the cause for this development lies in the past.

Gilbert White already stated in his UNESCO book *Ecology of Irrigation Systems* in 1978 some of the most necessary items which could help:

- Avoid irrigation projects of less than 1000 ha
- Divide large irrigation projects into lateral units of between 2000 and 6000 ha, depending on topography
- Let each lateral unit contain a number of rotational units, the size of which should vary between 70 and 300 ha, depending on topography
- Operate main, lateral and sublateral canals on a schedule of continuous flow
- Within a rotational unit, organise the rotation of water supply to farm inlets or group inlets independently of the distribution in adjacent units
- On large irritation projects of more than 10,000 ha, decentralise the project management so that each lateral unit has its own staff

Sometimes it could even help to reduce the size of the irrigated lands. It is better to have good maintenance of 50% of the fields than 100% in bad shape. In India, experiments on the effectiveness of irrigation were conducted and showed that sometimes less than 5% of the possible production on farm land was reached (White 1978).

Summary

There is a chance that irrigation and river diversion do not inevitably lead to salinization and desertification, but political action has to be asked for.

Economical problems are obviously always the trigger of misuse and problems. Therefore a sound financial base must be present not only during installation but also during the sustainable functioning of the system.

To assure this, commercial knowledge and information have to be used in connection with the local possibilities.

Measures to fight moving soils are known, but due to misunderstanding and underestimating the importance of combating the starting point is ignored. This has to be altered by involving and educating the local farmers.

More activity should be directed on using local rain-fed agriculture of natural plants for fodder or building material within or adjacent to the irrigation schemes.

Also more attention has to be given to local specialities in soil, climate, flora and fauna, to open a special market for these goods.

The chances of keeping the unique flora of natural deserts are widely neglected, but they open possibilities in using them as a genetic reservoir of unknown value.

Education and integration of local people in the decision processes will have a wide effect on the sustainable stability of the ecology.

It must be realized that a changing in attitude to the environment and to ecological problems also means a change in the attitude of the people involved. This includes, first, their economic basis and, second, their education.

People seem to realise that irrigation is far more than just water to grow crops. It is a network of possibilities and dangers to almost all relevant human and environmental factors. The results of scientific research have always to be viewed under the light of such a comprehensive task.

Acknowledgements. This study was initiated and financially supported by the BMBF/UNESCO project on the Rehabilitation of the Aral Sea. Thanks are due to all my colleagues working in this project and especially to Dr. Vefa Moustafaev from UNESCO, who helped to pinpoint many of the stated problems.

References

Aladin NV (1998) Some paleolimnological reconstruction and history of the Aral Sea basin and its catchment area. In: UNESCO Aral Sea project. Ecological research and monitoring of the Aral Sea deltas. UNESCO, Paris, pp 3–12

Asimov DA, Mirabdullayev IM, Golovanov VI, Kuzmetov AR, Shakarbeyev E, Turemuratova GT (1998) Aquatic invertebrates as intermediate hosts of parasites. In: UNESCO Aral Sea project. Ecological research and monitoring of the Aral Sea deltas. UNESCO, Paris, pp 263–271

Bogdasarov V, Bolshakova L, Mollaev T, Kuturina N, Sergeeva, I (1998) Migration of Pesticides in time and their distribution in the soil–water–vegetation system. In: UNESCO Aral Sea project. Ecological research and monitoring of the Aral Sea deltas. UNESCO, Paris, pp 81–91

Brown LR (1990) State of the world. World Watch Institute. Norton, New York, 253 pp

Ghezae N (1998) Irrigation water management. Uppsala University Library, Uppsala, Sweden, 296 pp

Islam MM (1997) Irrigation, agriculture and the Raj. Punjab, 1887–1947. South Asia Institute, New Delhi Branch Heidelberg University, South Asian Studies No XXX. Manohar, New Delhi, 180 pp

Novikova NM, Kust GS, Kuzmina JV, Trofimova GU, Dikariova TV, Avetian SA, Rozov SU, Deruzhinskaya VD, Safonicheva LF, Lubeznov UE (1998) Contemporary plant and soil cover changes in the Amudarya and Syrdarya river deltas. In: UNESCO Aral Sea project. Ecological research and monitoring of the Aral Sea deltas. UNESCO, Paris, pp 55–80

Sydykov ZS, Poryadin VI, Vinnicova TN, Oshlakov GG, Dementiev VC, Dzhakelov AR (1998) Estimation and forecast of the state of ecological-hydrogeological processes and systems. In: UNESCO Aral Sea project. Ecological research and monitoring of the Aral Sea deltas. UNESCO, Paris, pp 159–177

White GF (1978) L'Irrigation des terres arides dans les pays en dévélopement et des conséquences sur l'environnement. UNESCO, Paris , 67 pp

Part I:

The Aral Sea Crisis

Part I.

The Aral Sea Crisis

The Aral Sea Crisis Region

Siegmar–W. Breckle, Walter Wucherer, Okmir Agachanjanz and Boris Geldyev

Keywords. Aral Sea crisis, Small Aral Sea, Great Aral Sea, Aralkum desert, ecological situation

Abstract. The term Aral Sea crisis region covers the lower stretches of the rivers Amudarya and Syrdarya, where the effects of the diversion of the river water from the Amudarya and Syrdarya are a blatant case. The drying of the Aral Sea, according to the Scientific Board for Global Environmental Changes of the German Federal Republic (WBGU), was declared the greatest environmental disaster caused by mankind in this century by changing the regional water budget. The disintegration into smaller water bodies is reality. Meanwhile the knowledge and information basis on the status of land degradation and on current activities in the crisis region is still insufficient. Up-to-date information on monitoring of land use and of the situation of the ecosystems in the crisis region is important, since the ecological situation is very dynamic and unstable. Degradation and desertification are threatening the whole population of the area and endangering their basis of living.

Introduction

The huge irrigation projects in Central Asia during Soviet times are the main cause for the catastrophic desiccation of the Aral Sea. The luxury use of the water resources of the Aral Sea basin have led to a severe loss of equilibrium between the natural water sources in ecosystems and the water use in agricultural irrigation. This predominantly unintentional and unreflected water use has led to an agricultural system characterized by extensive use of water and thus to steadily increasing damage to and impact on the ecosystems of the whole Turan Basin.

The Aral Sea can be regarded as the centre, where all the political and other interests of the new Central Asian states are connected to each other. The northeastern half of the Aral Sea belongs to Kazakhstan, the southwestern half to Uzbekistan. The catchment area of the two large rivers, the Amudarya and the Syrdarya, are part of the mountain republics Tadzhikistan and Kirgizistan, but the main water consumers are Kazakhstan, Turkmenistan and Uzbekistan (Kuznetsov 1986; Agachanjanz 1988; Glazovskii 1990; Letolle and Mainguet 1996; Walter and Breckle 1994; Klötzli 1997; Micklin and Williams 1996; Breckle et al. 1998).

The term Aral Sea crisis region covers the lower stretches of the rivers Amudarya and Syrdarya, the Aral Sea and adjacent surroundings, where the effects of the diversion of the river water from the Amudarya and Syrdarya are a blatant case (Giese 1997). This is mainly the case for the districts of Khorezm and Karakalpakistan in Uzbekistan, the districts Kzyl–Orda in Kazakhstan and Tashauz in Turkmenistan. The area of this region is about 500 000 km^2, the population about 3 800 000. It is likely that especially Karakalpakistan is severely affected, along the lower stretches of the Amudarya. In September 1995 the Nukus Declaration was adopted by the Central Asian States and by international organizations, declaring as an important aim the solution of the problems of the region by a development of a sustainable environment in the Aral Sea Basin. Since then, some activities in the region have started: attempts to supply the towns and cities with clean drinking water; projects to improve land use; construction of dams for better water use in agriculture etc.

Meanwhile the knowledge and information basis on the status of land degradation and on current activities in the crisis region is still unsufficient. Up-to-date information and the monitoring of land use and of the situation of the ecosystems in the crisis region are important, since the ecological situation is very dynamic and unstable. Especially the process of water loss of the Aral Sea has now become crucial. The salt concentration of the Aral Sea water is now higher than sea water; the new dry sea floor is strongly saline. Salt desertification in the Aralkum covers huge areas, thus forming salt deserts which are the main source of salt dust and sand storms. The delta areas are suffering from lack of water, the remaining tugai forests have died and were cut down, the oases are suffering secondary salinization. Degradation and desertification are threatening the whole population of the area and are endangering their basis of living.

The Aral Sea

Formerly (before 1960) the Aral Sea was the fourth largest inland lake on the globe with a surface area of about 68 000 km^2. Since 1960 the surface area has decreased drastically (see Table 1). About 80% of the water body and about 60% of the water surface have been lost (Fig. 1). The area of the dry sea floor, the Aralkum, is about 40 300 km^2 (level 1999).

The drying up of the Aral Sea according to the Scientific Board for Global Environmental Changes of the German Federal Republic (WBGU) was declared the greatest environmental disaster which mankind has caused by changing the regional water budget. The developing new land was called the Aralkum desert (see also Walter and Breckle 1994). It is obvious that the speed of drying up and the increase in the dry sea floor area in the 1980s was especially fast. The water supply by the rivers to the Aral Sea in this time dropped to about 4 km^3 a^{-1}, before 1960 it was on average about 60 km^3 a^{-1}. During the past years, in the 1990s, the average water input to the Aral Sea has increased again to about 14 km^3 a^{-1}. This is a very positive sign, which in future will slow down the drying out of the Aral Sea.

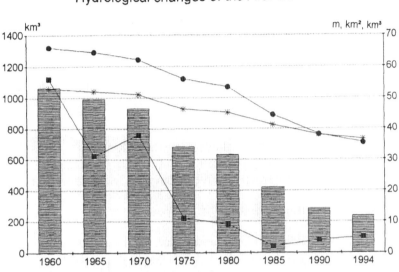

Fig. 1. Dynamics of the water volume and sea level of the Aral Sea

Table 1. Dynamics of the water surface area of the Aral Sea and dry sea floor. Data from 1960, 1970, 1980 according to Aralskoe More (1983), for 1990 from Ressl (Aral Sea Homepage 1999) and for 1999 from B. Geldyev

Year	Water level (m NN)	Water surface (km^2)	Area of dry sea floor (km^2)
1960	53.4	68 000	0
1970	51.5	61 000	7 000
1980	46.0	52 000	16 000
1990	38.5	38 800	29 200
1999	[a]	27 700	40 300

[a] In 1999 the water level in the Small and the Great Aral Seas were different.

However, the situation has reached a status where the dynamics of the northern and the southern sea basin have started to be different and the formation of separate sea basins started: the northern Small Aral Sea and the southern Great

Aral Sea; the Great Aral Sea is almost divided north–south by the island Vozrozhdenie into the deeper Western Aral Sea and the shallower Eastern Aral Sea. This disintegration into smaller water bodies is almost reality, as well as the new terms, Aralkum and Aral Sea Syndrome, as complex phenomena of various desertification processes.

The Ecological Situation of the Great Aral Sea Area

The lowering of the sea level of the Great Aral Sea is still continuing. The water level of the Great Aral Sea in 1999 was about 33.8 m NN and the surface area of the sea was about 24 400 km^2. The drying out of the Great Aral Sea causes four main environmental problems:

- the threat to the existence of the nature reserve Barsa–Kelmes;
- the opening of the former military experimental site by a land bridge;
- the formation of a huge new salt desert between the island Vozrozhdenie and the eastern coast;
- loss of the remnant fish fauna.

During the years 1996 to 1998 a land bridge was formed between the island Barsa–Kelmes, which was originally in the centre of the Aral Sea and then the eastern coast of the Aral Sea. The area between the island Barsa–Kelmes and the peninsula Kokaral is now in a terrestrial developmental phase. The island Barsa–Kelmes is one of the most precious nature reserves within the Central Asian deserts, which was established in 1939. The zonal ecosystems of the island with *Artemisia* and *Anabasis* vegetation complexes are amongst the best-preserved ecosystems of the northern Kazakho–Dsungarian deserts. The flora of the island comprises about 257 species. Some antilopes (*Gazella subgutturosa* and *Saiga tatarica*) as well as the onager were introduced (*Equus hemionus*) on the island years ago. The animal populations of *Gazella subgutturosa* and *Equus hemionus*, which are part of the Red Faunal List, are especially to be mentioned. The natural isolation of this Nature Reserve gave it a perfect protection, thus the status of this area corresponds better with the international rules of IUCN for strict nature reserves than any other protected area in Central Asia. However, in 1999 the island became accessible from the mainland. Geologists from Kazakhstan and from have reached the island, coming from the east coast by four-wheel trucks and have made trial drillings on the island. Native people have started hunting. Without special additional means of protection this former island will rapidly lose its role as one of the most important nature reserves.

It is obvious that by continuous drying out of the Great Aral Sea, it will be soon divided into two separate water bodies. The island Vozrozhdenie will mark the watershed between the Western Aral Sea and the Eastern Aral Sea. The island Vozrozhdenie was formerly the experimental military site of the Soviet army. Remnants of arms, technique and military supplies are still left on the island, probably also remnants of biological and chemical weapons. It is not known what kind of military experiments were made here. In the very near future the island Vozrozhdenie will be accessible from the mainland as was Barsa–Kelmes. Immigration of animals and people from the mainland might result in unexpected biological effects or toxic damage in the area. Thus, drastic means to prevent uncontrolled access to the area by civilians have to be put into effect, as long as the epidemiological, toxicological and ecological situation of the island is unclear.

Caused by the drying out of the eastern part of the Eastern Aral Sea, the formation of another huge open salt desert will occur. The colonization of the dry sea floor on the areas from the 1960s and the 1970s is rather dense, their salinity is low, but the areas from the 1980s and the 1990s are already covered by salt deserts with very isolated single plants only (Fig. 2). In continuation of the increase in salt desert areas, salt and dust storms will become more common and more frequent.

Fig. 2. The salt desert on the dry sea floor at the east coast of the Aral Sea

The increase in salt concentration (to more than 50 g l^{-1}) in the Great Aral Sea has already led to the extinction of almost all native fish species as well as of the plaice (summer flounder), which was introduced several years ago. At the recent southwest coastline of the Kokaral peninsula the authors have seen salt marshes covered by dead plaices.

The Ecological Situation of the Small Aral Sea and the Dam

The water level of the Small Aral Sea in the 1990s exhibited large fluctuations. In 1990 to 1993 the sea level was about 38–39 m NN; the water surface of the Small

Aral Sea varied between 2600 and 3100 km^2 and the volume of the water body was about 17–22 km^3 (Bortnik 1996). In the second half of the 1990s the water level of the Small Aral Sea rose rather rapidly and the water body increased correspondingly (Fig. 3). According to data from the ecological station at Kasalinsk in August 1997, the water level was about 40.3 m NN and in August 1998 at about 41.8 m (according to nivellements at the transect Karabulak by Wucherer). The surface area then had increased to about 3700 km^2. Fishing on a small scale had restarted. The introduced plaice flourished. According to local fishermen, the daily haul reached about 1000–1500 kg. This overall positive development was caused by a drastic decrease in the irrigation area of the paddy fields of the Ksyl–Orda-district, as well as by the construction or reconstruction after damage of a dam between the Kokaral and Kosaral peninsulas, separating the Syrdarya water discharge totally from the Great Aral Sea. Bortnik (1991) has given a pessimistic and an optimistic prognosis for the development of the Small Aral Sea (Fig. 4). It is obvious that the real rise in water level of the Small Aral Sea exceeded even the optimistic variant of the prognosis. The hydrological water budget of the Small Aral Sea and the Syrdarya discharge would enable a recovery of the Small Aral Sea within 5 to 7 years. However, this would need a strong and durable dam construction. The present dam is rather provisionary and primitive (Fig. 5). The material used is solely loose sand and loam; but the local people and the local administration tried to do their best when the situation at the dam became alarming during the spring tides; the whole available technical vehicle fleet was involved. Despite the lack of money and a severe economic situation, the district government of Ksyl–Orda and the Aralsk administration tried to improve the situation of the Small Aral Sea by erecting and repairing the dam. However, in April 1999 the water level rose very rapidly and a storm caused the total breakage of the dam. Figure 6 shows how the water flood wrecked the vehicles. The flood caused huge damage on the northeastern dry sea floor of the Great Aral Sea. The sea level dropped by about 2.5 m. The great efforts of the whole decade (the 1990s) to stabilize the Small Aral Sea were useless.

The Aral Sea Crisis Region

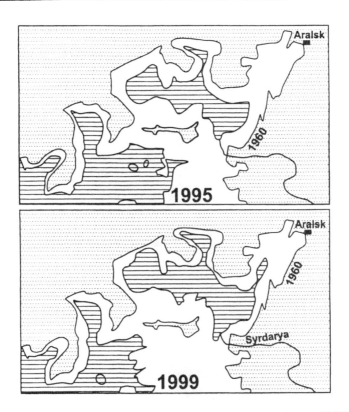

Fig. 3. Increase in the water body of the Small Aral Sea in the past 5 years (1995–1999)

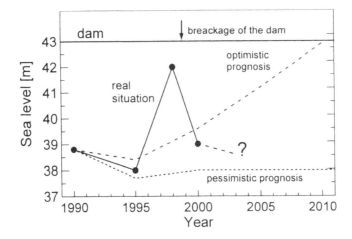

Fig. 4. Dynamics of the sea water level of the Small Aral Sea

Fig. 5. The dam between the Small and Great Aral Sea

Fig. 6. Damaged and abandoned technology on the breakage of the dam

A solid dam cannot be realized without sufficient international help. A sound technical solution to control the sea level of the Small Aral Sea by solving the construction of a solid dam with electric generators etc. will most probably have a whole set of positive ecological and social effects. Maintenance and support of various small villages around the Small Aral Sea, reviving the fishery and support of the city of Aralsk, development of salt marshes, enhancement of a more sustainable economy based on camel grazing, less salt and sand storms etc.

The Aralkum Desert

The Aralkum desert with the remnant water bodies and the recent and older terraces is part of the centre of the Aral Sea crisis region. The dry sea floor is a complicated mosaic of solonchak and sandy ecosystems. The psammophytic ecosystems were formed mainly during the first phase of drying out until the middle of the 1970s, around the islands of Barsa–Kelmes and Vozrozhdenie until the end of the 1980s. This means that the landscapes dominated by sandy substrates are the typical areas along the former coastline. On the sandy substrates, perennial psammophytic plant communities have developed, mainly *Stipagrostis pennata*, *Eremosparton aphyllum*, *Calligonum* and *Astragalus* species (Fig. 7). On the new dry surface new sand dunes developed rapidly, mainly barchan-shaped. One can distinguish two types of dune systems on the dry sea floor: open dune systems with single dunes, covering much less than 50% of the area; and those with a dense pattern of dunes, covering more than 50%.

Fig. 7. The psammophytic plant community on the southeast coast of the Aral Sea

Since the 1980s on the dry sea floor almost exclusively solonchak deserts have formed. The open dry sea floor of the Aral Sea today is a huge salt flat and a source of salt dust in the larger vicinity. The direct influence of aerosols (salt particles, dust) on the adjacent agricultural areas and their salinization as well as on the natural ecosystems and on the health of the people is still under dispute. The huge salt deserts which have developed since the 1980s are characterized by a rather strong deflation and denudation by wind at a speed of 2 mm a^{-1} (Semenov 1990). This means that the upper 4-cm layer has been blown off during the past 20 years. This investigation takes into account the saline dry sea floor deserts which date from the 1980s. An increasing drying of the sea floor will create even more and greater saline flats, and thus a more enhanced salt dust regime by storms in the near future. This is a threat to the agricultural areas adjacent and beyond the southern and eastern former coast line. It is to be suspected that the salt desertification is spreading in the whole Aralkum and the surrounding areas. Several questions have to be solved: will there be a limit to natural plant colonization? If so, where it will be? How long will the primary and secondary salt deserts prevail? Are there other species which will colonize the dry sea floor? Is it possible to use phytomelioration (Meirman et al., this Vol.) as a means to effectively combat sand movement, salt dust storms and thus salt desertification?

Deltas and Oases

The drastic cut in discharge of the river waters in the delta areas has caused severe degradation and desertification of soils and dense shrub and river meadow vegetation of the river deltas. In the 1990s the BMBF-UNESCO project studied the delta regions. The main problem of the delta regions again is the unintentional wastage of water. A realistic and sound, sensible and reasonable planning of water use for both nature conservation and overall use in the areas for a sustainable grazing regime is totally lacking. The agriculture of the oases depends totally on irrigation water and again is wasting water. Irrigation certainly has to be kept as the basis for the main agricultural systems in Central Asia. Since the water needs and the consumption will increase in future, the possibilities of water-saving irrigation techniques have to be applied as well as possible, since the extensive increase of irrigated areas is limited. The only solution for the Central Asian states can be to increase the efficiency of water use, to use recycling techniques for water use in order to avoid severe water shortages.

Conclusions

The Aral Sea no longer exists. The separation into smaller water bodies has taken place and will continue. The Great Aral Sea will split into the Western and the Eastern Aral Sea by the continuation of drying out. The Western Aral Sea is rather deep and will remain for a longer time, the shallow Eastern Aral Sea will remain only in small fractions (west of the island Vozrozhdenie). The Small Aral Sea will remain despite the breakdown of the dam. Technical solutions for an improved construction of a new dam for the control of the water level of the Small Aral Sea are necessary.

The nature reserve Barsa–Kelmes is under severe threat. The former island Vozrozhdenie with its military remnants is a great potential danger unless the conditions and the hazardous situation are clarified. For the oases regions and the river deltas, a realistic planning strategy for a sustainable development and nature protection is lacking, but urgently needed. Since the 1980s the development of open salt deserts has increased tremendously, but the natural colonization by

The Aral Sea Crisis Region 37

plants remains slow in comparison with increased desertification. The increasing by severe effects of salt dust and sand storms from the open dry sea floor and the secondary salinization in vast areas of the whole region, in oases and deltas, is an excellent example of the salt desertification syndrome, which, with all its various threatening aspects for the local economy, is very characteristic for this Central Asian region.

Acknowledgements. This research is funded by the German Federal Ministry for Education and Research (BMBF grant 0339714).

References

Aralskoe more (1983) Nauka, Alma–Ata, 236 pp

Agachanjanz O (1988) Wasserbilanz und wasserwirtschaftliche Probleme in der mittleren Region der UdSSR (Mittelasien und Westsibirien). Petermanns Geogr Mitt 132, H2:109–116

Bortnik VN (1996) Changes in the water-level and hydrological balance of the Aral Sea. In: Micklin PP, Williams WD (eds) The Aral Sea basin. Springer, Berlin, Heidelberg, pp 25–32

Bortnik VN, Kuksa VI, Tsytsarin AG (1991) Sovremennoe sostoyanie i vozhmozhnoe budushchee Aral'skogo morya (In Russian). Izv Akad Nauk SSSR Ser Geogr 4:62–68

Breckle SW, Agachanjanz O, Wucherer W (1998) Der Aralsee: Geoökologische Probleme. Naturwiss Rundsch 51:347–355.

Giese E (1997) Die ökologische Krise der Aralseeregion. Geogr Rundsch 49:293–299

Glazovskii NF (1990) Aral'skii krizis. Prichiny vozniknoveniya i puti vykhoda. Nauka Moskva (in Russian)

Klötzli S (1997) Umweltzerstörung und Politik in Zentralasien – eine ökoregionale Systemuntersuchung. Europ Hochschulschriften, Reihe IV, Geographie, vol 17, Lang, Bern, 251pp

Kuznetsov NT (1986) O nekotorykh aspektakh Problemy Aral'skogo morya i Priaral'ya. Izv Akad Nauk SSSR Ser Geogr 3:56–62

Letolle R, Mainguet M (1996) Der Aralsee: eine ökologische Katastrophe. Springer, Berlin Heidelberg New York 517pp

Micklin PP, Williams (1996) The Aral Sea basin. Springer, Berlin, Heidelberg, 186 pp

Ressl (1999) Aral Sea Homepage

Semenov OE (1990) Otsenka ob'emov vynosa peska i solei s osushayushcheisya chasti dann. In: Gidrometeorologicheskie problemy Priaralya. Gidrometizdat, Leningrad, pp 200–216

Walter H, Breckle SW (1994) Ökologie der Erde. Band 3: Spezielle Ökologie der gemäßigten und arktischen Zonen Euro–Nordasiens. (2. Aufl. by Breckle SW, Agachanjanz OE) Fischer, Stuttgart, 726 pp

Flora of the Dry Sea Floor of the Aral Sea

Walter Wucherer, Siegmar–W. Breckle and Liliya Dimeyeva

Keywords. Chenopodiaceae, plant-geography, psammophytes

Abstract. In 1999, the dry sea floor of the Aral Sea had an area of about 40 000 km². The new land is being steadily invaded by plants. The list of the flora consists of 266 species. It is a young immigration flora (40 years old) that established under harsh habitat conditions. Adjacent to the former coastline there are mostly sandy soils which already are free of salt. They are replaced on younger sea floor areas by mixed sandyloamy or loamy coastal and marsh solonchaks. The family Chenopodiaceae was dominant on the solonchaks of the dry sea floor from the 1970s, 1980s and 1990s. The share of the Chenopodiaceae in the flora amounts to 28.2% (75 species). Along the former coastline on the sands of the dry sea floor from the 1960s the Polygonaceae (*Calligonum*), Fabaceae (*Astragalus*), psammophilous Chenopodiaceae and some representatives of the Poaceae play a major role. The composition of the flora and of the important families is discussed.

Introduction

The vegetation stretching over vast distances of the flat dry sea floor at the eastern side of the Aral Sea is rather heterogenous and varying. Huge areas are covered by therophytic communities, sand and salt deserts and even perennial woody vegetation. The flora and vegetation of the surrounding belts of the Aral Sea has been until today poorly investigated, except for the delta areas.

First knowledge about the flora and vegetation of the direct surrounding of the Aral Sea comes from Borshchov (1865), who studied the northwest and southwest coasts of the Aral Sea. Smirnov (1875) collected plants on the southeast coast of the Aral Sea between and from the deltas of the Syrdarya and Amudarya. Based on the plant collection of Berg (1900–1902), Litwinov (1905) published a list including 219 species. Sherbaev (1982) investigated the flora of the islands on the southeast coast of the Aral Sea and gave a list of the flora with ca. 235 species. These works give a fragmentary idea of the flora of the direct surroundings of the Aral Sea before the drying process started, without exact geographical and habitat differentiation. The Pedagogical Institute of Saint Petersburg pursued an autonomous research with the help of students in the State Reserve of the island of

Flora of the Dry Seafloor of the Aral Sea

Barsa–Kelmes. There is a reference to the flora of this island with 257 species (Kuznetsov 1995).

Our Own Floristic Investigations

The effects of the diversion of the Amu- and Syrdarya drain and the retreat of the Aral Sea were explored in the framework of a complex programme under guidance of the Geographical Institute of Moscow in 1977–1985. During this programme the ecosystems of the dry sea floor were investigated for the first time. Dimeyeva (1990) and Dimeyeva and Kuznetsov (1999) published on the basis of the research of the 1980s a list of the flora of the Kazakh Aral Sea coastal region including about 300 species. From these lists, 154 species also occurr on the dry sea floor. One should distinguish clearly between the Aral terraces and the dry sea floor. The Aral terraces are areas which dried out already 800–3000 years ago. The new marine Aral terrace is 800 years old, the old marine terrace 3000 years. They were flooded inregularly by rather old high water levels of the Aral Sea. The Aral terraces lie between 53 and 60 m NN. Their flora and vegetation are very complex. The flora of the Aral terraces is still insufficiently investigated.

The surface of the actual dry sea floor (Aralkum desert) lies below 53 m NN and is rather recent— only 40 years old. Here, we take only the flora of the dry sea floor into consideration, because it is clear to define by spatial and temporal parameters. Wucherer investigated (DFG project/Kassel) the north, east and south coasts of the Aral Sea in 1994. At that time the list of the flora of the dry sea floor included 201 species (Breckle et al. 1998). In 1998–1999 an 8-month expedition was carried out (BMBF Project Successional Processes on the Dry Sea Floor of the Aral Sea and Perspectives of the Land Use). Special attention was given to the investigation of the family Brassicaceae and Calligonaceae. The total species list became enlarged by 65 additional species. The plant collection from the coast and dry sea floor of the Aral Sea comprises ca. 1000 specimens (Herbaria BIEL, KAS). This plant collection of the dry sea floor is the largest from this area. The investigation of the Chenopodiaceae is time-consuming, especially of the therophytes from the genera *Suaeda, Salsola* and *Petrosimonia*.

This list of the flora records the plant species of the whole dry sea floor area of the Aral Sea, including Kazakhstan and Uzbekistan, with the exception of the archipelago Akpetki in the southeast of the Aral Sea. This is a special region which consists of hundreds of little islets, between which small sections of the dry sea floor are enclosed. This area is difficult to approach and until today it is floristically almost not investigated. Theoretically, one could assume that the small areas of this archipelago become populated faster and could contribute considerable to the flora of the dry sea floor. This reason and the lack of knowledge is forcing us to exclude this area. One more white floristical spot is the dry sea floor around the island Vozrozhdenie (a former experimental site of the Soviet armed forces in the Aral Sea region). Since the ferry services on the Aral Sea were given up, this island is not approachable; it might be accessible in coming years after the formation of a mainland bridge between this island and the south coast.

The species concept for the Chenopodiaceae was taken from *Flora Iranica* (Hedge et al. 1997), except for the genera *Salsola* and *Climacoptera*. As long as there is no complete revision of the genus *Climacoptera*, we keep the taxa *Climacoptera* separated from *Salsola*. The species names for the other families are used according to Cherepanov (1995), except for the genus *Calligonum*. The species concept of this genus has been very controversial up to the present. Therefore we use the more detailed species concept from the *Flora of Kazakhstan*. The tribes were accepted by Takhtadzhan (1987) and *Flora Iranica*.

Results

Physicogeographical Description of the Dry Sea Floor

The Aral Sea is located within the zone of the temperate deserts of Central Asia. The climate is continental. The geological and geomorphological structure of the dry sea floor is relatively complicated. The depression of the Small Aral Sea is formed differently from that of the Great Aral Sea. The north–south gradient also plays a definite role. The seepage, the particle size distribution and the salinity of the soils is of special importance for the composition of the flora and the colonization by plants. From the former coastline to the present one we can observe a typical catena: the substrate becomes finer and the salinity increases. Accordingly, the soil mosaic and the vegetation often exhibit a rather striped pattern. Important soil types are the marsh solonchaks, the coastal solonchaks, the degraded coastal solonchaks and a variety of arenosols.

Composition of the Flora

The flora of the dry sea floor is an immigration flora that has developed since 1960. It consists of 34 plant families, with 134 genera and 266 species (see Appendix), and is a typical desert flora. Almost two thirds of the species belong to the four most frequent plant families. Table 1 gives a list of the ten most important plant families of the Aralkum flora. The remaining 24 plant families are only represented by one to three species per family. In most genera the number of species per genus is rather low (one to three), except in *Artemisia, Calligonum, Salsola, Suaeda* and *Tamarix*, which is an indication of a flora type which is not yet saturated.

Flora of the Dry Seafloor of the Aral Sea

Table 1. Floristic data on the vascular plants of the dry sea floor of the Aral Sea: the dominant ten plant families and percentage of total flora

Family	Genera	Species	Species (%)
Chenopodiaceae	27	75	28.2
Polygonaceae	5	36	13.5
Brassicaceae	23	31	11.7
Asteraceae	16	28	10.5
Poaceae	14	18	6.8
Fabaceae	9	17	6.4
Boraginaceae	6	8	3.0
Tamaricaceae	1	7	2.6
Cyperaceae	3	5	1.9
Zygophyllaceae	3	5	1.9
Other families	27	36	13.5
Total	134	266	100.0

Discussion

Family Chenopodiaceae

The Chenopodiaceae are a cosmopolitan plant family. They play a special role in the dry areas of Eurasia (Il'in 1936; Lavrenko 1962), but also in the Nearctic and in other deserts, e.g. in Australia. The representatives of this family may often dominate the vegetation cover of the Irano–Turanian deserts including Central Asia. Chenopodiaceae are mainly xerophytes and halophytes. All three subfamilies – Chenopodioideae (8 genera and 26 species), Salicornioideae (4 genera and 5 species) and Salsoloideae (15 genera and 44 species) – are plentyfully represented in the Aralkum. Almost half of the genera which occur on the dry sea floor are mediterranean–Irano–Turanean or Irano–Turanean. The cosmopolitan genera *Atriplex*, *Salicornia* and *Suaeda* are also of considerable importance for the composition of the flora and vegetation of the dry sea floor. The deserts and semideserts in Irano–Turan exhibit 18 endemic or almost endemic genera. Out of these, 11 genera (*Agriophyllum, Halostachys, Bienertia, Climacoptera, Ofaiston, Horaninovia, Girgensohnia, Arthrophytum, Haloxylon, Nanophyton, Halimocnemis*) have already reached the Aralkum desert. The species richness of the family is also relatively high (Table 2).

The flora of the Chenopodiaceae on the dry sea floor is not yet saturated, as can be concluded from the data of Table 2. It could well be assumed that in future about 30–40 more species could enrich the flora of the Aral Sea floor. This is even more accentuated if we compare the flora with a larger region. On the dry sea floor (area 40 000 km²), for example 57% of the genera and 30% of the species of the Chenopodiaceae of Kazakhstan (2 700 000 km²) are found. For Central Asia (GUS space) altogether 52 genera and 318 species of the Chenopodiceae are recorded (Pratov 1987). The delta areas of Amu- and Syrdarya with the Aral terrasses and the Akpetki archipelago in the south, and the dry sea floor are part of the large evolutionary centre of the Chenopodiaceae in Eurasia. One should additionally pay attention to the fact that a great variety of vegetation types dominated by Chenopodiaceae occur in this area. They form plant communities with a highly varying composition.

Table 2. Number of Chenopodiaceae in different desert areas bordering the dry sea floor. Data for different districts are taken from Bakhiev (1985)

Districts	Dry sea floor	Karakalpakistan (Erezhepov 1978)	East Ustyurt (Sarybaev 1973)	Lower part of Amudarya valley (Bakhiev 1985)
No of species / % whole flora	75 / 28.2	140 / 15.9	52 / 17.6	113 / 18.0

The life-form composition of the Chenopodiaceae is represented by small phanerophytes (trees, shrubs, half-shrubs), chamaephytes and therophytes, the latter amounting to 54 (72%). The following genera are exclusively therophytes: *Ceratocarpus, Bassia, Corispermum, Agriophyllum, Bienertia, Climacoptera, Ofaiston, Horaninovia, Girgensohnia, Petrosimonia, Halimocnemis, Halogeton.* They form their own plant communties or synusia within perennial plant communities. Only the *Haloxylon* species and *Halostachys caspica* are shrubs higher than 1 m or even up to 4 m. On the dry sea floor *Haloxylon aphyllum* occurs only in shrub form, not as small trees, as elsewhere. The other perennial Chenopodiaceae are prostrate-growing shrubs or half-shrubs.

In the subfamily Chenopodioideae the tribe Atripliceae, Camphorosmeae and Corispermeae are floristically and phytocenologically significant. The genus *Atriplex* is represented by 11 species, but only 4 of them (*A. fominii, A. pungens, A. tatarica, A. aucheri*) are widespread and thus exhibit some dominance in relevant communities. The northwest Turanian *Atriplex fominii* (Fig. 1) forms widespread communities in the whole area of the dry sea floor. The north Turanian *Atriplex pungens* forms narrow vegetation stripes along the former coastline of the Small Aral Sea. *Atriplex aucheri* and *Atriplex tatarica* occur more sporadically. The ecology of the *Atriplex* species differs considerably. *Atriplex dimorphostegia* is a real psammophyte, but *Atriplex verrucifera* is a real halophyte. From the Camphorosmeae the old mediterranean (old Tethys area) *Bassia hyssopifolia* has the richest diversity in many plant communties, especially in areas bordering the river deltas. This species often forms its own alliances and occurs regularly as a component of several other plant communities. The *Corispermum* species have spread exclusively on sandy soils. The Caucasian and north Turanian *Corispermum aralo-caspicum* is the most typical psammophyte. *Agriophyllum squarrosum* as a pioneer species covers coarse sand degraded solonchak areas on the north coast of the Aral Sea.

Fig. 1. *Atriplex fominii*, an example of an abundant annual plant species

The species from the subfamilies Salicornioideae and Salsoloideae are the dominant species of the halophytic communities on marsh and coastal solonchaks. The Salicornioideae are represented by five species, but all five are important dominants in the whole area of the dry sea floor: *Salicornia europaea, Halocnemum strobilaceum, Kalidium caspicum, Kalidium foliatum, Halostachys caspica*. All five species are eu-halophytes and widespread in the whole Irano-Turanian region. The Salsoloideae are represented by many *Suaeda* and *Salsola* species. They are also often important components of distinct plant communities. On the dry sea floor eight *Suaeda* species are reported. The species *Suaeda crassifolia* and *S. salsa* are rarer and occur singly or in small groups on the marsh solonchaks. *Suaeda heterophylla* was found only once in the Bozkol bay on the east coast on the dry sea floor. The species *Suaeda acuminata* and *S. microsperma* prefer more crusty and crusty-puffy coastal solontchak stands. Both pairs of species form geographical and ecological vicariances. *Suaeda acuminata* and *S. salsa* are very widespread, they are Pontic–Mongolian–Irano–Turanian species; their centre of distribution is in steppes and cold deserts. In contrast, *Suaeda crassifolia* and *S. microsperma* are thermophilous Irano–Turanian desert species. This vicariance is visible also in the area of the Aral Sea. *Suaeda acuminata* occurs, for example, in the northern part of the dry sea floor. On the south and southeast coast it is replaced by *Suaeda microsperma*. The tribe Salsoleae is likewise marked by some peculiarities: the eight *Salsola* species (four therophytes and four shrubs and half-shrubs) prefer sandy soils and moderately saline stands. The *Climacoptera* species are found mostly on very salty crusty-puffy coastal solonchaks. The *Petrosimonia* species are widespread on the north coast. *Halimocnemis* species occur one by one on the dry sea floor of the 1960s and 1970s.

The plant-geographical analysis reveals that only a few widespread species are considerably involved in plant communities. From the 11 old mediterranean and south palearctian species, only *Salicornia europaea, Halocnemum strobilaceum* and *Bassia hyssopifolia* are widespread. Some 19 species are widespread in steppes and deserts with a distribution centre in the cold Anatolian, Kasakhian and Central Asian semideserts and deserts, and even spreading to the Pontic and West Siberian area. To this group belong *Atriplex verrucifera, Suaeda acuminata, Suaeda salsa, Kalidium foliatum, Ofaiston monandrum, Atriplex cana* etc. The remaining 40 species of the Chenopodiaceae are typical desert plants. Some of them are typical Irano–Turanian species and play an important role on the dry sea floor: *Kalidium caspicum, Halostachys caspica, Haloxylon ammodendron, Salsola arbuscula, Salsola nitraria, Salsola orientalis, Anabasis aphylla.* Strictly Turanian species are *Climacoptera aralensis, Halimocnemis longifolia* and *Arthrophytum lehmannianum.* The following species exhibit a north Turanian distribution: *Atriplex fominii, Atriplex pungens, Corispermum aralo-caspicum, Climacoptera affinis.* It is very obvious that the Chenopodiaceae form the broad basis of the very diverse halophyte flora on the dry sea floor of the Aral Sea, so to say the Aralkum.

Polygonaceae

The life form of shrubs dominates in this family (in comparison to the families Chenopodiaceae and Brassicaceae, where annuals prevail). The genus *Calligonum* (Colour Plate 1) consists of 30 species. It often forms plant communities on the eastern former coastline and on the desiccated sea floor of the 1960s. The taxonomy of the genus is still uncertain. There are four sections in the genus *Calligonum: Calliphysa, Pterococcus, Pterygobasis, Eucalligonum.* The latter three are represented on the dry sea floor. The section *Pterococcus* is the most controversial. Alexandrova and Soskov (1968) have combined 83 species from this section to only 4. Here, we use the concept of taxa according to the *Flora of Kazakhstan* (1956–1966). The distribution of *Calligonum* and the formation of plant communities is very important for fixation and stabilization of the moving sands on the dry sea floor areas.

Brassicaceae

The representatives of this family are widespread, especially in the area of the Small Aral Sea along the northern and western coasts. They grow in different ecological conditions, preferably with non-saline and weak saline upper horizons of soils. Most of the species are spring annuals. Only three species are perennials: *Lepidium latifolium, L. obtusum* and *Cardaria pubescens.* The genus *Lepidium* (Lepidieae) is represented by four species. *L. perfoliatum* is widespread in degraded coastal solonchaks in the whole area of the dried sea bed. *L.latifolium* and *L.obtusum* were found close to the present coast in the Small Aral Sea area. *Cardaria pubescens* is a typical synantropic species, growing in salt meadows in the area of the Butakov Bay. *Isatis* violascens (Sisymbrieae) often occurs in degraded coastal solonchaks of the northern coast. *Sameraria bidentata and Pachypterigium multicaule* are rare on the northern and eastern coasts of the Aral Sea. *Strigosella circinnata* and *S.africana* (Hesperideae) are widespread in the whole coastal area. *Euclidium syriacum* forms microphytocenoses on the foothills of Butakov Bay and on the south coast of Koktyrnak Peninsula. *Matthiola stoddartii* (Colour Plate 2) forms beautiful aggregations on the northern coast of the Aral Sea.

The species from the tribe Alysseae are typical elements of the zonal vegetation, and colonize the area of the exposed sea bed relatively slowly. *Meniocus linifolius* and *Alyssum turkestanicum* are widespread species. *Draba nemorosa* is one of the few species represented on the dry sea floor with a circumpolar distribution.

Asteraceae

One of the largest families in the world is represented by 28 species on the dry sea floor. The life-form diversity in comparison with other families is more diverse: semishrubs, perennials, spring and summer annuals, geophytes. The subfamily Cichorioideae is represented by *Chondrilla* and *Lactuca* species. The north Turanian–Dzhungarian *Chondrilla* species have a low coverage in plant communities; however, they are permanent components of sandy habitats of the northern part of the dry sea floor. The south Palearctic *Lactuca tatarica* and *L. serriola* are components of the saline meadow vegetation in delta areas or form aggregations on the *Zostera* carpets. *Artemisia* species are the most widespread in subfamily Asteroideae. The species of the section *Seriphidium*, which are dominant in zonal semidesert vegetation, have a restricted distribution in the Aralkum. Only north Turanian–Dzungarian *A. scopaeformis* and more rarely the *A. pauciflora* form plant communities or can be subdominant in the halophytic communities on coastal or takyr-like solonchaks. *A. arenaria*, from section *Dracunculus*, is present in all plant communities in the sandy area of the whole northern coast of the Aral Sea as well as in the Aralkum. In May, the late spring annual *Senecio noeanus* often occupies huge areas of the coastal solonchaks, creating an impressive green-yellow aspect. The Asteraceae in most cases are components of different plant communities but never dominate.

Poaceae

The Poaceae play a subordinate role on the dry sea floor, the Aralkum. However, some species are important for the understanding of the floristic and vegetational structure. The tribe Triticeae comprises five species. *Eremopyrum triticeum* and *E. orientale* have an Irano–Turanian distribution. For these species the new land is a suitable environment to be colonized. *Agropyron* species rarely occur in the northern and eastern coasts of the Aral Sea and on the island Barsa–Kelmes.

Leymus racemosus is widespread on the northern and eastern coasts of the Aral Sea especially in Butakov Bay. *Anisantha tectorum* (Bromeae) occurs by aggregations and single plants on *Zostera* carpets. The tribe Poaeae comprises *Poa* and *Puccinellia* species. *Puccinellia distans* is distributed in the whole area of the Aral Sea, *P. gigantea* only along the southern coast. *Phragmites australis* (Arundineae) forms thickets in Bozkol and Akkol bays and the delta areas of the Amu- and Syrdarya. Remnants of red are common on various parts of the dry sea floor, documenting huge reed areas during the retreat of water on shallow parts of the Aral Sea. *Stipagrostis pennata* (Aristideae) is the most widespread species of the psammophytic communities in a sandy belt around the Aral Sea. Hemicryptophytes and spring annuals are typical lifeforms of most grasses. The Poaceae form a basic composition of saline meadow vegetation (*Phragmites australis, Puccinellia* species, *Aeluropus littoralis*) and psammophytic communities (*Stipagrostis pennata, Leymus racemosus*).

Fabaceae

The representatives of this family occupy mostly sandy, less saline areas. *Ammodendron* species are widespread on the northern and northeastern coasts. Their occurrence near villages, e.g. Bugun, Karateren, Akespe, is caused by anthropogenic impact. *Eremosparton aphyllum* forms plant communities on the desiccated sea floor of the 1960s and the 1970s. *Astragalus brachypus* can occupy slightly saline as well as non-saline environments. It is one of the pioneer species growing on degraded coastal solonchaks of the 1970s and forms communities especially on the eastern Aralkum.

Other families

The Tamaricaceae are widespread in delta areas, where they form a tugai type of vegetation. Their distribution on the dry sea floor includes the areas from the 1960s, 1970s and partially1980s.

The Limoniaceae are represented by four species, three of which (*L. otolepis, L. gmelinii, L. suffruticosum*) occur around the whole Aral Sea.

The Nitrariaceae (closely related to Zygophyllaceae) are represented by *Nitraria schoberii* which forms a phytogenic type of relief (typical nepka) along the eastern coast on the exposed marine surface of the 1970s.

Conclusions

The flora of the dry sea floor (Aralkum) is an integrated floristic complex with 266 species. The dominance of the Chenopodiaceae is to be expected and is well documented. This family plays a leading role in the Irano–Turanian and Central

Asian deserts. The prevalence of Brassicaceae and Polygonaceae is obviously connected with the active fast-colonizing plant strategy of the species from these families. The contribution of Asteraceae and Fabaceae will increase with time. Thus, the flora of the dry sea floor of the Aral Sea is a new immigrant flora in a stage of transformation, steady new developments and high dynamic processes.

Acknowledgements. The research project was funded by BMBF (No. 0339714). This is very much appreciated as well as the help and advice of Dr. Joachim Kutscher. Thanks are due to Dr. Vefa Moustafaev (UNESCO/Paris) and to the local people in the villages in the area, in Aralsk, Saksaulsk, Kazalinsk and Kzylorda.

Thanks are due to Prof. Helmut Freitag and Dr. S. Rilke (Kassel) for advice and cooperation.

References

Alexandrova LA, Soskov YuD (1968) Khrosomosomnye chisla vidov roda *Calligonum* L. v *svyazi* s sistematikoi roda. Bot Zh 54:196–201

Bakhiev A (1985) Ekologiya i smena rastitel'nykh soobshchestv nizov'ev Amudar'yi. FAN, Tashkent

Borshchov IG (1865) Materialy dlya botanicheskoi geografii Aralo–Kaspiiskogo kraya . Prilozheniye k sed'momu tomu sapisok Imper. Akad Nauk Sankt–Petersburg

Breckle SW, Agachanjanz O, Wucherer W (1998) Der Aralsee: geoökologische Probleme. Naturwiss Rundsch 9:347–355

Cherepanov SK (1995) Sosudistye rasteniya SSSR. Nauka, Leningrad

Dimeyeva LA (1990) Flora i rastitelnost' poberezhii kazakhstanskoi chasti Aralskogo morya. Nauka, Almaty, 20 pp

Dimeyeva LA, Kuznetsov N (1999) Flora poberezii Aralskogo morya. Bot. Zhurn. 8:39–52

Erezhepov SE (1978) Flora Karakalpakii, selskokhosyaistvennaya charakteristika, ispolsovanie i okhrana. FAN, Tashkent

Flora of Kazakhstan (1956–1966) Vol 1–9

Hedge I, Akhani H, Freitag H, Kothe–Heinrich G, Podlech D, Rilke S, Uotila P (1997) Chenopodiaceae. In: Rechinger KH (ed) Flora Iranica 172. Akademische Druck- und Verlagsanstalt, Graz, Austria, 371 pp

Il'in MM (1936) Chenopodiaceae. In: Komarov VL (ed) Flora SSSR 6. Moskva, AN SSSR, Leningrad

Kuznetsov LA (1995) Flora of the island Barsa-Kelmes. In: Alimov AF, Aladin NV (eds) Biological and environmental problems of the Aral Sea and Aral region. Proc Zool Institute, Saint Petersburg, pp 106–128

Lavrenko EM (1962) Osnovnye cherty botanicheskoi geografii pustyn' Evrasii i severnoi Afriki. AN SSSR, Moskva

Litwinow D (1905) Rasteniya beregov Aralskogo morya. Izv Turk Russk Geogr Obshch 5: 1–41

Pratov U (1987) Marevye (Chenopodiaceae Vent.) Srednei Azii i Severnoi Afriki (Systematika, Filogeniya i botaniko-geograficheskii analiz). Kurzfassung der 2. Doktorarbeit. Leningrad

Sarybaev B (1982) Flora i rastitelnost' Vostochnogo chinka Ustyurta. FAN, Tashkent

Sherbaev B (1982) Sostav flory yuzhnogo poberezhya Aralskogo morya. Bot Zh 67:1372 –1377

Smirnov SM (1875) Botanicheskie issledovaniya v Aralo–Kaspiiskom krae. Izv Russk Geogr Obshch 11, 3:190–219

Takhtadzhan A (1987) Systema Magnoliofitov. Nauka, Leningrad

Appendix

List of species

Alliaceae J. Agardh. 1. *Allium sabulosum* Stev.

Amaryllidaceae Jaume 1. *Ixiolirion tataricum* (Pall.) Herb.

Apiaceae Lindl. 1. *Ferula caspica* Bieb., 2. *F.nuda* Spreng

Asclepiadaceae R. Br. 1. *Cynanchum sibiricum* Willd.

Asparagaceae Juss. 1. *Asparagus inderiensis* Blum ex Pasz., 2. *A.persicum* Baker

Asteraceae Dumort. 1. *Acroptilon repens* (L.) DC., 2. *Artemisia arenaria* DC., 3. *A.pauciflora* Web., 4. *A.santolina* Schrenk, 5. *A.semiarida* (Krasch. et Lavr.) Filat., 6. *A.scopaeformis* Ledeb., 7. *A.scoparia* Wldst. et Kit., 8. *A.songarica* Schrenk, 9. *A.terrae-albae* Krasch., 10. *Chondrilla ambigua* Fisch. ex Kar. et Kir., 11. *Ch.brevirostris* Fisch. et Mey., 12. *Cirsium arvense* (L.) Scop., 13. *Epilasia hemilasia* (Bunge) Clarke, 14. *Hyalea pulchella* (Ledeb.) C. Koch, 15. *Inula caspia* Blum, 16. *Karelinia caspia* (Pall.) Less., 17. *Koelpinia linearis* Pall., 18. *K.tenuissima* Pavl. et Lipsch., 19. *K.turanica* Vass., 20. *Lactuca serriola* L., 21. *Lactuca serriola* L., 22. *L.tatarica* (L.) C. A. Mey., 22. *Scorzonera pusilla* Pall., 23. *Senecio noeanus* Rupr., 24. *Sonchus oleraceus* L., 25. *Taraxacum bessarabicum* (Hornem) Hand.–Mazz., 26. *Tragopogon marginifolius* Pavl., 27. *T.sabulosus* Krasch. et S. Nikit, 28. *Tripolium vulgare* Nees.

Boraginaceae Juss. 1. *Argusia sibirica* (L.) Dandy, 2. *Heliotropium argusioides* Kar. et Kir., 3. *H.dasycarpum* Ledeb., 4. *Heterocarium subsessile* Vatke, 5. *Lappula semiglabra* (Ledeb.) Guerke, 6. *L.spinocarpos* (Forssk.) Aschers. O. Kuntze, 7. *Nonea caspica* (Willd.) G. Don., 8. *Suchtelenia calycina* (C. A. Mey) A. DC.

Brassicaceae Burnett 1. *Alyssum dasycarpum* Steph., 2. *A.turkestanicum* Regel et Schmalh., 3. *Cardaria pubescens* (C. A. Mey.) Yarm., 4. *Chorispora tenella*

Flora of the Dry Seafloor of the Aral Sea   49

(Pall.) DC., 5. *Descurainia sophia* (L.) Webb ex Prantl, 6. *Diptychocarpus strictus* (Fisch. ex Bieb.) Trautv., 7. *Draba nemorosa* L., 8. *Erysimum sysimbrioides* C. A. Mey., 9. *Euclidium syriacum* (L.) R. Br., 10. *Goldbachia laevigata* (Bieb.) DC., 11. *Isatis minima* Bge., 12. *I.violascens* Bge., 13. *Lachnoloma lehmannii* Bge., 14. *Lepidium latifolium* L., 15. *L.obtusum* Basin., 16. *L.perfoliatum* L., 17. *L.ruderale* L., 18. *Leptaleum filifolium* (Willd.) DC., 19. *Litwinowia tenuissima* (Pall.) Woronow ex Pavl., 20. *Matthiola stoddartii* Bge., 21. *Meniocus linifolius* (Steph.) DC., 22. *Octoceras lehmannianum* Bge., 23. *Pachypterigium multicaule* (Kar. et Kir.) Bge., 24. *Sameraria bidentata* Botsch., 25. *Strigosella africana africana* (L.) Botsch., 26. *S.circinnata* (Bunge) Botsch., 27. *S.turkestanica* (Litw.) Botsch., 28. *Syrenia montana* (Pall.) Clock., 29. *Tauscheria lasiocarpa* Fisch. ex DC., 30. *Tetracme quadricornis* (Steph.) Bge., 31. *T.recurvata* Bge.

Caryophyllaceae Juss. 1. *Gypsophyla paniculata* L., 2. *G. perfoliata* L., 3. *Silene nana* Kar.et Kir.

Chenopodiaceae Vent. 1. *Agriophyllum minus* Fisch. et Mey., 2. *A.squarrosum* (L.) Moq., 3. *Anabasis aphylla* L., 4. *Anabasis salsa* (C. A. Meyer) Benth. ex Volkens, 5. *Arthrophytum lehmannianum* Bge., 6. *Atriplex aucheri* Moq., 7. *A.cana* C. A. Mey., 8. *A.dimorphostegia* Kar. et Kir., 9. *A.fominii* Iljin, 10. *A.littoralis* L., 11. *A.micrantha* C. A. Meyer, 12. *A.pungens* Trautv., 13. *A.saggitata* Borkh., 14. *A.sphaeromorpha* Iljin, 15. *A.tatarica* L., 16. *A.verrucifera* Bieb., 17. *Bassia eriophora* (Schrad.) Aschers. 18. *B.hyssopifolia* (Pall.) O. Kuntze, 19. *B.sedoides* (Pall.) Aschers., 20. *Bienertia cycloptera* Bunge, 21. *Chenopodium glaucum* L., 22. *Ceratocarpus arenarius* L., 23. *Corispermum aralo-caspicum* Iljin, 24. *C.hyssopifolium* L., 25. *C.laxiflorum* Schrenk, 26. *C.lehmannianum* Bunge, 27. *C.orientale* Lam., 28. *Climacoptera affinis* (C. A. Meyer) Botsch., 29. *C. aralensis* (Iljin) Botsch., 30. *C.brachiata* (Pall.) Botsch. 31. *C.ferganica* (Drob.) Botsch., 32. *C.lanata* (Pall.) Botsch., 33. *Girgensohnia oppositiflora* (Pall.) Fenzl, 34. *Halimocnemis karelinii* Moq., 35. *H.longifolia* Bge., 36. *H.sclerosperma* (Pall.) C. A. Mey., 37. *Halocnemum strobilaceum* (Pall.) Bieb., 38. *Halogeton glomeratus* (Bieb.) C. A. Mey., 39. *Halostachys caspica* (Bieb.) C. A. Meyer, 40. *Halothamnus subaphyllus* (C. A. Mey.) Botsch., 41. *Haloxylon ammodendron* (C. A. Mey.) Bge. ex Fenzl [Syn. H. aphyllum (Minkw.) Iljin], 42. *Haloxylon persicu*m Bge. ex Boiss. & Buhse, 43. *Horaninovia anomala* (C. A. Mey.) Moq., 44. *H.minor* Schrenk, 45. *H.ulicina* Fisch. et Mey., 46. *Kalidium caspicum* (L.) Ung.– Sternb., 47. *K.foliatum* (Pallas) Moq., 48. *Kochia prostrata* (L.) Schrad., 49. *K. stellaris* Moq., 50. *Krascheninnikovia ceratoides* (L.) Gueldenst., 51. *Nanophytum erinaceum* (Pall.) Bge. 52. *Ofaiston monandrum* (Pall.) Moq., 53. *Petrosimonia brachiata* (Pall.) Bge., 54. *P.glaucescens* (Bge.) Iljin, 55. *P.sibirica* (Pall.) Bunge, 56. *P.squarrosa* (Schrenk) Bge., 57. *P.triandra* (Pall.) Simonk., 58. *Salicornia europaea* L. s. l., 59. *Salsola arbuscula* Pallas, 60. *S.australis* (R.) Br., 61. *S.dendroides* Pall., 62. *S.foliosa* (L.) Schrad., 63. *S.nitraria* Pall., 64. *S.orientalis* S. G. Gmel. 65. *S.paulsenii* Litv., 66. *S.richteri* (Moq.) Kar. ex Litv., 67. *S.tamariscina* Pall., 68. *Suaeda acuminata* (C.A. Mey.) Moq.,69. *S.altissima* (L.) Pall., 70. *S.crassifolia* Pall., 71. *S.heterophylla* (Kar. et Kir.) Bunge, 72. *S.microphylla* Pall., 73. *S.microsperma* (C. A. Mey.) Fenzl., 74. *S.physophora* Pall., 75. *S.salsa* (L.) Pall.

Convolvulaceae Juss 1. *Convolvulus subsericeus* 2. *C.erinaceus* Ledeb.

Cyperaceae Juss. 1. *Bolboschoenus maritimus* (L.) Pallas 2. *Carex pachystylis* J. Gay, 3. *C.physodes* Bieb. 4. *Scirpus lacustris* L., 5. *S.tabernaemontani* C. C. Gmel.

Elaeagnaceae Juss. 1. *Elaeagnus oxycarpa* Schlecht.

Ephedraceae Dumort. 1. *Ephedra distachya* L., 2. *E.intermedia* Schrenk et C. A. Mey., 3. *E.strobilacea* Bunge

Equisetaceae Rich. Ex DC. 1. *Equisetum ramosissimum* Desf.

Euphorbiaceae Juss. 1.*Euphorbia seguierana* Neck.

Fabaceae Lindl. 1. *Alhagi pseudalhagi* (Bieb.) Fisch., 2. *Ammodendron bifolium* (Pall.) Yakovl., 3. *A.conollyi* Bunge, 4. *A.karelinii* Fisch. et Mey., 5. *Astragalus amarus* Pall., 6. *A.ammodendron* Bunge, 7. *Astragalus testiculatus*, 8. *A.brachypus* Schrenk, 9. *A.lehmannianus* Bunge, 10. *A.vulpinus* Willd.,11. *Halimodendron halodendron* (Pall.) Voss., 12. *Eremosparton aphyllum* (Pall.) Fisch. et Mey., 13. *Glycyrrhiza glabra* L., 14. *Psedosophora alopecuroides* (L.) Sweet, 15. *Sphaerophysa salsula* (Pall.) DC., 16. *Trigonella arcuata* C. A. Mey., 17. *T.orthoceras* Kar. et Kir.

Frankeniaceae S. F. Gray 1. *Frankenia hirsuta* L.

Geraniaceae Juss 1. *Erodium oxyrhrinchum* Bieb.

Iridaceae Juss. 1. *Iris tenuifolia* Pall.

Juncaceae Juss. 1. *Juncus gerardii* Loisel.

Lamiaceae Lindl.1. *Chamaesphacos ilicifolius* Schrenk

Liliaceae Juss. 1. *Tulipa biflora* Pall., 2. *Rhinopetalum karelinii* Fisch. ex Alexand.

Limoniaceae Lincz. 1. *Limonium caspium* (Willd.) Gams., 2. *L.gmelinii* Willd. O. Kuntze, 3. *L.otolepis* (Schrenk) O. Kuntze, 4. *L.suffruticosum* (L.) O. Kuntze

Papaveraceae Juss. 1. *Hypecoum parviflorum* Kar. et Kir.

Poaceae Barnhart 1. *Aeluropus littoralis* (Gouan) Parl., 2. *Agropyron desertorum* (Fisch. ex Link) Schult., 3. *A.fragile* (Roth) P. Candargy, 4. *Anisantha tectorum* (L.) Nevski, 5. *Calamagrostis dubia* Bunge 6. *Catabrosella humilis* (Bieb.) Tzvel., 7. *Crypsis schoenoides* (L.) Lam., 8. *Eremopyrum orientale* (L.) Jaub. et Spach., 9. *Eremopyrum triticeum* (Gaertn.) Nevski, 10. *Leymus racemosus* (Lam.) Tzvel., 11.

Flora of the Dry Seafloor of the Aral Sea

Phragmites australis (Cav.) Trin. ex Steud., 12. *Poa bulbosa* L., 13. *Puccinellia distans* (Jacq.) Parl., 14. *P.gigantea* (Grossh.) Grossh., 15. *Schismus arabicus* Nees, 16. *Stipa caspia* C. Koch, 17. *S.sareptana* Beck., 18. *Stipagrostis pennata* (Trin.) de Winter

Polygonaceae Juss. 1. *Atraphaxis replicata* Lam., 2. *A.spinosa* L., 3. *Calligonum acanthopterum* Borszcz. 4. *C.alatiforme* Pavl., 5. *C.alatum* Litv., 6. *C.androssovii* Litv., 7. *C.aphyllum* (Pall.) Guerke, 8. *C. borszczowii* Litv., 9. *C.cancellatum* Mattei, 10. *C.caput-medusae* Schrenk, 11. *C.colubrinum* Borszcz., 12. *C.commune* (Litv.) Mattei, 13. *C.cristatum* Pavl., 14. *C.densum* Borszcz., 15. *C.dubjanskyi* Litv., 16. *C.elatum* Litv., 17. *C.erinaceum* Borszcz., 18. *C.humile* Litv., 19. *C.lamellatum* (Litv.) Mattei, 20. *C.leucocladum* (Schrenk) Bunge, 21. *C.macrocarpum* Borszcz., 22. *C.membranaceum* (Borszcz.) Litv., 23. *C.microcarpum* Borszcz., 24. *C.minimum* Lipsky, 25. *C.muravljanskyi* Pavl., 26. *C.palibinii* Mattei, 27. *C.platyacanthum* Borszcz., 28. *C.pseudohumile* Drob., 29. *C.rotula Borszcz.*, 30. *C.rubens* Mattei, 31. *C.squarrosum* Pavl., 32. *C.undulatum* Litv., 33. *Polygonum arenarium* Waldst. Ed Scit., 34. *P.monspeliense* Thieb. ex Pers., 35. *Rheum tataricum* L., 36. *Rumex marschallianus* Reichenb.

Ranunculaceae Juss 1. *Ceratocephala falcata* (L.) Pers.

Rutaceae Juss 1. *Haplophyllum perforatum* Kar. et Kir.

Scrophulariaceae Juss. 1. *Linaria dolichocarpa* Klok

Solanaceae Juss. 1. *Hyosciamus pusillus* L., 2. *Lycium ruthenicum* Murr.

Tamaricaceae Link. 1. *Tamarix elongata* Ledeb., 2. *T.hispida* Willd, 3. *T.hohenackeri*, 4. *T.karelinii* Bunge, 5. *T.laxa* Willd., 6. *T.litwiniwii* Gorschk., 7. *T.ramosissima* Ledeb.

Typhaceae Juss. 1. *Typha angustifolia* L.

Zygopyllaceae R. Br. 1. *Nitraria schoberi* L., 2. *Peganum harmala* L., 3. *Zygophyllum fabago* L., 4. *Zygopyllum eichwaldii* C. A. Mey., 5. *Zygophyllum oxianum* Boriss.

Vegetation Dynamics on the Dry Sea Floor of the Aral Sea

Walter Wucherer and Siegmar–W. Breckle

Keywords. Primary succession, succession stages, vegetation development

Abstract. The dry sea floor of the Aral sea is a new terrestrial surface, where plants (including seedbank) and animals have not existed before, and it is actively populated by organisms. The dry sea floor (Aralkum desert) is the largest area worldwide where a primary succession takes place. Unintentionally, mankind has created a huge experiment, an experimental set, a laboratory of nature with thousands of local events. The new knowledge on vegetation dynamics in the Aralkum desert, which is a mosaic of sand and salt desert ecosystems, is very important for the understanding of the ecosystem dynamics in the whole Central Asian area. The succession on the dry sea floor has continued for 40 years. We are able to determine the age of the drying and of the ecosystems. This is important to identify major mechanisms that determine the rate and direction of ecosystem changes on the dry sea floor. The distribution and dynamics of the vegetation and ecosystems were surveyed along some new transects. On the dry sea floor barchan and salt deserts have spread. The present and future development of the drying sea is characterized by the creation of salt desert flats. Along the former coast line the inhabitants of adjacent villages are using the dry sea floor more and more as grazing area and for hay production.

Introduction

The Aral Sea, in 1960 the fourth largest sea on the globe, is in a critical process of drying out. The development of separate smaller independent water bodies is now a reality. A new dry surface area of about 40 000 km^2 is exposed. This is the new desert called Aralkum. The dry sea floor is a new area, where land plants (including seedbank) and animals have not existed before. It is actively invaded by organisms. The formation of plant communities, soils, a new groundwater level, aquifers, all components and processes of ecosystems is occurring simultaneously. The dynamics on the drying sea floor is unique. The dry sea floor is the largest area worldwide where a primary succession is taking place. For our purposes here we will define succession as the replacement of one ecosystem by another or the colonization of the new surface. Unintentionally, mankind has created a huge experiment, an experimental set, a laboratory of nature with thousand of local events with an open end. There are two questions to ask:

- What is this new desert?
- What does this mean for the people in the Aral Sea Region?

In 1977 a complex research program was started in Central Asia to study the negative effects of the drying process of the Aral Sea and the change in the discharge of the Amu- and Syrdarya (under the leadership of the Geographical Institutes of Moscow). Transects were set up at different coastal areas, stretching from the former coast to the present coastline. The distribution of the soils and the vegetation was surveyed along these transects. One of the authors had participated in this program (Wucherer 1979, 1984, 1990; Kabulov 1990). This program was stopped in the mid 1980s. At present, the ecological situation on the dry sea floor is different compared to 15 years ago:

- The diversity of landscapes and plant communities has increased enormously.
- Barchanes and salt deserts spread out on the dry sea floor.
- Salt dust is blown into the atmosphere and hundreds of km into adjacent areas.

In 1990 the UNESCO program started with special emphasis on the delta areas. The problems of the Aralkum desert itself have not been touched since the middle of the 1980s. Therefore a new research program as the logical continuation of the BMBF and UNESCO initiatives in the Aral Sea region resulted in an international BMBF project: Successional Processes on the Dry Aral Sea Floor and Perspective of Land Use. This project has two important aims:

- The study of the ecosystem dynamics in the Aralkum desert. The research and evaluation of the ongoing processes and a prognosis for future scenarios are vital for this area.
- What measures can be taken to accelerate the natural colonization by plants? Are experimental plantings on the dry sea floor successful?

At present, the dry sea floor of the Aral Sea is a huge salt flat. According to several estimations, it is the source of many million tons of salt and dust blown out by wind annually and transported to rather distant adjacent areas with irrigated fields and settlements. The present and future development of the drying sea is characterized by the creation of salt desert flats. It is high time for strict measures to minimize the salt dust output.

Studies along Transects

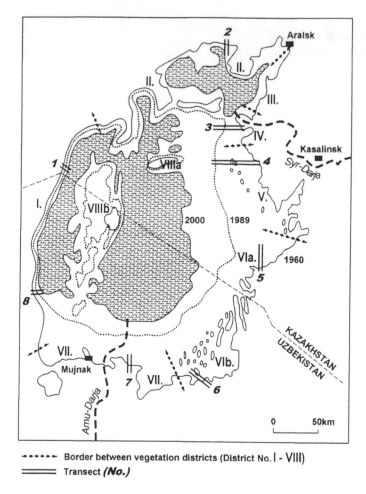

••••• Border between vegetation districts (District No. I - VIII)
═══ Transect *(No.)*

Fig. 1. Transects, vegetation and succession districts on the coast of the Aral Sea

The new knowledge of vegetation dynamics in the Aralkum desert, which is a mosaic of sand and salt desert ecosystems, is very important for the understanding of the ecosystem dynamics in the whole Central Asian area. The dry sea floor is a huge open area, open for invasion by plants and animals. For such an invasion process two different terms were proposed: autotropic succession (Begon et al. 1986) and syngenetic succession (Sukachev 1938, 1954). Sukachev emphasizes syngenesis as being the main mechanism for a succession, driven by the vitality and competitive force of the plants. Certainly, succession is a complex process, interrelated with the exogenesis, endoexogenesis and syngenesis of plant communities. Succession is a process controlled by tolerance, inhibition and other successional mechanisms (Drury and Nisbet 1973; Connell and Slatyer 1977; Glenn–Lewin et al. 1992). In the first few years after the drying process of the sea floor, the exogenic factors dominate the ecosystem dynamics. The study of the mechanisms of the ecosystems dynamics and the ecological attributes of the dominant species is of great importance for clarifying the following open questions:

• Which geological, geomorphological and edaphical processes are affecting the present ecosystem development?
• Which mechanisms are governing the development of salt desert and sand desert (barchanes)?
• Will there be an ecological limitation to colonization by plants, and where it will be?

Figure 1 surveys the distribution of the old and new transects for studying the dynamics of the vegetation and ecosystems. The transects stretch from the old coast line to the present one. Along the transects, groundwater level, distribution of soils and vegetation types were studied. The long-term transects were set up at the end of the 1970s and at the beginning of the 1980s. The data set now covers 19–23 years. The main vegetation types are: halophytic, psammophytic, tugai and salt meadow communties. The halophytic plant communities are formed by eu- and hemihalophytic vegetation on saline soils. Plant species are mainly shrubs, semishrubs, perennials and annuals from the Chenopodiaceae (*Halostachys, Halocnemum, Salicornia, Suaeda*), some Tamaricaceae, Limoniaceae and Zygophyllaceae. The cover percentage is often rather high, 10–100%. The great variability of this vegetation is a matter of the very varying salinity and water availability. The typical soils are marsh and coastal solonchaks, degraded and sandy coastal solonchaks as well as takyr-like solonchaks. The halophytic vegetation is present on the dry sea floor of the 1970s, 1980s and 1990s. The psammophytic vegetation consists of eu- and mesoxerophytic shrub and semishrub vegetation on desalinized sands. The specific canopy of these plant communties is determined by vigorous species of *Haloxylon* and *Salsola* (Chenopodiaceae), *Calligonum* (Polygonaceae) and some species of Fabaceae and Poaceae. The main canopy is about 0.5–2.0 m but can reach 5 m. The plant cover percentage is about 20–60%. The sandy soils are preliminary with no or only very slight formation of horizons. Wind erosion causes the development of a complicated relief. The psammophytic vegetation is most dominant on the dry surface areas from the 1960s, partly from the 1970s and rare from the 1980s. The typical tugai biome is represented by the shrubs and small woods of the delta regions and lake bays. The characteristic species of this vegetation type are mainly the *Populus* species from the subgenus *Turanga, Elaeagnus oxycarpa* and shrubs from the Tamaricaceae. They depend on available ground-water and are only slightly resistant to salinity. The salt meadow vegetation is mainly composed of reed communities with many perennial hemicryptophytes (*Puccinellia* and *Limonium* species, *Aeluropus littoralis, Karelinia caspica* etc.), which also can withstand soil salinity. The typical pattern of landscapes, vegetation types and soils is striated. This banded pattern is most characteristically seen along the east coast of the Aral Sea (Ishankulov and Wucherer 1984).

Factors Influencing Vegetation Development

Primary succession can be considered from the viewpoint of time, habitat and the plants (Bradshaw 1993).

Time and Age

The succession on the dry sea floor has been active for about 40 years. The time factor plays no active role, but influences indirectly the vegetation development. Species which arrived rather late are often not able to establish themselves in reasonable numbers. Either the ecological conditions have changed and became unfavourable or other species have already occupied the relevant ecological niche. The momentaneous stabilization of the sea level of the Small Aral Sea has demonstrated that a whole set of species has established on the coastal solonchaks very rapidly, especially under conditions of a stabilized groundwater level. The change in the ecological factors leads to a change in colonization rate and thus to a change in the time sequence of the succession. We know the annual sea water level since 1961, so that we are able to determine almost exactly the age of the drying process of a distinct portion of the sea floor and thus the development of the relevant ecosystems. This is important to identify major mechanisms that determine the rate and direction of ecosystem changes on the dry sea floor.

Plant Sources

On the dry sea floor alone, to date 266 plant species have been reported (Wucherer et al., see this volume). The precondition for colonization by plants is the surrounding flora and its seed production. It is mainly the flora of the old Aral terraces along the old coast line and the adjacent mainland all around the Aral Sea. The given scheme (Fig. 1) indicates the main directions and specialities of the vegetation development on the dry sea floor. It depends on the geomorphological and landscape pattern as well as on the pattern of the vegetation units along the coast (Lymarev 1969; Ishankulov 1980; Kurochkina et.al. 1983). The districts I–II (Fig.1) on the north and west coasts of the Aral Sea lack the old Aral terraces almost completely, or these are very narrow belts (only 10–20 m up to some 100 m). The azonal flora of the Aral terraces and the steep slopes of the Chinks are the main sources for the seed bank (Colour Plate 3). The zonal plant communities of the former mainland are hardly relevant as a source for colonization with plants and for formation of vegetation units on the dry sea floor. It is remarkable that on the dry sea floor area from the 1960s there is an essential proportion of Brassicaceae and Chenopodiaceae but a sparse representation of Fabaceae and Polygonaceae. In the districts III and VI, the belt of the Aral terraces is continuous and relatively broad (some 100 m to 3–5 km). The azonal vegetation of the Aral terraces and the zonal psammophytic plant communities of the sand deserts Kyzylkum and Priaralskii Karakum are the sources for the plant species for the dry sea floor. The Polygonaceae and Fabaceae are significantly richer on those areas of the dry sea floor from the 1960s and 1970s than the Brassicaceae and Chenopodiaceae. In the delta districts IV and VII the irregular floods of the rivers are an important factor. These inundations led to an extensive distribution of seeds and an activation of the seed bank already present on the dry sea floor. Mainly the Tamaricaceae, Limoniaceae, Asteraceae and Chenopodiaceae are conspicuous. Within district V only the halophytic communities and the zonal communities (*Artemisia terrae-albae, Anabasis salsa, Haloxylon aphyllum*) are present on the Aral terraces. Here, mainly Chenopodiaceae dominate the dry sea floor from the 1960s and 1970s. Within the island district VIII the belt of the Aral terraces is interrupted. Again, here the zonal vegetation units are of no importance for the colonization processes. The flora of the Aral terraces and of the distinctly lower Chinks is poor in comparison with the mainland. The composition of flora of the former coast is different in each district and influences the course of the succession on the dry sea floor.

Exogenic Factors

These are the sea and river floods, the geological and geomorphological, climatical, edaphical and eolian factors. The areas adjacent to the coastline are subject to inundations. The pattern of sedimentation, as well as the former geological history of the Aral Sea determine the particle size distribution and sedimentation layers of the new soils of the dry sea floor. The geomorphological structure of the Aral Sea basin is complicated. Plains predominate in the eastern part of the depression with an inclination of 0.2° to 0.6°. Therefore the present coast line is situated up to 100 km away from the former eastern coastline. On the west coast, between the plateau Ustyurt and the islands Barsa–Kelmes and Vozrozhdenie, the inclination of the plain is steeper and amounts to 2°–5°. Correspondingly, the dry sea floor belt is only 4–10 km wide.

The new alluvial deposits of the retreating Aral Sea cover 1–6-m older layers. The salinization of the substrate varies to a great extent, causing a wide variety of saline soil types: marshy solonchaks, crusty and puffy solonchaks, solonchaks slightly covered by sand, degraded coastal solonchaks, takyr solonchaks and takyr soils, sandy soils.

On sandy or sandy loamy soils deflation of soil particles takes place. Barchans or complexes of barchans and salt deserts develop. They are widespread on the east coast. Since the middle of the 1980s, open salt deserts have developed on a large scale. Some plant populations have decreased according to the salt desert formation. The fast increase of the salt desert areas, the changing of soil texture and increase in salt content of the soils has caused the absolute dominance of halophytes as pioneer plants, mainly species from the Chenopodiaceae, to the exclusion of most other life forms.

Succession

The development of vegetation units is different on sandy and loamy or even clay soils. Some typical examples will be mentioned.

Succession on Loamy Soils (Fig. 2)

On marsh solonchaks at the present coast line (stage I), dominant plant communities are found with the following species: *Salicornia europaea, Suaeda crassifolia* and *Tripolium vulgare (Aster tripolium)*. These are pioneer plants that directly follow the retreating sea water level. During the vegetation period, this area is often flooded, creating ecological conditions favourable for the development of annuals (Colour Plate 4). The plants normally are only 30–50 cm high. Figure 3 illustrates the development of the *Salicornia* unit on the marsh solonchaks from year to year on the Bajan transect on the east coast. The density of the plant population varies tremendously from year to year. When the water level sinks, the pioneer plants die and the coastal solonchaks develop (stage II). The crusty and puffy coastal solonchaks are formed when the groundwater level goes down to about 1–2 m and no further inundations occur. The strong capillary upwards movement of water leads to an extremely strong salinization of the upper most soil surface and the formation of a salt crust, containing up to 16–20% of salt. The ecological conditions for colonization and establishment of plant species are then very severe in comparison with marsh solonchaks.

Only few species are able to colonize such stands. These are the pioneer plants of the second generation (Colour Plate 5). colonizing the dry sea floor (*Climacoptera aralensis, Petrosimonia triandra, Bassia hyss*opifolia) or developing successional salt deserts. Those successional deserts can be defined as stands, where for lack of a seed bank, colonization by plants has not yet taken place. Thus, during further succession salt deserts are often formed, and can exist for 10–20 years and more. They are the main area and source of the salt dust output from the dry sea floor. The third stage of succession is a loose mixture of annuals and perennials like *Halocnemum strobilaceum, Halostachys caspica* etc. on degraded solonchaks.

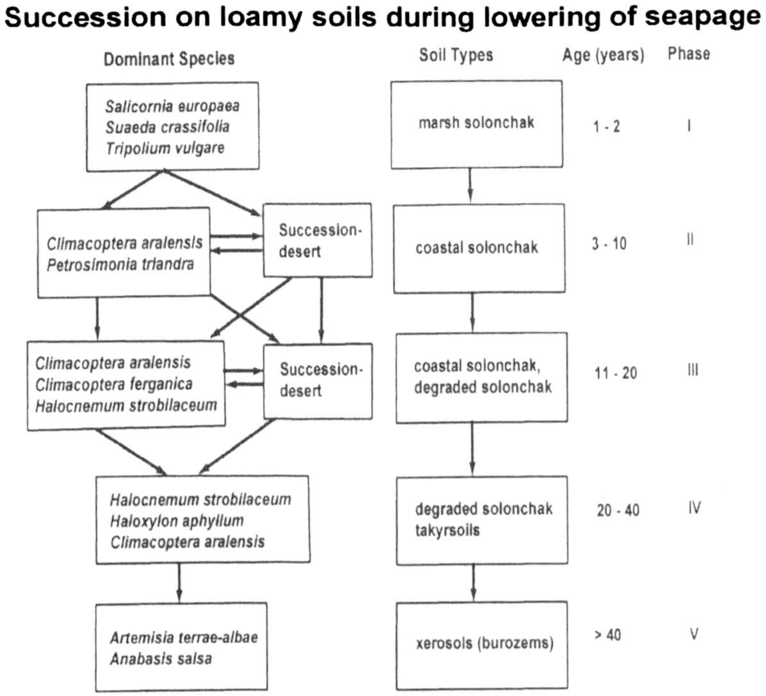

Succession on loamy soils during lowering of seapage
Dominant Species
Soil Types
Age (years)
Phase
Salicornia europaea
Suaeda crassifolia
Tripolium vulgare
marsh solonchak
1 - 2
I
Climacoptera aralensis
Petrosimonia triandra
Succession-desert
coastal solonchak
3 - 10
II
Climacoptera aralensis
Climacoptera ferganica
Halocnemum strobilaceum
Succession-desert
coastal solonchak, degraded solonchak
11 - 20
III
Halocnemum strobilaceum
Haloxylon aphyllum
Climacoptera aralensis
degraded solonchak takyrsoils
20 - 40
IV
Artemisia terrae-albae
Anabasis salsa
xerosols (burozems)
> 40
V

Fig. 2. The main direction of succession on loamy soils

Cover (%)
Number of species
Cover (%)
Number of species
0
20
40
60
80
100
0
1
2
3
4
5
6
1981
1982
1983
1984
1986
1987
1989
1990
1994

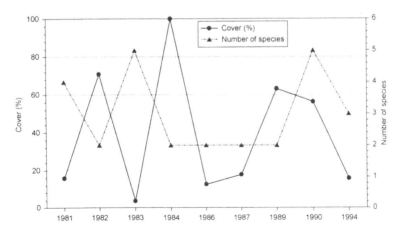

Fig. 3. Development of *Salicornia europaea* community on the marshy solonchaks from year to year (transect Bayan)

Degraded coastal solonchaks develop from coastal solonchaks by further lowering of the groundwater table and thus drying of the surface. The capillary threads no longer reach the surface, but end below the surface at a depth of about 10–30 cm. With time communities with a cover percentage up to 20% and more are formed (stage IV). On the Kaskakulan transect, permanent communities of *Halocnemum strobilaceum* and *Haloxylon aphyllum* are widespread throughout the east coast (Fig. 4).

Fig. 4. Permanent communities of *Haloxylon aphyllum* with some *Halocnemum strobilaceum* (stage IV) on the dry sea floor at the east coast of the Aral Sea

Succession on Sandy Soils (Fig. 5)

The psammoseries of vegetation are most perfectly developed around the islands of Barsa–Kelmes and on the southeast coast of the Aral Sea. The soil profile is sandy up to at least 2 m in depth. The succession is influenced also by the sinking groundwater level as well as by the desalinization by means of a good water percolation along the soil profile. On the marsh solonchaks (stage I) the floristic composition is slightly different. Together with *Salicornia europaea*, *Suaeda acuminata* and *Atriplex fominii* form pioneer units. On the coastal solonchaks (stage II) the species compositions and the vegetation units are richer and more variable. Already during this stage deflation events start. The nanorelief with small hills (5–10 cm high) is already distinct. Plants accumulate sand around them and the small hills increase (15–20 cm high). These stands are optimal for the development of *Atriplex fominii* units. Further decrease of the groundwater table to 1.2–1.7 m and desalinization of the top soil (down to 50 –60 cm depth) is the precondition for the first colonization with perennial psammophytes (stage III, Fig. 6) an. With time psammophyte communities with *Stipagrostis pennata, Calligonum* and *Astragalus* species thus develop on the sandy soils.

Succession on sandy soils
during lowering of seapage and increasing eolian activity

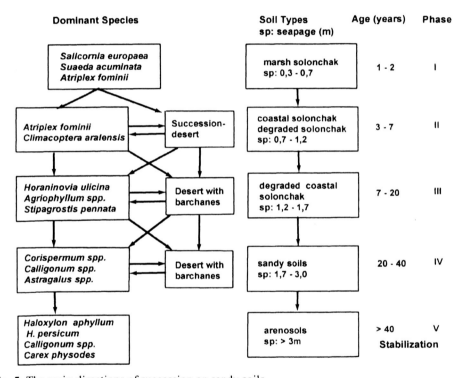

Fig. 5. The main directions of succession on sandy soils

On average, the succession on loamy stands can be described by two to four stages, and on sandy soils by three to five stages. The existence of a distinct stage is a matter of the ecological conditions and their lability or stability, and thus might vary between 2 to 30 years. In the course of further drying out, ecosystems with new characteristics arise: the granulometric composition, salt content and microrelief may be different. Particularly striking is the contrast between the dry areas of the 1960s and 1970s and those of the 1980s and 1990s. Accordingly, the temporal change of the vegetation and of the soils do not match the spatial position of plant societies.

Fig. 6. Beginning of first colonization with perennial psammophytes of *Stipagrostis pennata* on sandy soils (stage III)

Fragments of the Vegetation Development around the Small Aral Sea

In 1998 the dry sea floor around the Small Aral Sea (12 transects) was surveyed. The new hydrological situation, in contrast to that of the Great Aral Sea, has influenced the colonization of the dry sea floor in two ways. The therophytic vegetation with *Salicornia europaea* and *Suaeda* species was inundated and is almost lacking now on most parts of the present coastline because of the rapid rise of the water level. On some coastal parts also perennial and woody vegetation was inundated. First observations show that the resistance of the various species to inundation varies greatly.

The Pioneer-Species

Pioneer species like *Salicornia europaea, Suaeda crassifolia, Suaeda salsa* and *Suaeda acuminata*, which form plant associations close to the coast line, entered a new ecological situation. With the rise of the water level the therophytic belt also drifts backwards to the older former coastline. The rise of the water level occurs so rapidly that some pioneer species from the first colonization wave simply disappear. The distribution of the pioneer species around the Small Aral Sea is very interesting (Table 1).

Table 1. Distribution of the pioneer species around the Small Aral Sea arranged according to successional sequence of localities

Locality/ Species	1	2	3	4	5	6	7	8	9	10	11	12
Salicornia europaea	++	++	++	++	++	++	++	+				
Suaeda acuminata	++	++	++	+	-	+	-	+				
Suaeda crassifolia	+	+	+	++	+	+	+	+				
Suaeda salsa				+	-	-	-	-				
Petrosimonia triandra					-	+	+		+	+	+	+
Climacoptera aralensis									+	++	-	+
Atriplex fominii									-	+	+	+
Phragmites australis									+	-	+	-

++ dominant species; + sparse; - missing

The small bights (Butakov, Dzhideli, Table 1: 1–4), where wave activity is low and salinization slightly enhanced, exhibit a well-developed plant cover with typical pioneer species like *Salicornia europaea*. Around the larger, more open bights (Shevchenko, Table 1: 5–8) the typical pioneer species are only subdominant or members in other plant associations. The third group (Table 1: 9–12) are the open coast lines, where the halosucculent therophytes, like *Salicornia* and *Suaeda* species have disappeared and are replaced by pioneers from the second generation (*Climacoptera* and *Petrosimonia* species). On the remaining dry sea floor of the Small Aral Sea conditions are favourable for a second flush of therophytes (with several *Climacoptera, Salsola* and *Petrosimonia* species) as well as for some woody species from the tugai vegetation. The overlapping of the ecological amplitudes of several species seems to have become wider. This may result in future in more stochastic processes during formation of vegetation units and a higher interspecific competition.

Land Use

On the northeast coast of the Butakov bight (Table 1, transects 3, 8) the small village of Tastjubek and the camel farm Akkuduk are located. The camel flock consists of 50 animals. Along this coast line the inhabitants are using the dry sea floor as grazing area and for hay production. The annual *Atriplex* and *Suaeda* species are ideal, monotonous salty meadows. The salinization of the upper soil surface, grazing mowing are prevent the perennial species from spreading out.

Ecological Peculiarities of the Plant Colonization of the Dry Sea Floor

Flora

The dry sea floor and the coastline of the Aral Sea are one of the centres of diversity for Chenopodiaceae and Polygonaceae (Calligonum) in Central Asia.

The Chenopodiaceae are a cosmopolitan plant family. They play a special role in the arid areas of Europe and Asia. This family determines the appearance of the vegetation cover in the Central Asiatic deserts and on the dry sea floor. At present, about 75 species of Chenopodiaceae and about 30 species of *Calligonum* are recorded from the dry sea floor (Wucherer, in this volume).

Expansion of Plant Species

Hundreds of species had the opportunity to fill out or expand their distributional area. The primary seed bank is on the Aral terraces at the former coastline. The second seed bank is being formed today on the dry sea floor. By nivellements, with the help of bathymetric maps and GPS measurements, we were able to determine relative precisely the age of the relevant dry sea floor stand and thus the speed of colonization by various species. The fastest are the halosucculent therophytes which directly follow the retreating coast line, at a speed of about 70 km within 30 years.

Unique Plant Communities and Ecosystems

There are many examples of unique composition by various species. We will discuss a few examples from the psammophytic, halophytic and hygrophytic units. On the sandy substrate around the island Barsa–Kelmes and at the south-eastern coast a plant unit either with *Stipagrostis pennata* and *Halocnemum strobilaceum* or *Stipagrostis pennata* and *Phragmites australis* (Colour Plate 6) can often be found. Such a combination is peculiar and needs explanation. The sandy coastal solonchaks first are colonized by *Halocnemum strobilaceum*. With the further decrease of the ground-water table, the upper sandy layer over 8–10 years becomes slowly desalinized down to about 1 m depth. The main active roots of

Halocnemum strobilaceum are at a depth of 1–2 m. *Stipagrostis pennata*, however, has main roots in the upper sandy soil. This vegetation unit is an intermediate vegetation type. With the stabilization of the sea water level, it might be possible for such vegetation unit to become rather permanent.

Development of Monotonous Widespread Vegetation Units

The dry sea floor of the Aral Sea is a huge open flat plane. This is a perfect condition for wide-ranging plant dispersal. The therophytes may cover hundreds of km^2 within a very short period of less than 2 or 3 years. *Petrosimonia triandra* is a very typical anemochorous plant; it is cut off at the base by wind and then the whole plant is dispersed like the steppe runners by wind. Very dense stands were formed by *Petrosimonia triandra* in 1998 on the northeast coast; *Atriplex fominii* was extremely dense in 1994 on the southwest coast. As a rule, those units are often poor in species and over vast distances only three to seven species are found. The perennials can also form uniform, monotonous stands. *Halocnemum strobilaceum*, as a succulent chamaephyte, forms extensive stands along the east coast of the Aral Sea. Such units are ideal for studies on the genetic and biological variability of these plant populations.

Development of New Ecological Attributes of the Species

The variability of the stands on the dry sea floor is, in comparison with the adjacent mainland, much higher. There are various sandy, loamy and clayey soils, also with rather varying horizontation. The salt content varies over a wide range between 0.3 and 20% of dry matter. The groundwater table lies between 1 and 5 m depth. Atypical ecological conditions may force plant species to specific adaptations. As an example, the annual *Salicornia* species normally develop a root system not deeper than 30–40 cm. The ecological conditions of the marsh solonchaks of the dry sea floor may change very fast. During early spring the tiny seeds of *Salicornia* germinate rather close to the water level or in standing water, especially after a rainy day. The groundwater level is less than 20 cm deep. However, during the vegetation period it may recede rapidly and the groundwater level may go down to about 2–2.5 m depth at these stands and, accordingly, the water availability in the top soil decreases. Since the density of plants is extremely high, there is a water deficiency quite soon. The reaction of the strongest individuals of *Salicornia* is, accordingly, to produce roots which follow the capillary availability of water down to 1 m. The maximum root length found was about to be 1.06 m.

Because of the variability of stands with extreme conditions, the dry sea floor is an ecological challenge even for halophytes. Due to such habitat variability and rapid changes it is possible to see and to become familiar with the ecological amplitudes and the overlapping of species adaptations as well as the ecological limits of distinct species in a better way.

Unexpected Combinations and Interrelationships

Species which under natural conditions cannot be found together with other species despite an apparently similar ecological behaviour, here form new vegetation units. As an example, *Climacoptera aralensis* occurs on crusty and puffy solonchaks as well as on secondary solochaks in fields in the Kzyl–Orda district. It forms isolated, monotonous units, sometimes together with a few salt meadow halophytes. *Petrosimonia triandra* is distributed very locally only and prefers moderate marsh conditions or slightly saline meadows. Both species do not occur together at the same localities. However, on the dry sea floor they meet and form extensive stands. *Petrosimonia triandra* is dispersed faster and colonizes the coastal solonchaks first. One to two years later, *Climacoptera aralensis* follows. Due to the winged fruits the latter species has a high potential for wide range dispersal, but in comparison with *Petrosimonia* remains always only subdominant. Both species form mixed stands with a high cover percentage of about 60–80%. If salinity of the top soil increases, then *Petrosimonia* frequency decreases, but *Climacoptera* remains. The ecological range of both species apparently overlapps broadly, but some of their life strategies are different. *Petrosimonia triandra* has a more rapid dispersal ability, *Climacoptera aralensis* is more salt-resistant.

Contrasting Fluctuating Dynamics

The great annual variability in cover percentages of spring annuals in deserts is well known. In most cases in the Aral Sea floor we have halophytic therophytes which have a life cycle of about 6–8 months. Again, their density varies tremendously from year to year. The greatest contrast could be observed between the years 1981 and 1982 and 1998 and 1999. The first case is described from the north coast of the Aral Sea (transect Karabulak) and was mapped by Wucherer and Galieva (1985). The second case will be illustrated by an example from the transect Bayan. In 1998 along the transect therophyte units were developed with a percentage cover of about 80–100% with many species, mainly *Suaeda acuminata, Suaeda crassifolia, Petrosimonia triandra, Climacoptera aralensis* etc. One year later on the very same areas an absolutely barren salt desert could be found without any plants. The soil surface, however, was covered by an incredible amount of seeds, which had no chance to germinate. About 30–450 seeds per 100 cm^2 on the surface of the soil below dead *Suaeda acuminata* plants from the previous year could be counted. The reasons for such a regressive dynamics are mainly too dry spring months and too warm winters, poor in snow.

Conclusions

The successional processes on the dry sea floor are diverse and peculiar. The geomorphological, sedimentological and pedological factors determine the vegetation and succession patterns. The biological characteristics (plant strategy) of the plants are also very important for colonization of the dry sea floor. By the start of the succession, especially in the first 5 years, abiotical factors control the development of vegetation. Later, biotical factors intervene. The importance of facilitation and inhibition in early succession is only generally known. To what degree is physical amelioration required for invasion? The ecological situation on the dry sea floor is very changeable and the dry sea floor is a large dynamic ecosystem complex. The development of vegetation in the area of the Small Aral Sea is different from the Great Aral Sea. The actual trend in dry sea floor characteristics is the formation of salt deserts and spreading of halophyte vegetation. Long-term studies on vegetation development on the dry sea floor will contribute greatly to our understanding of primary succession and ecosystem dynamics in Central Asia.

Acknowledgements. This research is funded by the German Federal Ministry for Education and Research (BMBF grant 0339714).

References

Begon M, Harper JL, Townsend CR (1986) Ecology. Individuals, populations and communities. vols. 1–2. Blackwell, Oxford

Bradshaw AD (1993) Introduction: understanding the fundamentals of succession. In: Miles J, Walton DWH (eds) Primary succession on land. Blackwell, Oxford, pp 1–3

Connell JH, Slatyer RO (1977) Mechanisms of succession in natural communities and their role in community stability and organisation. Am Nat 111, 982: 1119–1144

Drury WH, Nisbet ITC (1973) Succession. J Arnold Arbor Harv Univ 54, 3: 331–368

Glenn–Lewin DC, Peet RK, Veblen TT (1992) Plant succession. Theory and prediction. Chapman & Hall, Cambridge

Ishankulov MSh (1980) The classification of the landscapes of the actual coasts of the Aral Sea. Probl Osvo Pustyn 5: 18–24 (in Russian)

Ishankulov MSh, Wucherer W (1984) Nature complexes of the east coast of the Aral Sea in 1982. Probl Osvo Pustyn 1, 52–58 (in Russian)

Kabulov SK (1990) Izmenenie fitotsenozov pustyn pri aridizatsii. FAN, Tashkent (in Russian)

Kurochkina LYa (1979) Botanical survey in the Aral Sea basin. Probl Osvo Pustyn 2: 25–33 (in Russian)

Kurochkina LYa, Makulbekova GB, Wucherer W, Malaisarova A (1983) Vegetation of the dry sea floor of the Aral Sea. In: Aralskoe more. Nauka, Alma–Ata, pp 91–144 (in Russian)

Lymarev VI (1969) The coasts of the Aral Sea. Nauka, Moskva (in Russian)

Sukachev VN (1938) The main definitions in the vegetation science. In: Rastitelnost' SSSR, vol 1. AN SSSR Moskva, pp 15–37 (in Russian)

Sukachev VN (1954) Some general theoretical problems in vegetation science. In: Problems of botany, vol 1. AN SSSR Moskva, pp 291–309 (in Russian)

Wucherer W (1979) Primary colonisation of the dry sea floor of the Aral Sea. Probl Osvo Pustyn 3: 9–18 (in Russian)

Wucherer W (1984) Vegetation formation on the dry sea floor of the Aral Sea. In: Problemy Aralskogo morya i delty Amudar'i. FAN, Tashkent, pp 155–161 (in Russian)

Wucherer W (1990) Vegetation development on the new habitats in the desert. Gylym, Alma

–Ata (in Russian)

Wucherer W, Galieva L (1985) Vegetation mapping of the formed dynamic ecosystems. Isv AN KasSSR Ser Biol 4:3–13 (in Russian)

Methods of Conservation and Restoration of Vegetation Cover on the Aral Sea Coast

Liliya A. Dimeyeva

Keywords. Biodiversity, ecosystem, *Haloxylon*, protected areas, *Tamarix*

Abstract. The formation of desert plant associations of the Aral shore has had a long history. This chapter describes the contemporary coastal vegetation. It was determined that main threat of loosing biodiversity in the region lies in extending anthropogenic desertification. Two methods of biodiversity conservation are discussed: (1) restoration of the lost ecosystems and (2) establishment of a network of regional protected areas.

The formation of desert landscapes, ecosystems of the Aral Sea region similar to modern ones, can be assigned to the Miocene (Fedorovich 1946). According to pollen analyses, development of desert plant associations with *Haloxylon*, *Tamarix* and *Nitraria* in the region started in the Eocene (Zaklinskaya 1954). By the time of origin of the Aral Sea in the late glacial age, sand and saline landscapes and clay deserts with various plant communities had formed on the Turan lowland. Their development took millions of years, exstinction can ocour within a few years.

The present Aral Sea appeared about 10 000 years ago. Its level was always distinguished by high dynamism (Rubanov et al. 1987). Several periods of transgression and regression occurred during the Holocene, caused by climate change, solar activity change and the dynamics of the Syrdarya and Amudarya runoff regime. We can find preserved ancient trunks of *Haloxylon* on the exposed sea bed, which testify to former regressions. The coastal lines (Aral marine terraces) correspond to the transgression periods. Old Aral and New Aral marine terraces are marked by absolute height and age (56–59 m, 3000 years old and 53–56 m, 1000 years old respectively). Vegetation formation of coastal ecosystems has a history of 3000 years.

Modern vegetation on the Aral Sea coast is composed of three types: psammophytic, halophytic and meadowtugaic. Phytocenotic diversity and type of vegetation depend on the age of the Aral Sea marine terraces and the type of dominant succession (Dimeyeva 1995 1998).

Sand desert vegetation prevails in the Old Aral marine terrace. It is represented by zonal plant communities of *Haloxylon aphyllum* (Minkw.) Iljin, *H. persicum* Bunge ex Buhse, *Artemisia terrae-albae* Krasch. and seral plant communities [*Ammodendron bifolium (Pall.)* Yakovl., *Eremosparton aphyllum* (Pall.) Fisch. et Mey., *Calligonum* spp.]. Intrazonal communities [*Tamarix* spp., *Halocnemum strobilaceum* (Pall.) Bieb., *Nitraria schoberi* L., *Aeluropus littoralis* (Gouan) Parl.) are represented on the territory by less than 10%. Successions on the Old Aral marine terrace are secular (phylocenogenesis, sensu Sukachev 1942) and endodynamic. Woodlands of *H. aphyllum* and various plant communities of different stages of sand invasion [*Astragalus brachypus* Schrenk, *Stipagrostis pennata* (Trin.) de Winter, *Leymus racemosus* (Lam.) Tzvel., *Ephedra distachya* L. etc.] are dominant on the New Aral marine terrace. Intrazonal phytocenoses [*Tamarix, Lycium ruthenicum* Murr., *Nitraria, Anabasis salsa* (C.A.Mey.) Benth. ex Volkens] occupy about 25% of the area. Succession types are endodynamic and exodynamic against the background of hologenic successions (Rabotnov 1978). The leading role on the desiccated Aral Sea bed belongs to halophytic ecosystems; they are formed by perennial and annual saltworts [*H. strobilaceum, Halostachys belangeriana* (Moq.) Botsch., *Salicornia europaea* L., *Climacoptera aralensis* (Iljin) Botsch). Woody, shrub tugai and grass-forb meadow vegetation grows in the Syrdarya river mouth and as fragments in the sea bed [*Phragmites australis* (Cav.) Trin. ex Steud., *Elaeagnus oxycarpa* Schlecht., *Tamarix*]. Dominant successions are neogenic (singenetic and exodynamic). These temporal changes of vegetation are natural. Nowdays, anthropogenic successions have begun to play a leading role in the coastal ecosystems. They are only one of the exodynamic changes, but the most important one.

The main threat to biodiversity in the region is the processes of desertification caused by human activity. These are mainly overgrazing, clearing trees and bushes, technogenic factors related to background of the Aral crisis. Anthropogenic habitats differ in structure and composition from natural biotopes.

The vegetation of the Aral Sea coast is usually used as pasture, of which the most valuable are *A. terrae-alba - Agropyron fragile* (Roth) Gandargy - *Kochia prostrata* (L.) Schrad. on sandy soils and *A. terrae-albae-Anabasis aphylla* L. – Ephemeroidae on brown desert soils. Human impact on the rangelands is leading to the gradual disappearance of dominants and species replacement. *Artemisia arenaria* DC. and *A. scoparia* Waldst. et Kit. are species suited for medium grazing pressure. An abundance of *Peganum harmala* L., *Euphrbia seguieriana* Neck., *Iris tenuifolia* Pall., *Heliotropium dasycarpum* Ledeb. indicates processes of overgrazing. Zonal vegetation under strong grazing pressure responds by a wide distribution of *A. aphylla* and ephemers. Rangelands on saline soils are not overgrazed, but woodlands of *Haloxylon* and *Tamarix* have been cut down for fuel. The disappearance of trees and shrubs results in change and loss of biodiversity. It is to be noted that these ecosystems are characterized by high floristic and phytocenotic diversity.

There are ten species of *Tamarix* on the Aral Sea coast: *T. elongata* Ledeb., *T. laxa* Willd., *T. litwinowii* Gorschk., *T. hohenackeri* Bunge. *T. hispida* Willd.,

---

[1] Latin names of plants are given according to S. Cherepanov (1995)

T. aralensis Bunge, T. leptostachys Bunge, T. ramosissima Ledeb., T. karelinii Bunge, T. androssovii Bunge, and two species of Haloxylon: H. persicum, H. aphyllum. Coastal Haloxylon ecosystems are formed mainly by H. aphyllum. They are distinguished by high biological diversity and composed of 62 plant species. Nineteen plant associations have been described for these ecosystems: H. aphyllum - E. distachya; H. aphyllum - Atriplex fominii Iljin; H. aphyllum – A. terrae-albae - Eremopyrum orientale (L.) Jaub. et Spach and others. Each association is characterized by its own species composition and ecological conditions. Ecosystems of Tamarix in the Aral Sea region are distinguished by high floristic and phytocoenotic diversity. They consist of 84 plant species and 40 associations, among which the most widespread are: T. hispida, T. laxa - H. strobilaceum; T. ramosissima – H. aphyllum; T. elongata – A. brachypus and others. The high biological diversity of the ecosystems is of significant scientific and practical value and requires restoration and conservation.

A palpable loss of biodiversity in the Aral Sea region began in the 19th century. Woodlands of *Haloxylon* were cut to provide the Aral steamship flotilla with firewood. Clearing became heavier after the construction of the railway at the beginning of the 20th century and as a result, *Haloxylon* woodlands were almost totally cleared on the islands and coast of the Aral Sea. The boundary of shrubland distribution shifted to the south. Nowadays, all this has been repeated with another species – *Tamarix*.

Two methods for biodiversity conservation are suggested: (1) restoration of lost ecosystems, (2) establishment of a network of regional protected areas. Phytoamelioration and afforestation have been carried out periodically on the Aral Sea coast. Plantations by the Kazak Institute of Agroforestry in the dried sea bed have successfully demonstrated a better way for increasing vegetation coverage. Restoration of natural ecosystems could be realised by artificial phytocenoses. The natural vegetation remaining on the sea coast can serve as a model for the creation of man-made ecosystems, as well as a source of seeds. Reintroduction of the vanished wild vegetation in suitable habitats will provide the formation of seed banks and the natural dissemination of vegetation. Thanks to grant GEF/SGP, a test area of 4 ha and several plots of 100 m^2 were established at the southwest edge of the town of Aralsk in 1998 to revive *Haloxylon* and *Tamarix* communities. Seedlings of *H. aphyllum*, *T. laxa* and *T. hispida* were planted. Seeds of *H. aphyllum*, *Krascheninnikovia ceratoides* (L.) Gueldenst, *Salsola orientalis* S.G.Gmel. and several annuals were sown in the test areas. Three small protected areas of 400 m^2 with a buffer zone around them will be established in the project. Each area represents a rare ecosystem of *Tamarix* as a pattern of coastal vegetation.

1. *Atraphaxis* Ecosystem on the New Aral marine terrace. Vegetation coverage is 50%, floristic composition 25 species: *Atraphaxis spinosa* L., *T. laxa*, *Ammodendron*, *Limonium suffruticosum* (L.) O.Kuntze, *Scorzonera pusilla* Pall, *Euphorbia*, *Alhagi pseudalhagi* (Bieb.) Fisch., *Asparagus persicus* Baker, *Atritplex sagittata* Borkh, *Eremopyrum orientale* (L.) Jaub. et Spach, *Anisantha tectorum* (L.) *Nevski*, etc. The *Atraphaxis* community with such species composition is rare since it was formed on the New Aral terrace under the indirect

influence of the sea. The community is composed of both psammophytic and halophytic elements, which rarely happens on inland continental sands. The plot has a slight anthropogenic transformation; however; the presence of spurge indicates grazing.

2. *Tamarix* ecosystem on dunes. Vegetation coverage is 60%, height of dunes 3–5 m. Floristic composition is 15 species: *T. laxa, Atraphaxis, Limonium otolepis* (Schrenk) Kuntze, *Leymus, Peganum, Allium sabulosum* Stev. ex Bunge, *Heliotropium argusioides* Kar. et Kir., *Agriophyllum squarrosum* (L.) Moq, *Coryspermum aralo-caspicum* Iljin, *Salsola paulsenii* Litv, *Horaninovia ulicina* Fisch. et Mey., *Nonnea caspia* (Willd.) G.Don fil., etc. The ecosystem is unique by the fact that it was formed under the direct influence of the sea; high phytogenic hills were created by wind and sea splash activity. The sea receded and the hills remained as relicts. Clearing the tamarisk will damage the vegetation and disperse large sand massifs.

3. *Tamarix-Ammodendron-Artemisia* ecosystem on the New Aral marine terrace. Vegetation coverage is 55–60%. There are 30 higher plants in the species composition: *Artemisia santolina* Schrenk, *Ammodendron, T. laxa, Atraphaxis, Astragalus scabrisetus* Bong., *A. brachypus, Limonium. gmelinii* (Willd.) O.Kuntze, *L. suffruticosum, Agropyron desertorum* (Fisch. ex Link) Schult., *Leymus, Peganum, Euphorbia, A. sabulosum, Hyalea pulchela* (Ledeb.) C.Koch., *Corispermum orientale* Lam., *Koelpinia linearis* Pall., etc. The ecosystem was formed under the indirect influence of the sea. Lately, the anthropogenic pressure has become stronger. According to a survey in 1995, the extent of desertification was low. At present, *Tamarix* is cleared and this results in the formation of centres of wind erosion.

The Protected areas described above are fragments of remaining relict ecosystems. They were formed at the coastal line of the Aral Sea under the direct influence of the sea; no such conditions exist at present.

A network system of protected areas has not been developed in the region. There is an island – the State Reserve Barsakelmes – which represents a standard for vegetation in the region and thus of unchanged land. The vegetation of the plateau is to zonal grey-brown desert soils. Predominant communities are formed by *A. terrae-albae* and *Anabasis salsa*. Subdominants in plant associations are *Eremopyrum orientale, Agropyron desertorum* (Fisch. ex Link) Schult., *Anabasis aphylla, H. aphyllum*. Halophytic communities grow on saline soils [*H. strobilaceum, Climacoptera aralensis, Halimocnemis sclerosperma* (Pall.) C.A.Mey., *Bienertia cycloptera* Bunge, *Limonium suffruticosum* (L.)O.Kuntze]. The vegetation of sands has links with the Aral marine terraces, and is represented by communities of *Atraphaxis spinosa, Calligonum, Ephedra distachya, H. aphyllum, Salsola arbuscula*, etc. Communities of *Caragana grandiflora* (Bieb.) DC., *Rosa persica* Michx. ex Juss. and *Stipa lessingiana* Trin. et Rupr. have less significance in the island ecosystems. Thus, in the territory of the island Barsakelmes we can find the main types of ecosystems of the Aral Sea region.

The island was joined to the original eastern coast after the Aral Sea level lowered by 18 m. The exposed sea bed with the former islands Kaskakulan, Uzynkair and Akbasty is an unique example of primary successions and the formation of rare ecosystems of *H. aphyllum, Nitraria* and an endemic species — *A. brachypus.* It is very important to expand the protected area and include the territory of exposed sea bed. Urgent measures can protect the woodlands from clearing. In the most accessible places *Haloxylon* has been cut down. Setting up a network of regional protected areas at the local authority level will conserve biological diversity in its natural surrounding as a natural monument. A legal mechanism of official registration of local protected areas in the Aral regionhas not yet been organized, but it will be our future task.

Acknowledgements. The present study is a result of longterm investigations carried out in the Institute of Botany of the Kazak Academy of Sciences. It is now supported by the UNESCO project Evaluation of Desertification Processes in the Syrdarya River Delta, and a project UNDP-GEF Small Grants Program Oasis.

References

Cherepanov S (1995). Vascular plants of Russia and surrounding states. Mir i sem'ya, St-Peterburg, 992 pp

Dimeyeva L (1995). Ecologo-historical stages of Priaralie vegetation forming. Bull Mosc Soc Nat Biol Ser 100:72–84

Dimeyeva L (1998) Floristic and phytocenotic diversity of coastal ecosystems of the Aral Sea: present-day state and tendencies of change. In: Proc Int Conf on Environmental problems of the Aral Sea and surrounding areas. September 9–11, 1997. Almaty, Kazakstan, pp 107–111

Fedorovich B (1946) Questions on paleogeograghy of the Central Asia plains. In: Issues of Institute of Geography AS USSR, Moscow, pp 152–174

Rabotnov T (1978) Phytocenology. MSU, Moscow, 384 pp

Rubanov I, Ishniyazov D, Baskakova M, Chistyakov P (1987) Geology of the Aral Sea. Fan, Tashkent, 247 pp

Sukachev V (1942) The idea of development in phytocenology. Sov Bot 1–3:5–17

Zaklinskaya E (1954) Pollen specters of North Priaralie in the Eocene. Moscow, Papers of AS USSR, 49:621–624

Vegetation Dynamics on the Syrdarya Delta and Modern Land Use

Natalia P. Ogar

Keywords. Vegetation, ecosystem, succession, Syrdarya Delta, Aral Sea, desertification

Abstract. The vegetation dynamics of the Syrdarya Delta has been studied according to its main ecosystem types. The decisive factors in anthropogenic transformation of vegetation are the irrational utilization of water, irrigated agriculture and overgrazing. Aridization of the climate caused by the shrinkage of the Aral Sea speeding up the anthropogenic successions. Vegetation dynamics on all relief positions goes in the direction of halophytization and is chaotic in character. Desertification processes are in progress and expressed in biodiversity loss, convergence of plant communities and simplification of the spatial structure of vegetation cover. Stages of desertification of the meadow vegetation have been shown.

The vegetation of the modern Syrdarya Delta is characterized by a complicated spatial structure and rapid dynamics. The reason has to do with the physical -geographic peculiarities of the region, consequences of centuries-old economic use and the active impact of modern anthropogenic and anthropogenically stimulated processes.

An analysis of vegetation dynamics and modern land use was made for the main ecosystem types, which we determined for the ecosystem map of the modern Syrdarya Delta at a scale of 1:500000 scale (Ogar and Evstifeev 1998).

The peculiarity of the territory lies in the combination of zonal (automorphous) and intrazonal (hydromorphic and semihydromorphic) types of ecosystems which occupy certain positions in the mesorelief.

Long-term studies of the flood lands and river deltas of Central Asia (Ogar 1999) allowed us to determine the typological dimension of hydromorphic ecosystems. The elementary ecosystem corresponds to the biogeocoenose theory (Sukachev 1947). Further typification is made according to the type or subtype of the soils. Soil type integrally reflects the character of alluvial processes (thickness, composition and age of alluvium, expression of fluvial relief, dynamics of water -salt regime). In addition, significant differences in composition and structure of vegetation are expressed on this taxonomic level. The phytocoenotic constituent of the ecosystem is described by phytocoenotic dynamics as the elementary unit of vegetation cover. Thus the ecosystem, as a territorial unit at mesostructural level, is characterized by one type or subtype of soil, a set of plant communities with common ecobiomorphes (ecological and life forms) and typical species, an equal range of fluctuation of a floristic composition and productivity, similar successional tendency, and a similar reaction to natural and anthropogenic influences and resistance to them. An ecosystem is characterized by a certain type of relationship between its components and exchange of substances and energy with the outer environment. An ecosystem within a territorial unit is convenient for studying and mapping of the highly dynamic vegetation of river flood lands and deltas where exogenic successions are often of catastrophic character.

Powerful factors in the modern dynamics of the Aral ecosystems are the shrinkage of the Aral Sea, decline and contamination of the runoff of the Syrdarya and Amudarya Rivers and irrigated agriculture.

Character, pace and direction of vegetation successions under modern conditions are determined by hydrodymanic and halogeochemical processes against the background of an arid climate. Most noticeable changes in vegetation are observed in intrazonal ecosystems.

Intrazonal hydromorphic and semihydromorphic ecosystems are located on delta alluvial and primary marine plains. They differ by origin, formation conditions and evolution, and are characterized by the particular additional moistening regime. The resource potential of intrazonal vegetation is 10—30 times higher than that of surrounding desert vegetation types.

Ecosystems of the alluvial delta plain are formed as a result of erosion -accumulation activity of the Syrdarya River. The particular hydrological and hydrogeological regimes of the territory strongly influence the formation and dynamics of delta ecosystems. Highly mineralized groundwater lies close to surface and the relief of microdepressions of the water confining layer makes it difficult for groundwater to flow towards the Aral Sea. Redistribution and flow of groundwater are substantially affected by river flow and modern irrigation.

The aridity of the climate, combined with the absence of outflow from this area, induces progressive soil salinization. A positive salt balance is caused by progressive salt accumulation within the delta as well as by additional salt supply (impulverization) from the desiccating bed of the Aral Sea (Mozhaitseva 1979).

The ecosystems of the delta plain are characterized by a complicated spatial structure and high dynamics. They are highly productive and valuable for agriculture (hay lands, pastures) (Plisak et al. 1989).

The ecosystems of grass bogs are formed on soils of bog series under conditions of excessive moistening in deep depressions between river courses. They are represented mainly by reed thickets [*Phragmites australis* (Cav.) Trin. ex Steud.] with participation of *Bolboschoenus maritimus* (L.) Palla, *Juncus gerardii* Loisel and *Chenopodium rubrum* L.. Last 20 years area of reed thickets on bog soils has decreased by 60–70%.

Modern reed thickets are of secondary nature: they were formed within the past 30 years as a result of development of irrigated agriculture, particularly the cultivation of rice. They differ by insufficient formation of floristic composition and the structure of communities from natural reed thickets which develop in depressions on the soils of bog series. Their spatial distribution is of belt (around collector lakes) and linear (along canals) character. The vitality of the dominant species and the composition of accompanying species depends on the pattern of flooding or wetting conditions of the current year. These conditions vary significantly from year to year and result in weak intraphytocoenotic links between species and instability of the community in space and time. Suspension of surface flooding of habitats by collector waters leads to a succession from reed thickets to hylophytic shrubs. The succession series is represented by the following stages: reed community [*Phragmites australis* (Cav.) Trin. ex Steud.] on bog soils → annual saltwarts → reed community [*Phragmites australis* (Cav.) Trin. ex Steud., *Climacoptera aralensis* (Iljin) Botsch., *Chenopodium rubrum* L.] on bog solonchak drying soils → *Halostachys caspica* C.A. Mey. community on crust-puff solonchacks → sparse *Anabasis* L. community [*Anabasis salsa* (C.A. Mey.) Benth. ex Volkens, *A. aphylla* L.] on takyr-like solonchaks. The successions are of catastrophic character. Reed communities are replaced by *Anabasis* L. within 5—7 years (Table 1).

The reason for this is that the collector waters are clarified, do not contain silt deposit, and soils are not enriched by nutritive elements. Species and phytocoenotic diversity of reed thickets is very low.

Ecosystems of true mesophytic meadows [*Calamagrostis epigeios* (L.) Roth, *Elytrigia repens* (L.) Nevski, *Glycyrrhiza uralensis* Fisch., *G. glabra* L.] were widely distributed under conditions of natural flooding prior to the regulation of the Syrdarya runoff. They were replaced by halophytic meadows as a result of the influence of long-standing irrigation.

Modern meadows of the delta are secondary and formed by fragments on the areas underflooded by canals and irrigated fields. Their natural habitats are occupied currently by cultivated crops. They are characterized by heterogeneity of composition and structure and by chaotic dynamics. Phytocoenotic links within communities are very weak and floristic composition is of accidental character. Plant communities are serial, short-living and unstable in space and time.

Meadow vegetation dynamics is of a chaotic character due to the influence of irrigation, and depends on the condition of the land in the current year. Halophytization processes are in progress in these habitats. Ecosystems of halophytic meadows with a dominance of *Aeluropus littoralis* (Gouan) Parl., *Puccinellia distans* (Jacq.) Parl., *Hordeum bogdanii* Wilensky *and Leymus multicaulis* (Kar. & Kir.) Tzvel. develop on meadow solonchaks and alluvial soils of meadow series. As the water table lowers, they are replaced by communities of halophytic shrubs, *Halostachys caspica* C.A. Mey., *Kalidium caspicum* (L.) Ung. –Sternb., *K. foliatum* (Pall.) Moq., *Tamarix hispida* Willd. on secondary, common and crust-puff solonchaks.

Vegetation Dynamics on the Syrdarya Delta and Modern Land Use

Table 1. Stages of desertification of meadows

Stages	Communities and dominated species groups	No. of species	Productivity Overground / Underground	Overground Underground ratio
I.	Xero- and halomesophytic (grasses, phreatophytic forbs)	15–25	2884.0 / 3267.1	1:1
II.	Meso- and halomesophytic (phreatophytic forbs, saltworts, grasses, seedlings of shrubs)	20–40	848.0 / 1963.6	1:3
III.	Haloxerophytic (saltworts, forbs, shrubs: *Tamarix* L., *Halostachys* C.A. Mey., *Kalidium* Moq.)	15–25	665.2 / 3839.2	1:5
IV.	Xerophytic and haloxerophytic (saltworts, shrubs: *Eurotia* L., *Haloxylon* Bunge, *Lycium* L.)	5–15	179.1 / 1221.1	1:8
V.	Sub-climax xerophytic (desert dwarf semishrubs: *Artemisia* L., *Anabasis*)	3–7	103.2 / 1229.5	1:12

Stages	Soils	Water table (m)	Duration of the stage (years)	Time of restoration (years)	Economic use
I.	Exsiccating hydromorphic non-saline and saline	2.0–3.5	3–5	2–3	Hay land
II.	Exsiccated hydromorphic non-saline and meadow solonchaks	3.5–5.0	5–8	3–5	Haymaking in places, pasture
III.	Desertificating hydromorphic and common solonchaks	5.0–7.0	5–8	5–6	Pasture
IV.	Desertificating hydromorphic	6.0–10.0	5–6	7–9	Pasture
V.	Takyr-like non-saline and solonchakous	7.0–15.0	Long period of time towards zonal type	10–15	Pasture

The encroachment of meadows by woody vegetation is a characteristic feature of river deltas with regulated runoff, and is caused by the decrease in the volume of runoff and by change in the flooding regime (duration and dates) (Gorbachev and Lutsenko 1973; Novikova 1983; Bakhiyev 1985). This process increases with the development of desertification and drying of meadows.

The phytocoenotic structure of secondary meadows in the central part of the delta, flooded and underflooded by collector waters, is totally different. Plant communities are dominated by grasses [*Phragmites australis* (Cav.) Trin. ex Steud., *Aeluropus littoralis* (Gouan) Parl] along with xerophytic and mesoxerophytic forbs [*Karelinia caspica* (Pall.) Less., *Alhagi pseudalhagi* (Bieb.) Fisch., *Cirsium arvense* (L.) Scop., *Inula britanica* L., *Limonium otolepis* (Schrenk) O. Kuntze] with an abundance of shrubs [*Halostachys caspica* C.A. Mey., *Tamarix hispida* Willd., *Nitraria sibirica* Pall., *Kalidium caspicum* (L.) Ung.–Sternb.].

The vegetation succession goes in the direction of halophytization on all types of relief. Communities of *Nitraria sibirica* Pall. and *Tamarix hispida* Willd. are formed on soils of light mechanical composition in areas of former forb-grass meadows, and communities of *Halostachys caspia* C.A. Mey. on soils of heavy mechanical composition. Communities of *Halostachys caspia* C.A. Mey. also formed on the dried areas of former paddy fields.

Thus, desertification of the delta plain is accompanied by the convergence of vegetation cover and loss of biodiversity on the species, phytocoenotic, ecosystem and landscape levels.

The composition and structure of communities become simplified, meadows and forb-bogs are replaced by halophytic shrubs, which, in turn, are replaced by subclimax zonal communities when the ground-water level declines below 5 m.

Tugai (floodland forest) communities have remained on the delta as fragmentary vegetation. Their area has decreased by 80–90%. Their composition and structure have significantly changed: species abundance of *Salix* L. and *Glycyrrhiza* L. genera has strongly decreased, and abundance of weed species *Lepidium seravschanicum* Ovcz. & Junussov, *Cirsium arvense* (L.) Scop., *Zygophyllum fabago* L. and halophytic shrubs *Nitraria sibirica* Pall., *Halostachys caspica* C.A. Mey., *Tamarix ramosissima* Ledeb., *T. hispida* Willd. has increased.

Woody tugais (*Elaeagnus oxycarpa* Schlecht., *Salix alba* L., *Clematis orientalis* L., *Tamarix ramosissima* Ledeb., *Lycium ruthenicum* Murr.) on alluvial meadow-tugai soils is replaced under desertification by desert plant communities on takyr-like soils dominated by black saxaul [*Haloxylon aphyllum* (Minkw.) Iljin].

Vast areas in the Syrdarya Delta are covered by ecosystems of alluvial -proluvial delta plains. They are desertified areas that lost the influence of floods long ago. Vegetation cover is composed of subclimax communities with dominance of dwarf semishrubs, *Artemisia terrae-albae* Krasch., *Anabasis aphylla* L. and black saxaul [*Haloxylon aphyllum* (Minkw.) Iljin] formed on takyr -like solonetz-solonchak soils, solonchaks and saline sands.

Ecosystems of the primary marine plain are formed on the dry Aral Sea bed. The boundaries of their distribution change with changes in the sea level.

The spatial structure of ecosystems of the dry sea bed is not complicated and has a pattern of belts caused by the yearly decline of the sea level. Thus, ecosystem groups of different ages are distinguished corresponding to the belts within the dry Aral Sea bed.

Ecosystems of the 1st year of continental development, appearing in the open air, completely retain the features of submarine landscapes. They show up as vast areas of marsh solonchaks. The surface of the former sea bed dries, cracks, is covered by a thick salt crust, and at the end of the 1st year of continental conditions wind starts to break it down. The dessicating part of the sea bed is annually fixed at the splash zone by an algae layer and permanently influenced by up-and-down flows. Pioneer plants of succulent annual saltworts [*Suaeda salsa* (L.) Pall., *S. prostrata* Pall., *Salicornia europaea* L.] follow the receding coastline.

Ecosystems of the 2nd–10th years of continental development are characterized by drying up of the soil profile accompanied by the formation of deep polygonal cracks. Vegetation cover is strongly rarefied in the 2nd–3rd year and dies completely in the 4th–5th year. Vast areas of solonchak wastelands are formed (abiotic ecosystems without vegetation), they are distributed on the dessicated sea bed in mosaic form or in the form of strips. By their ecological parameters, these ecosystems are the most advanced stage of desertification, with clearly expressed destruction processes (dessication and strong salinization of soils, movement of surface deposits by wind). Their duration and the consequent type of vegetation overgrowth are not predictable. Single plants of desert halophytic shrubs (*Tamarix hispida* Willd., *Halostachys caspica* C.A. Mey.) settle on the separate areas in the 6th–8th year of dessication.

The initial stage of inner landscape differentiation is observed under the influence of intensive halo-chemical and deflation-accumulation processes in ecosystems of the former sea bed with after 10–12 years of continental development. Formation of initial microforms of accumulative relief takes place in the form of sand ripple, plait-like hillocks and hillocks around shrubs (Geldyeva and Budnikova 1990).

Marsh solonchaks transform into maritime and then into maritime sand soils. The inner structure of the ecosystems of the dried sea bed becomes more complicated at this stage. Dominating forms of relief at this stage are eolian low hills and low barchans with single plants of psammophytic shrubs [*Tamarix hispida* Willd., *Eremosparton aphyllum* (Pall.) Fisch. & C.A. Mey., *Nitraria schoberi* L.] on maritime soils with wind blown mantle. They alternate with an aggregation of annual saltworts [*Climacoptera aralensis* (Iljin) Botsch., *Atriplex fominii* Iljin] on solonchacks.

Ecosystems of the dried sea bed at 12–25 years of continental development form a kind of threshold or buffer zone between the former submarine complex and continental natural complex. This stage is characterized by a relative stabilization of relationships between ecosystem components and the development of ecosystems under the influence of zonal processes, primarily eolian and halochemical (Geldyeva and Budnikova 1990). These ecosystems are of long duration, they have directional development towards the formation of zonal ecosystems and are distinguished by their higher degree of biodiversity; but their composition and structure still remain unstable at this stage.

If the tendency of the Aral sea level to decline remains unchanged, the formation and development of ecosystems of the dried sea bed will take place under conditions of severe desertification, and their dynamics will be primarily determined by the climatic pattern. The belt character of their spatial structure will remain in the future. However, the decline in the sea level will be followed by a shifting of their boundaries.

Ecosystems of the dry sea bed are the initial stages of formation of zonal and intrazonal types of ecosystems. They are not completely formed and therefore have weak and unstable relationships both between inner components (soils, vegetation, fauna) and between their different types (sands, solonchaks etc.).

The dynamics of communities of annual plants is dictated mainly by precipitation fluctuating from year to year and its seasonal distribution. The development of perennial plants depends on the depth and mineralization of groundwater as well as on the mechanical composition and salinity type of the soil horizons.

Zonal automorphous ecosystems in the Syrdarya River Delta are located on drained plains of deposition on remnant elevations.

They are represented by two subzonal types: grey desert soils within northern deserts and grey-brown soils within true deserts. The conventional boundary between them lies along the Syrdarya River.

Communities of desert dwarf semishrubs [*Salsola orientalis* S.G. Gmel., *Kochia prostrata* (L.) Schrad., *Anabasis salsa* (C.A. Mey.) Benth. ex Volkens, *Nanophyton erinaceum* (Pall.) Bunge] and semishrubs (*Anabasis aphylla* L.) prevail in the vegetation cover. *Artemisia terrae-albae* Krasch. is dominant in these plant communities. Ephemerous plants and ephemeroids are abundant in the spring [*Poa bulbosa* L., *Eremopyrum orientale* (L.) Jaub. & Spach, *Carex physodes* Bieb.].

The dynamics of zonal ecosystems is determined by the influence of climatic and anthropogenic factors. Groundwater decline, increase in its mineralization and worsening of water supply caused by change of climatic parameters lead to degradation of the vegetation cover.

Other consequences of these changes are loss of biodiversity by all live organisms (soil microflora and fauna, insects, birds, plants, and animals), loss of humus, decrease in soil fertility, aggravation of desertification, and loss of ecosystem integrity.

Ecosystems of the eolian plains. A particular type of automorphic ecosystems is formed on eolian plains. These plains are formed as a result of deflation of pre -Quartenary and Quartenary deposits. Predominant plains are those with medium

Vegetation Dynamics on the Syrdarya Delta and Modern Land Use 81

-size hills and desert sand soils. Biodiversity and spatial distribution of ecosystems depend on the absolute heights of hilly relief.

Vegetation cover is formed by communities, dominated by psammophytic shrubs [*Haloxylon persicum* Bunge ex Boiss. & Buhse, *Calligonum* L. species, *Ammodendron argenteum* (Pall.) Kryl., *Lycium ruthenicum* Murr., *Atraphaxis spinosa* L., *Krascheninnikovia ceratoides* (L.) Gueldenst.], grasses [*Agropyron cristatum* (L.) Beauv., *A. fragile* (Roth) P. Candargy, *Elymus giganteus* Vahl] and sagebrushes [*Artemisia terrae-albae* Krasch., *A. santolina* Schrenk, *A. arenaria* DC.]. These ecosystems are characterized by a high level of floristic diversity.

Dynamic factors here are the same as for zonal ecosystems, but the processes induced by them (loss of biodiversity, desertification) are faster because of the low resistance of sands to anthropogenic impact. Sand massifs near settlements have no vegetation, have completely lost biodiversity, and are the source of deflation.

Semihydromorphic ecosystems on low plains. Automorphous ecosystems of drained plains of deposition and eolian plains form complexes and combinations with semihydromorphic ecosystems which are distributed on negative forms of relief and sand massifs with close groundwater. These ecosystems are solonchaks, takyrs, saladas and saline sands. Composition and structure of vegetation depend on relief, soil and hydrogeological conditions.

Communities of black saksaul [*Haloxylon aphyllum* (Minkw.) Iljin] with participation of semishrubs (*Artemisia terrae-albae* Krasch., *Salsola orientalis* S.G. Gmel.) prevail on tekyr-like soils. Ecosystems with a dominance of *Haloxylon aphyllum* (Minkw.) Iljin are of the highest resource value among them. These ecosystems are distinguished by the high biodiversity of all biota components; but as a consequence of the fuel-energetic problems of the past years, and the related merciless cutting of saksaul trees by the local population for firewood, these ecosystems have lost their resource and ecological potential.

Low plains of deposition are covered by algae and lichen takyrs which often have no vegetation or have single plants of dwarf semishrubs [*Anabasis salsa* (C.A. Mey.) Benth. ex Volkens, *Nanophyton erinaceum* (Pall.) Bunge].

Salada depressions also have no vegetation. Halophyte-dwarf semishrub monodominant communities [*Halocnemum strobilaceum* (Pall.) Bieb., *Nanophyton erinaceum* (Pall.) Bunge, *Anabasis salsa* (C.A. Mey.) Benth. ex Volkens] are formed on remnant solonchaks.

In semihydromorphic ecosystems of eolian plains, on «churot» sands, there is abundant cover of perennial herbs, namely, *Karelinia caspica* (Pall.) Less., *Alhagi pseudalhagi* (Bieb.) Fisch and *Phragmites australis* (Cav.) Trin. ex Steud., indicating close groundwater.

Because of their low economic value, these ecosystems are practically everywhere in a satisfactory condition. Shrinkage of the sea and decline of the water table have an indirect influence on them by strengthening salinization and aridization. Since the environmental conditions are already at a critical stage of desertification, and changes related to sea level decline proceed towards gradual loss of biotic components, these ecosystems with time will transform into abiotic (lifeless) ecosystems. Any economic activity speeds up these negative processes.

Change in the character of vegetation successions in semihydromorphic and automorphic habitats is influenced by increasing aridization of climate, first of all by an increase in temperature (by 0.5–0.7 ^{0}C) and decrease in humidity (by 10 –20%) at ground surface level as well as by the reduction of breeze from the sea. This is evident from the loss of biodiversity in plant communities, the simplification of their structure, the loss of certain plant species and synusia and the redistribution of the composition of ecobiomorphes.

The Syrdarya Delta is characterized by unstable land use under conditions of Aral Sea shrinkage. The Area of irrigated lands and hayfields has been continuously decreasing since the beginning of the 1990s and is gradually being transformed into pastures. Grazing pressure on hydromorphic ecosystems has sharply increased as a result of severe degradation of pastures of adjacent deserts, eolian and remnant (deluvial-proluvial) plains. The vegetation of hydromorphic habitats degrades very rapidly under conditions of water deficiency. Successions are of catastrophic irreversible character and accompanied by the convergence of plant communities and loss of their resource potential.

The integrity of delta ecosystems is highly disturbed by agriculture. The consequences of its influence are contamination of surface waters by pesticides and fertilizers, disturbance of structure and physical-chemical properties of soil cover and development of erosion and deflation. The natural vegetation cover of this areas is irreversibly transformed. Agricultural ecosystems (paddies, irrigated and watered fields, abandoned fields) are characterized by the dominance of species of adventitious flora.

As studies have shown (Chalidze 1974; Mozhaitseva 1979; Novikova 1983; Bakhiyev 1985; Plisak et al. 1989; Vukhrer 1990), successions of vegetation communities on delta plains and dessicated sea bed are rapid and of irreversible nature. For instance, according to our study, meadow and tugai types of vegetation are replaced by desert within 7–10 years; coenotic diversity declines 20–30 times, and on the species level, floristic composition is completely or only partly replaced. Total biological productivity decreases 15–20 times, and economic productivity 20–30 times.

Stages of vegetation and ecosystem transformation are determined, in general, by complex ecological and phytocoenotic criteria, most important of which is the loss of biodiversity by plant communities.

The total cumulative effect of the natural-anthropogenic dynamics of vegetation is the progressive desertification of the Syrdarya Delta accompanied by the decline in the natural potential of ecosystems, with loss of biological diversity and the possibilities for the reproduction of resources.

References

Bakhiyev AB (1985) Ecology and successions of plant communities of Lower Amudarya. Tashkent, 196 pp

Chalidze FN (1974) Vegetation succession in the areas between river courses of the ancient and modern Syrdarya River Delta. Ecology 3: 20–29

Geldyeva GV, Budnikova TJ (1990) Landscapes of the Kazakhstan region of Priaralie. In: Aral today tomorrow. Kainar, Alma–Ata, pp 144–183

Gorbachev BN, Lutsenko AI (1973) Vegetation change of the Lower Don after establishment of regulated runoff. In: News of North Caucasian Scientific Center of Higher School, Natural Sciences, pp 126–129

Mozhaitseva NF (1979) Landscapes evolution in the course of desiccation of the eastern coast of the Aral Sea. Problems of desert lands development 3: 18–25

Novikova HC (1983) Influence of anthropogenic change of river runoff on ecosystems of modern delta plains in arid regions. Problems of desert development, 3: 18–24

Ogar NP (1999) Vegetation of river valleys in semiarid and arid regions of continental Asia. Abst of the doctoral dissertation, Almaty

Ogar NP, Evstifeev YG (1998) Map of ecosystems of the modern Syrdarya Delta. Scale 1 : 500000. In: Ecological research and monitoring of the Aral sea. A basis for restoration. UNESCO. Paris, pp 15–41

Plisak RP, Ogar NP, Sultanova BM (1989) Productivity and structure of desert zone meadows. Nauka, Alma–Ata, 186 pp

Sukachev VN (1947) Principles of biocoenology theory. Moscow.

Vukhrer VV (1990) Vegetation formation on a new land in desert. Gylym, Alma–Ata, 215 pp

Zaletaev VS, Novikova NM (1995) Changes in the biota of the Aral region as a result of anthropogenic impact between 1950 and 1990. Geojournal 35 (1): 23–27

Ecological Basis for Botanical Diversity Conservation Within the Amudarya and Syrdarya River Deltas

N.M. Novikova

Keywords. Plant species, communities, groundwaters, soil salinization, desertification, landscape dynamics

Abstract. Botanical diversity (BTD) is an important part of biodiversity and a goal for real action for conservation. BTD includes, according to our understanding, plant species, plant communities and their spatial combinations (symphytotaxones). We differentiate between the potential, the registered and the actual BTD.

Potential species richness (local flora) of the Amudarya and Syrdarya River Deltas includes 774 species. 295 species are presented in bight deltas, 307 species are only in the Amudarya and 154 species are only in the Syrdarya Delta.

Potential and registered plant community richness includes 71 associations and 21 formations. Actual community richness has 2 associations less of *Saliceta songooricae* formation, because these have disappeared within the two deltas.

The ecological positions as a frequency of distribution within the scales of the groundwater table and soluble salts in the soils were studied for the main plant community.

An ecological interpretation of the scheme of landscape desertification within the Amudarya Delta indicated the area where different conditions allow tugai communities to survive and possible water management to support existing tugai communities and renew wetland conditions.

Background

Conservation of the tugai flora and vegetation under conditions of regional desertification is one of the major problems of the Aral ecological crisis. The steps undertaken in this direction have an arbitrary character, and are not based on the theoretical positions established by the ecological sciences, so they are doomed on failure.

According to our understanding, for a solution of this problem, it is necessary to unite on one hand an understanding of an ecology of species and communities, with the peculiarities of their spatial-temporal dynamics, and on the other, a

Ecological Basis for Botanical Diversity Conservation 85

knowledge of deltaic landscape evolution and its modern dynamic condition with the purpose of maintaining or reconstructing landscape processes which can the save flora and vegetation as a unique ecologic-dynamic system.

To gain new information on the ecology of communities and landscapes, the tasks are:

- To understand botanical diversity within the two deltas at the landscape level.
- To analyze the ecological diapason of vegetation communities of formation range and to reveal the optimum conditions, based on the frequency of communities.
- To determine the current conditions for plant communities growing in different kinds of landscapes.

Material and Methods

As a main methodological position, we improve our understanding of botanical diversity as an important element of biodiversity and an object for protection. Botanical diversity we understand as a complicated integrated system, resulting from florogenesis and phylogenesis and characterized by a certain set of taxa, syntaxa and chorological units of vegetation, supported by a system of exo- and endodynymical successions.

For this purpose, the ecological-geographical database system (DBS) was worked out (Novikova and Trofimova 1993). The empirical data mainly involved in this DBS were gathered during field expeditions from 1947 to 1996.

The DBS contains a set of dynamic information for the main components of landscapes gathered at one site: plant species, plant communities, description of soil profiles, and results of chemical analysis for soil samples, groundwater table depth and chemical analysis data for groundwater samples. All the information is gathered in geographical sets according to the landscape units, into which the area of the Amudarya and Syrdarya River Deltas was divided.

An inventory of floristical diversity was obtained as a local flora for the two deltas on the basis of the DBS and data from regional floras (Bondarenko 1964; Eredjepov 1978; Korovina et al. 1982; Conspect of the Flora of the Middle Asia in 12 volumes; Flora of Kazakchstan in 9 volumes; the latter with Latin names of plants after S. K. Cherepanov (1995).

Analysis and Discussion

We divide three types of floristic diversity: potential – known from Floras, registered – known from geobotanical field descriptions throughout the investigations, and actual – obtained in field work over the last years. According to our data, the potential species richness for the two deltas is 774 species, common species (spreading in bight deltas) are 295; 308 species are only in the Amudarya and 154 species are only in the Syrdarya Delta.

At the Amudarya River Delta, potential floristical richness includes 601 species, registered (from more than for 50 years field) is 230 species, and actual (from 10 years) 192 species.

For the Syrdarya Delta the potential floristical richness of the hydromorphic landscapes consists of 436 species. Unfortunately, we have only a small data set of fieldwork for checking and actual richness. The 156 registered species are according to data over nearly 40 years (F. Chalidze from 1956 to 1962; N. Mojaitzeva from 1968 to 1973; N. Novikova 1979, 1985, 1992). The actual one come from different data: from 1956–1962 122 species; for 1968–1973 94 species and for 1979–1992 96 species.

To evaluate the floristical diversity within landscape units, we obtained gamma the diversity for ten kinds of landscapes (Novikova 1998). It is the highest for oasis (153 species), tugai forests (113), levees (113) and dry sea bed (92), and smaller within zonal landscapes (near 50) and beds of dry lakes (49); but these data are insufficiently correct as they depend on information gathered from a number of descriptions. Landscapes with the most data have a greater number of species.

A more correct characteristic for landscape flora diversity is the number of species in one geobotanical site (10×10 m^2 for herb and bush communities and 20×20 m^2 for trees). This unit is known as α-diversity. For the whole region around the Aral Sea it is 3 on 10×10 m^2, the least in Central Asia. In the Amudarya River Delta it is higher, nearly by 5 or more, and it is highest in landscapes of lakes, tugai forests and oases – from 10 to 8; it is 7 in most of the landscapes of the delta and reaches 8 in desert landscapes of the inselbergs. The least number of species – 5, is in the landscapes of the dry Aral Sea and in the Holocene part of the delta.

As can be seen, for conservation the landscapes with high gamma and α-diversity: levees, oases and tugai forests hold more promise. There appears then the new problem – how to organize biodiversity conservation within an irrigated area.

Classification of the plant communities divides 21 taxonomy units at the level of formations and 71 associations within the two deltas as community richness (Table 1). There are 26 associations common for bight deltas, 26 are present only in Amudarya and 16 are only in Syrdarya. There is no high level of similarity (less than 50%) in syntaxones at the association level. At the level of formation the similarities are greater, 67%, and within the whole 27 units 18 are common to the two deltas. So for plant community preservation we need to be especially attentive, because many associations are unique for each delta.

The next step was understanding the ecological positions of plant communities. For this an analysis of the frequency distribution of the plant communities was made, with the variation in groundwater table and soluble salts content in soils (Figs. 1, 2).

Ecological Basis for Botanical Diversity Conservation 87

Table 1. Botanical diversity within the landscapes of the Amudarya and Syrdarya river Deltas

Indicators	Total	Amudarya	Syrdarya	Common
Species, total	773	601	436	295
Species of tugai-type communities	308	259	154	94
Genera		178	149	
Families		62	49	
Formations	27	23	22	18
Associations	71	55	42	26

As can be seen, most plant communities are soil-tolerant and spread by means of soil salinization and groundwater table depth; but more often they are connected with classes of salinization 2–5 (Figs. 1, 2), which means the contents of soluble salts from 0.25% to 5% in 100 g of soil (from non-saline to solonchak). On the other hand, the frequency of soluble salts in different types of soil in the delta shows a high level of salinization. So such less saline soils as meadow tugai are now mostly very salty soils. At present more than 60% of the soil samples of this soil can be qualified as solonchaks; but the total amount of soluble salts is often in the crust (0–0.2 cm), and the deeper soil layers contain less than 0.3% of soluble salts. So these soils are non-toxic for trees, bushes and perennial herbs.

We found a correlation between the main growth factors of plant communities: groundwater table (gwt), groundwater mineralization (min) and percent of light fraction in granulometric composition (gc) of soils at 1 m depth. The correlation in the general selection (Table 2) shows slight positive links (k = 0.30) between gwt and min. The is gwt deeper, the more it is mineralized. Between gwt and gc is slight negative link (k = -0.23): the more percent of light particles in gc, the nearer is gw to the soil surface. Between min and gc is a medium negative link (k = - 0.37): the more percent of light particles (more sandy) in soil, the less is the mineralization of the groundwater.

We controlled this correlation in particle selections for various plant community formations, and found large variations in coefficients: from 1 to 0. This is possible because the characteristics are not for plant communities, but mainly for the different habitats.

Table 2. Correlation between the main ecological factors

Factors	gwt	min	gc
gwt	-	0.30	-0.23
min	0.30	-	-0.37
gc	-0.23	-0.37	-

gwt groundwater table, *min* groundwater mineralization, *gc* percent of the light fraction in granulometric composition

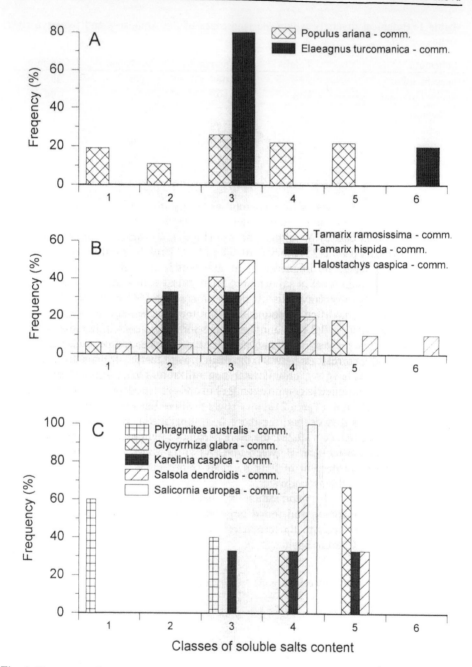

Fig. 1. Frequency of plant community (formation) distribution under different conditions of soil salinization; A trees, B bushes, C herbs. Classes of Soil salinization: 1 = <0.25% soluble salts in residual sec; 2 = 0.25–0.5%; 3 = 0.5–1%; 4 = 1–2%; 5 = 2–5%; 6 = >5%

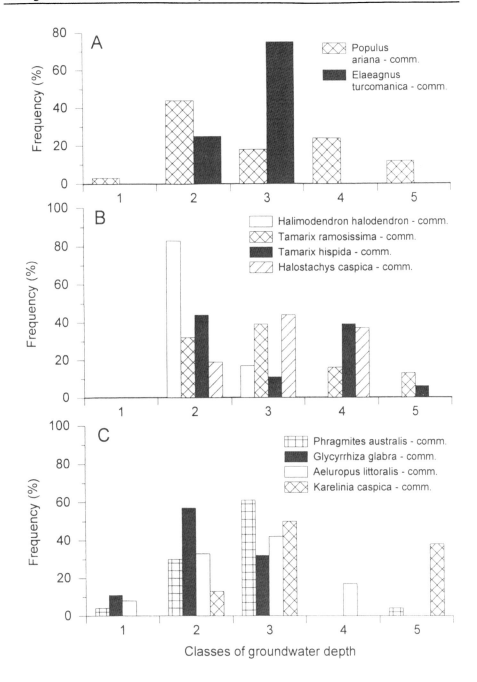

Fig. 2. Frequency of plant community (formation) distribution at different goundwater table depths; A trees, B bushes, C herbs; Classes of groundwater table: 1 = 0.00–0.50m, 2 = 0.51–1.50m, 3 = 1.51–3.00m, 4 = 3.01–5.00m, 5 = 5.01–15.00.

The inventarization of the objects as basis of the knowledge, gained by classification, permits us to control and analyze changes in plant communities and soils over time. For this purpose we divided the total time of desertification of the Amudarya River Delta environment into various intervals during which the watering of the delta differed considerably.

1st period 1947–1952, is considered to be stadard, water supply and development of delta were in the natural regime. Active irrigational cultivation of the deltaic plain was of local character.

1952–1965 is characterized by the development of irrigational construction in the river basins, but no reduction in water flow to the upper course of the deltas is detected.

2nd period 1972(74)–1977, the beginning stage of active drying of the coastal part of the deltas as a result of the construction of the Tahiatash and Kazalinsk hydroelectric stations in the upper course of the deltas and the cessation of natural flooding beginning in the Syrdarya Delta from 1972 and in the Amudarya Delta from 1978.

3rd period, 1978–1983, minimum water supply of delta, stage of automorphous development, groundwaters remain up to 5–10 m in non-irrigated area.

4th period, 1984–1989, beginning of the new water supply from the drainage water flow and construction of local irrigation systems. Groundwaters rise near lakes to 1.5–3 m.

5th period, 1990–1993, increase of flowing of river water to the upper ends of the deltas, vast filling of dried lake lagoons by river water, restoration of the previous water supply up to 30% of the deltaic plain that had been before 1960s. Groundwater table rises to 3–4 m in delta.

6th period, 1994, new decrease of river runoff and disappearance of many lakes again.

Composition of plant communities changes according to water conditions. Up to 1972 the most hydromorphic plant communities, two associations of *Saliceta songooricae* formation, disappeared in the Amudarya Delta. Later, until 1983, formations of *Populeta arianae, P. pruinosae, Elaeagneta turcomanicae, Halimodendreta halodendrii, Calamagrostideta dubiae* and some others died on the main area, and preserved patches are degraded. However, during this time, to 1994, new species appeared, mainly annual, salt-resistant ombrophytes, such as *Climacoptera aralensis, C. lanata*. In 1985 appear in the delta the first plants of takyr desert species *Haloxylon aphyllum*, in 1990 the first plants of the desert psammophyte *Ceratoides papposa*. As indicator of the development of the soil takyrization process, *Salsola dendroides* was widespread in the delta from 1985 to 1998. Since 1993 at the northeast of the delta within the area of desertified solonchaks, one further species, *Anabasis aphylla*, an indicator of the formation of takyr-like soils, has begun to develop.

The database shows us changes in soils (Fig.3). In the first period (1947–1952) there were mainly two classes of salinization – all swamp soils in the delta were not or only slightly saline. From 1978 to 1983 salty soils appear (Fig.3, class4) and during this time there were no non-saline soils. Rewatering of the delta show the appearance non-saline soils (Fig. 3, class 1) equal in percent to the 1st period, and other soils slightly salty (Fig. 3, classes 3, 4).

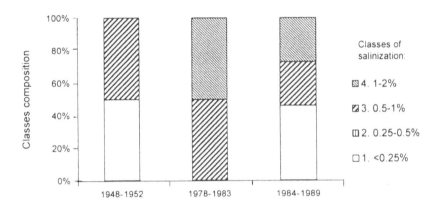

Fig. 3. Changes in average content of soluble salts in the 0–50-cm layer in swamp-type soil. Classes of soluble salts as in Fig.1

Control of the changes in space and in time is the most difficult. Our knowledge helps us again in this aim to understand the natural dynamics of landscapes and their dynamic positions by reviewing the ecological-genetic succession of plant communities which accompany the evolution of landscapes (Novikova 1996). The landscapes take on the features of progressing aridization as a result of the decrease in additional moisture, correspondingly to the hydromorphic, semihydromorphic and automorphic stages. This is reflected in the plant cover as a hydrogenic succession with a consequent change of communities which include species of different hydroecologic groups: hydrophytes, mesophytes, mesoxerophytes, mesohalophytes or mesoxerohalophytes. As usual, the initial stage is the communities of highgrass perennials, which change to the communities of meadows and meso-(mesoxero-)phylic shrub-woody tugai. At the next stage the mesoxero- or mesohalophylic shrubs become dominant, changing at the next stage to the final stage of haloxerophylic trees and subshrub communities. Each ecological group indicates changes in all landscape forming processes, components of landscapes and interlinks; but gamma and α-diversity becomes less.

This problem could be solved by connection of this "point" database with a geoinformation system. Such a step was made by interpretating the scheme of desertification by A. Ptichnikov (Ptichnikov et al. 1998) into the scheme of conditions for tugai community growth. There we can divide the area where are conditions for its support and renovation within wetland watered and near main river shore. The next kind of landscape area supports degraded plant communities near secondary river branches, from time to time with runoff. The areas with desertified and dead woody and bush plant communities are on the slopes of levees. The largest areas of the interrivers plains now have desert conditions and are being invaded by plants of desert ecology, such as annual and perennial bush and trees. On the irrigated area ruderal flora and open waters are also indicated.

Analysis has shown that considerable areas of the Sudochie, Toguz–Tore, Karajar and Mezhdourechenskoe lake systems and a water body in the Muinak Bay, etc. are now permanently flooded. These areas are a valuable genetic reserve for flora and fauna, typical for wetland ecosystems. Most of the rare, endemic and economically valuable types and communities are preserved here. They can spread from these sites to restored or newlyformed water bodies. Clearly, these areas should be preserved under permanent flooding.

Seasonal water level rises in lakes and inundation of land plots occupied by coastal vegetation are ecologically important. The depth of flooding should not exceed 100 cm, because flooding to a depth of between 10 to 100 cm caused riparian grasses to grow intensively, which ensures favourable conditions for spawning and shelter against predators. The recommended period of flood inundation is April–May.

However, it is impossible to preserve the entire diversity of wetland ecosystems, without implementing traditional measures such as the organization of reserves:

- The Sudochie and Kungrad lake systems, on the left of the Amudarya.
- The Tougus–Tore Lake and Akpetki sand solonchak massif, on its right bank.

Watering of tugai complexes is necessary for their preservation. It will be expedient to restore a hydrological regime, close to the natural one, located along the Amudarya main channel (the reach between the Kazakhdarya inflow and the Tachiatash dam). This can be done by using a cascade of spaced dams to increase the water level in the river. This will promote a rise in groundwater, intensification of alluvial process and improve the water supply of the coastal tugais. In this case it will be necessary to investigate closely variants of possible salinization of innerchannel depression slopes, because groundwater seeping from the river body will evaporate. It will also be necessary to determine the optimum height of the water level rise.

In the head part of the delta, near the Tachiatash dams, tugai communities are in better condition and renovation by seeds is taking place. Seed deposition and germination are promoted by active formation of islands, and tugai communities proliferate. The optimum regime of watering mature tugai is a period of up to 20

Ecological Basis for Botanical Diversity Conservation 93

days in spring and late in summer (July–August). The depth of flooding should be 50–100 cm. The groundwaters should rise to 1.5–3 m.

Moreover, the creation of a water conservation zone along the whole Amudarya channel downstream to the Tyuamuyn dam is needed to restore and protect the river bank by developing riparian forests. The forest belts stretching along the river channel should be 1.5–2 km wide. Land ploughing, cattle grazing, forest cutting, any construction works and wastewater discharge into the river should be prohibited within this belt.

In order to preserve the genetic and cenotic fund of tugai ecosystems on the territory of well-watered tugais, it is necessary to decrease the direct impact by restoring reserve regimes on the Nurumtubek and Nazarchan tugais, and also at the new mouth of the Raushan Canal and on the Kokdarya floodplains.

Conclusion

There is a great difference between natural desertification of the deltas and the processes we have recently seen in the Amudarya Delta. In nature, the processes of aridization are cyclic, finishing by forming a solonchak landscape in one part of the delta, while in an other part these processes begin from the water-bog stage. Thus, all the stages of the desertification processes develop in the area of a delta; many different types of successions support ecological variety and botanical diversity. At present all the area of the Amudarya and Syrdarya Deltas are under aridization, the first stages – water-bog systems and non saline soils are not complete. No new set of levees is being formed on its slopes, interriver downflow area and lakes; there are mainly artificial lakes with mixed drainage and river water, without fertilized deposits. So the delta forming processes cease – this is the meaning of an ecological crisis. Landscape degradation is the basis for biodiversity degradation.

The measures we recommend may be ordered in three groups:

1. Regime measures, without changes in the hydrological regime, to support existing ecosystems.
2. Measures aimed at preserving existing hydrological regimes to protect the conditions of ecosystems.
3. Measures to modify water supply or water quality
- to support existing ecosystems;
- to improve existing ecosystems;
- to restore ecosystems of the tugai type.

Acknowledgements. The author thanks the Federal Ministry of Education and Research of Germany (bmb+f) and UNESCO for financial support of the program 509/RAS/40, 41 – ARAL SEA. This gave the opportunity to organize field research and also the possibility to participate at the workshop Ecological Problems of Sustainable Land Use in Deserts.

References

Bondarenko ON (1964) Conspect of the flora of high plants of Karakalpakstan. FAN, Tashkent, 304 pp

Chalidze FN (1966) Vegetation as an indicator of the granulometric composition and relative ages of the deltaic deposits within the river Syrdarya. Thesis of the candidate dissertation. Moscow, 19 pp

Cherepanov SK (1995) Vascular plants of Russia and surroundings. St. Petersburg: Nauka. 992 pp.

Erejepov SE (1978) Flora of Karakalpakstan and its economy, medicine characterization. FAN, Tashkent, 298pp

Korovina ON, Bachiev AB, Tadjitdinov T, Sasybaev B (1982–1983) Illustrated definition for high plants in Karakalpak and Choresm. FAN, Tashkent, 2 vols.

Mojaitzeva NF (1979) Evolution of the landscapes of the eastern Aral see shore. Thesis of the candidate dissertation. Problems of the desert development.

Novikova NM (1996)Current changes in the vegetation of the Amu Dar'ya delta. The Aral Sea Basin. In: Micklin PhP, Williams WD(eds) NATO ASI Series 2. Environment – Vol12. Springer, Berlin Heidelberg New York, pp 69–78

Novikova NM (1998) Ways to preserve diversity of tugai (wetlands) Plant communities and species on the deserted deltas of the Aral Sea. Arid Land Stud 7S: 307–310

Novikova NM, Trofimova GY. (1994) Ecological-geographical database system composition and structure. Proc of the 1st Int Conf on Hydroinformatics. Delft, The Netherlands. September 19–23, 1994, pp 617–620

Novikova NM, Kust GS, Kuzmina JV, Dikareva TV, Trofimova GY (1998) Contemporary plant and soil cover changes in the Amudarya and Syrdarya river deltas. Ecological research and monitoring of the Aral Sea deltas. UNESCO Aral Sea Project. 1992–1996 Final Scientific reports. UNESCO, Paris pp 55–80

Ptichnikov AV, Glushko EV, Kapustin GA, Reimov P (1998) Electronic atlas as the first step toward a geographical information system of the Aral sea region. Ecological research and monitoring of the Aral Sea deltas. UNESCO Aral Sea Project. 1992–1996 Final Scientific reports. UNESCO, Paris, pp 293–300

The Tugai Forests of Floodplain of the Amudarya River: Ecology, Dynamics and Their Conservation

Sergey Y. Treshkin

Keywords. Biodiversity, communities, plant, soil, salinization, vegetation

Abstract. Terrestrial ecosystems are extremely vulnerable in arid regions. They respond even to insignificant changes of the environment, which may result in irreversible modifications of ecosystems and often in the complete loss of their scientific, social, and economical value. Intrazonal hydromorphic landscapes undergo the greatest transformation, the main component of which is floodplain vegetation.

The process of excessive use of river water for irrigation, and the appearance of the powerful waterfarm constructions in the States of Central Asia for a comparatively short period of time (1960–1995) caused the degradation of the environment, drying up the Aral Sea, Priaral desertification and serious social -economic consequences. All these are features of the environmental crisis in the Aral Sea basin which was assessed by the world community as a global ecological catastrophe.

One feature of the Aral environmental crisis is the fast degradation of the Amudarya River ecological system, especially of its delta. This degradation was caused by different kinds of anthropogenic interference mainly by the construction of dams and changing watersheds, changes in the hydronetwork, pollution etc. Amudarya is the largest river in Central Asia and supplies the most water, together with Syrdarya, it provided the viability of the large continental lake, the Aral Sea. Until recently it remained the only unregulated river among the large rivers of the former USSR. The rapid rates of development of land irrigation in the basin of the Aral Sea, in the 1960s, caused step by step the necessity of regulating its channel by constructing dams and diverting the greater part of its water resources for the needs of agriculture. As the result, about 90% of the total water use is used by agriculture, 3.6% by the municipal sector, about 2% by industry, 1.6% by village water supply, 0.8% by fishery and 1% by other consumers.

Before regulation, the Amudarya had a large annual inflow to the sea (average 50 km^3), formed vast floods in the delta, and created ideal conditions for the richest brushwood of water plants, fish and other animals. The period after regulation by dams (the Tahiatash dam was built in 1974 near Nukus, the Tuyamuyn dam in 1980) is characterized by a sharp reduction in water mass inflow to the delta (Fig. 1) and instability of rate regulations, sometimes reaching an input of zero. Furthermore, the flow through channels was stabilized, delta floods disappeared, a vast net of secondary channels with water-consuming irrigation channels and collectors was formed.

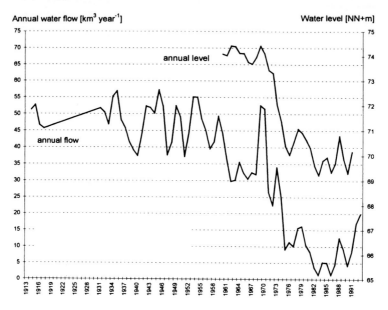

Fig. 1. Change in flow and water level in lower parts of the Amudarya for the past 80 years (at the hydrological station Chatly near Nukus)

At present, the water basin of the Amudarya may be characterized as a compound mutual waterfarm system, including, besides the river itself, a vast net of main and internal channels, collectors, lakes and reservoirs. About 17 km^3 of collector-drainage water is formed, about 30% of the average inflow of the river into the Aral Sea before the Aral catastrophe.

The chemical composition of the river water has changed under the impact of the discharges of drainage water from irrigated fields and from industries. First of all its mineralization increased from 0.4 g l^{-1} to 1.0-2.0 g l^{-1}. The river is contaminated by phenols, oil products, heavy metals, pesticides and nitrogen compounds. The Amudarya played an important role in the protection of biodiversity in floodplain ecosystems.

The problem of preservating biodiversity, in other words, the preservation of the whole biodiversity of the existent organisms on our planet at the maximal possible level under the present fast-changing conditions, is counted one of the present challenges, and requires an urgent solution. This problem acquires a particular actuality in the crisis districts, to which we can confidently attribute the floodlands of the Amudarya River, where intensive economic cultivation has continued for decades, as a result of which the largest tracts of tugai forests, once widely-distributed on this territory, have been almost completely destroyed. At the beginning of the 19th century its area in the floodplain of the Amudarya consisted more than 600000 ha (Fig. 2). We must note that the process of destroying the tugai forests is characteristic not only for the Amudarya floodplain but also for all the floodplains and river deltas of Central Asia.

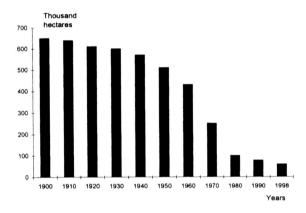

Fig. 2. Reduction of tugai forest area along the Amudarya River from 1990 to 1998

Floodplain vegetation is the most valuable kind of biological resource in Central Asia and is characterized by high biodiversity (more than 1000 species) and comparatively high productivity (190 t ha^{-1}).

The regulation of the river flow, together with the building of hydrotechnical constructions on the rivers, everywhere causes significant transformations of the floodplain ecosystems. Depending on concrete conditions, these changes can appear either as drying of floodplain landscapes or their flooding. In any case, regulation of flow causes a reduction in fluctuation amplitude in seasonal changes in river and groundwaters, significant changes in structure, diversity and biomass in floodplain ecosystems, destabilization of the moistening regime of soils and grounds in the floodplains during the year and reduction of seasonal moisture pulsation in the soil profile. The relatively unstable high or low groundwater table (GWT) that occurs, differs significantly from the natural one. This phenomenon causes landscape degradation and leads to transformation of soil and vegetation properties.

The genesis and natural development of tugai ecosystems are closely connected with the hydrological regime of rivers. The natural spread of seeds over floodplains and deltas in arid climates is possible only on essentially fresh alluvium, when natural (or anthropogenic) floods coincide with the fruit-bearing period of the main plant species in tugais.

Unfortunately, dynamics of natural tugai communities, was maintained only in the central part of the floodplains of the Amudarya River (Fig. 3). However, here again, necessary freshet flooding is possible only within catastrophic flooding, as for instance in 1998.

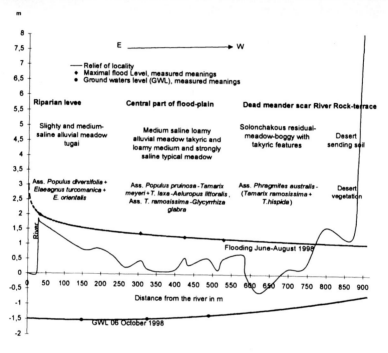

Fig. 3. Soil types, plant association and relief across the tugai forests of the Amudarya, Tugai Shabbas

Young tugais, which are 3–5 m in height at an age of 4 years, form monocenoses of poplar or mixed tugai forests with a closed upper canopy. By the age of 30–40 years, tree stands become thinned out and begin to die (Drobov 1950). With the growth of the community, the groundwater level (GWL) becomes lower.

There are several stages in the development of tree and bush tugais. At the initial stage of tugai formation on fresh deposits with high GWL (1–1.5 m), grass and also willow or oleaster poplar, willow or oleaster communities may occur. Soils are alluvial-meadow, non saline or slightly saline (Table 1). They are replaced by poplar and tamarisk and poplar communities on soils with lower GWL (2–4 m) and higher salinity. They gradually give way to bush tugais formed by tamarisk, which under natural conditions (lowering of the GWL without active salt

The Tugai Forests of Floodplain of the Amudarya River

accumulation in soils) usually degrade, forming desert takyr landscapes. The majority of investigators of plant cover explain the death of tree stands in the Amudarya Delta by the lowering of the GWL (Grave 1936). It is well known that the depth of the GWL in floodplains and deltas of arid regions is characterized by both frequent seasonal and yearly fluctuations, which depend on the alternating hydrological cycles of rivers.

Table 1. Soil and sediments under tugai vegetation in the lower reaches of the Amudarya River

	Plant association	Soils	Dry residue [%]	GWL [m]
1	*Phragmites australis – Typha laxmannii*	Medium-saline meadow-boggy alluvial	0.19–0.27	0.5–1
2	*Populus ariana – Salix songarica*	Non-saline sandy loam and loam weakly developed and alluvial meadow	0.05–0.12	1–1.5
3	*Elaeagnus turcomanica – Mixteherbosa E. turcomanica – Populus ariana*	Slightly saline and non-saline loamy alluvial meadow tugai	0.18–0.27	1.5–2
4	*Elaeagnus turcomanica –Glycyrrhiza glabra –Sphaerophysa salsola*	Typical clayey meadow solonchaks	1.20–1.75	1.5
5	*Populus ariana –Tamarix ramosissima –Halimodendron halodendron*	Saline loamy meadow	0.44–1.64	2–3
6	*Populus ariana –Tamarix ramosissima –Mixteherbosa*	Slightly and medium-saline alluvial meadow tugai	0.25–0.27	2.5–3
7	*Populus ariana –Tamarix ramosissima –Aeluropus littoralis*	Medium-saline loamy alluvial meadow takyric and meadow -takyric solonchaks	0.42–0.87	2–5
8	*Populus ariana*	Loamy alluvial meadow lakyric tugai	0.27–0.32	>4
9	*Tamarix ramosissima –Alhagi pseudalhagi –Aeluropus littoralis*	Loamy medium and strongly saline typical meadow and solonchaks	0.83–1.49	2
10	*Tamarix ramosissima*	Slightly- and medium-saline clayey and loamy meadow -takyric and takyric	0.23–0.90	5–7
11	*Phragmites australis –(Tamarix hispida)*	Solonchakous residual-meadow -boggy with takyric features	0.67–0.73	>2.5
12	*Halostachys belangeriana –Tamarix ramosissima –Tamarix hispida*	Residual-meadow clayey and loamy solonchaks	1.30–2.07	4–5

Consequently, the root systems of many tugai plant species are large in size, branching, have several levels, display high plasticity, and are capable, for example, of developing adventitious roots on trunks and stems (Mailun 1973). Because of these features, both poplar, the main edificator, and species of tamarisk can germinate on loamy, sandy loamy, and sandy deposits with different depths of GWL (Table 1). In the Amudarya delta groundwater may occur at a depth of 4–5 m near the river bed under poplar (*Populeta arianae*) groves, whereas in the irrigation zone it occurs at a depth of 1–1.5 m. The authors' investigations in tugais of the Amudarya Delta have shown that endogenic factors, which cause inter- and intraspecies competition in plant communities, are also very essential for the development and degradation of tugai communities (Treshkin and Kouzmina 1989).

Even 60–100 years ago, natural fluctuations of moistening in deltas and floodplains of desert rivers resulted in the extinction of tugai in some places and in the formation of new tugai on fresh deposits in others. These changes in pattern were the reason for N.P. Grave (1936) to define tugais as nomadic forests. At the present time, however, the environment in arid regions has fundamentally changed due to irrigation. Currently, because of regional anthropogenic influences, regulation of river flow – as well as local ones – cutting, overgrazing, fires, freight transport and others, tugai communities of the Central Asia region are undergoing processes of desertification. These processes are most pronounced in deltas, where the Amudarya River has become so shallow that its waters do not reach the Aral Sea. Nevertheless, anthropogenic salinization of soils remains the main trigger for degradation of tugai ecosystems. Anthropogenic salinization is caused by different factors and may result in the formation of both meadow (hydromorphic) solonchaks and meadow-takyrsolonchak soils, along with residual-hydromorphic takyr-like solonchaks.

Salinization of alluvial soils in the southern Aral Sea region violates the natural dynamics of tugai ecosystems. The following processes are observed there:

- Overall halophytization of tugais at different stages of series, which often results in the disappearance of typical floral differences of tugai communities in different regions
- Disappearance of typical tree and bush tugai communities
- Development of different grass tugais. which formerly were not widespread
- Accelerated irreversible transformation of typical tugais into communities of halophytes
- Considerable loss of species diversity in communities of grass tugais and solonchak deserts in comparison with typical tugais.

The degree of degradation of floodplain ecosystems and their changing composition are features of the critical condition of ecological problems of rivers of the Aral Sea basin. All this requires urgent actions to protect, maintain the genetic stock and support the population of rare plant and animal species which are under a threat of disappearance. However, biodiversity protection of floodplain ecosystems of the Amudarya and its basin is a complicated problem. It greatly depends on solving problems, which arise from the whole system of ways to reconstruct the nature of the South Priaral and overcome the Aral ecological crisis. The problem is also made more difficult by the different approach to decisions on the part of States that have become independent in the post soviet period in Central Asia.

At the beginning of the 1990s, Central Asia faced the serious social-economic and ecological problem of exhaustion of the water resources, the degradation of the environment, reduction in productivity of the irrigated areas and the desertification of the Priaral; the leaders of these States realized the necessity of unifying mutual efforts to overcome the ecological crisis. The regional water strategy of the Aral Sea basin was to be worked out to the present time by mutual actions, directed at elaborating ways and mechanisms to satisfy the national interests of every State in connection with the national problems of long-term management of water resources and the environment. The return of the Aral Sea to its former size was admitted to be unrealistic, because of the lack of water resources. The creation of the new nature-anthropogen complex in Priaral and especially in the delta of the river, which consists of wet zones, forest-protected areas and lakes rich in fish, as well as the area of river deltas and the floor of the dried Aral Sea. The ecological part of the strategy foresees supporting and developing the maximum possible useful biological diversity of the plant and animal populations.

The problem of preserving tugai forests at present can be solved only by government or non-government organizations, because it is connected with the redistribution of water resources, a problem, which has become an ecological -political and social-economical problem. Unfortunately, at present we should state that the problem is not being solved, but is being ignored. Not long ago the reserve Tigrovaya Balka located in Tadjikistan was the largest tugai massif in Central Asia, but military action in this republic led to the complete loss of this unique reserve. The fate of the numerous reserves in Turkmenistan is in question. So, the Convention of the Biological Biodiversity, signed by many countries of this region at the UN conference on Environment and Development in 1992 (meeting of the leaders on the Problems of the Earth in Rio de Janeiro), which came into effect in 1993 as an international statute, is being carried out only formally and has only a declarative character, although the Convention requires the government to take steps to provide consideration of the necessity of preservation and stable use of the biodiversity in the management and cultivation of natural resources. For this reason, the consideration and analyses of the modern condition of the tugai forests in the context of these problems are of great importance, as the tugai forests are related to the relict forests, they have a composition of distinctive flora and fauna and lie like an island of life standing out against the background of the desert landscapes transformed by humans. Moreover at present the floodlands of the River Amudarya have become the only place in Central Asia where the main area of the tugai forests are concentrated.

The most concrete complex of actions to reproduct and maintain the biological diversity of floodplain ecosystems in the Aral Sea basin must foresee:

- Phytomeliorative works in the river valleys and in the riparian zones of canals and reservoirs must be combined with regulating pasturing and other forms of agricultural activity in the floodplains, including ploughing, irrigation agriculture and laying-in of hay and firewood. On the patches that are not flooded, it is necessary to use water to form viable tree and bush springs of native species that are well adapted to the environmental conditions. On soils with low and average salinization it is advisable to use the following species: *Elaeagnus orientalis*, *E. angustifolia*, *Tamarix litwinowii*; and on the less salinized soils *Salix aegyptiaca*, *S. excelsa*, *Populus diversifolia*, *P. anana*, *P. alba*, *Tamarix florida*, *T. meyeri*, *T. ramosissima*.
- Taking into consideration the environment formation, the relict, scientific and practical meaning of tugai vegetation, it is necessary to protect tugais and their various ecotonal variants within the limits of international UN programs on the environment in all the countries of Near, Middle and Central Asia, where these valuable communities are still preserved, but are becoming extremely rare.

References

Bakhiev A, Treshkin SY, Kouzmina ShV (1994) Karakalpakia tugai's of modern state and their protection. Karakalpakistan Publishing House, Nukus

Drobov VP (1950) Tree and bush vegetation of tugais in Karakalpakistan. Works on productive forces of Uzbekistan, vol 1, Moscow, pp 55–107

Grave NP (1936) Tugai jungles in the lower part of the Amudarya River. Moscow

Kouzmina ShV, Treshkin SY (1997) Soil salinization and dynamics of tugai vegetation in the southeastern Caspian Sea region and in the Aral Sea coastal region. Eurasian Soil Science, vol 60, no 6, pp 642–649

Mailun ZA (1973) Tugai vegetation. Plant cover of Uzbekistan. Fan, Tashkent, pp 303–375

Novikova NM, Bakhiev A, Treshkin SY (1996) Dynamics of soil cover and vegetation in deltas of the Amudarya River. Problems of desert development. Ylym, Ashkhabad, no 6, pp 36–44

Treshkin SY, Kouzmina ShV (1989) Structure of tree and bush tugais in the lower part of the Amudarya River. Vestnik, Nukus, no 4, pp 35–39

Treshkin SY, Kouzmina, ShV (1993) Changes in floodplain forest ecosystems of the Amudarya and Sumbar rivers as a result of anthropogenic influence. Problems of desert development. Ylym, Ashkhabad, no 2, pp 14–19

Treshkin SY, Bakhiev A (1998) Present status of the tugai forests in the lower Amudarya basin and problems of their protection and restoration. Ecological Research and Monitoring of the Aral Sea deltas. UNESCO, Paris, pp 43–54

Soil Crusts in the Amudarya River Delta: Properties and Formation

Arieh Singer, Amos Banin, Lev Poberejzsky and Moshe Gilenko

Keywords. Salt crusts, solonchaks, takyr soils, crust micromorphology, soil development

Abstract. Particle size distribution, salt content and composition, mineralogy and micromorphology of a solonchak salt crust and a takyr crust from the Amudarya River Delta (ADRD) were determined. In the (salt-free) salt crust, 100-μm particles dominate, while silt and fine sand are present in minor quantities only and clay is absent. The soil below the crust has a similar particle size distribution. In the takyr crust, fine particles dominate. In both crusts, quartz dominates, and is accompanied by calcite and mica. Clay minerals (chlorite and kaolinite) appear in minor amounts only. Cl is the major anion among the soluble salts, followed closely by sulfate. The salt crust contains four times more salt than the takyr crust. The salt in the salt crust appears in the form of crystallites of halite and Na, Mg sulfates such as thenardite and epsomite, in addition to gypsum. The size of the interlocking crystallites is 5–10 μm. The takyr crust is highly porous, with fine pores on the upper surface, and coarse pores in the bottom part of the crust. It is proposed that particle size distribution of the sediment and groundwater are among the important factors in determining the development pattern of Amudarya River delta soils. A fine particle size composition and absence of groundwater (including drainage water) close to the surface will lead to takyr formation. A relatively coarser particle size distribution and the presence of groundwater close to the surface will lead to solonchak (+salt crust) formation.

Introduction

The soils of the Amudarya River Delta (ADRD) can roughly be divided into soils associated with and affected by the floodplain (former and present) of the river. These include wetlands consisting of hydromorphic meadow and bog soils with a relatively high clay content, that are slightly to moderately saline; these soils, that were closest to the river, were subjected to annual spring flooding and were therefore only in limited agricultural use; their extent is approximately 800 000 ha. Other soils are hydromorphic meadow and bog soil formed on alluvium, that are non-saline or only slightly saline; their extent is about 148 000 ha; finally, to this group of soils also belong hydromorphic meadow soils that are mildly to strongly saline and can consequently be termed solonchaks; their extent is over 460 000 ha. A second group of soils, more removed from the river bed, are the takyr soils; these soils are mildly to moderately saline, and partly also sodic; their extent is over 1 million ha; associated with them are gray-brown, sandy to loamy soils that are slightly saline; their extent is 306 000 ha. Finally, more removed from the river bed and on more elevated terrain to the west, are shallow, stony soils that are saline to varying degrees, from slightly saline when they are sandy, to more developed gypsic soils; their extent is 740 000 ha (Soil Survey Staff 1969).

This distribution of the ADRD suggests that with time and development, the river sediments develop either into strongly saline solonchak types of soils, or into less saline takyr types of soils. The most salient features of most solonchaks and takyrs are their crusts. In the following, some features of one solonchak crust and one takyr crust are given.

Sites and Methods of Examination

Sites

Sampling site I. Salt crust. Location: 43°30'/59°56'. The surface is covered by a nearly continuous whitish, 2–3-mm-thick salt crust (Fig. 1a). Small patches of reeds intervene between the salt covered surfaces. A core was taken down to 1 m depth. The profile is uniform, fairly wet, of a gray (10 YR 6/1, Munsell) color. Groundwater was encountered at 90 cm depth. The texture was silty at the surface, becoming more sandy (fine sand) with soil depth.

Sampling site II: takyr crust. Kasakhdaria. Location: 43°25'/59°35'. Cultivated fields are mainly on the northern side of the village. A sample was taken from an uncultivated area to the south of the village. Part of the surfaces was wet after some light rains. The dry surface portions were cracked into polygons, of roughly 5–10-cm diameter (Fig. 1b). The surface, partly occupied by bushes, was covered with a not very firm, undeveloped crust about 0.5 cm thick. Only the crust was sampled. The texture of the soil was silty clay.

Sampling site III. Muynak. Location: coord. 43°50'/59°05'. Stabilized surfaces, with a fairly dense cover of planted tamarisk-jangill. The sample was taken from an uncultivated field about 25 m east of a 10–12-m-wide ditch filled with water, that emptied a (storage?) lake near to Muynak into the Aral Sea. The sample of surface soil (up to 25 cm depth) was wet and had a sandy silt structure. No crusting could be observed. The wetness of the soil was due to rains that fell during the preceding night.

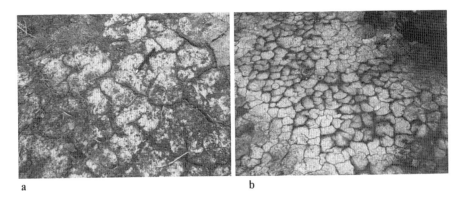

Fig. 1.a.,b. Soil crusts in the Amudarya River Delta. **a** Salt crust at site I; **b** Takyr crust at site II

Soluble salts were extracted and analyzed by ICP and ion chromatography. Particle size distribution was determined by the sedimentation method and also by laser. Mineralogy of both natural and salt-free material was determined by XRD. The micromorphology and chemical composition of the crusts was examined by scanning electron microscopy, to which an EDSA was attached.

Results and Discussion

Particle-Size Distribution

The salt-free particle-size distribution of the salt crust is given in Fig. 2a as determined by laser and in Fig. 2b as determined by sedimentation. Particles of 100 μm equivalent diameter absolutely dominate. Coarse sand and clay are negligible; silt and fine sand are present in minor quantities only. The soil below the crust has a very similar composition. In the takyr crust from site II, on the other hand, the maxima had shifted to lower sizes, with a major peak at about 65 μm, and a smaller peak at 8 μm. This corresponds to finer sand and a sizable silt fraction. The soil at site III resembles the takyr crust. That suggests that takyr formation is favored when the particle–size distribution of the material is relatively finer.

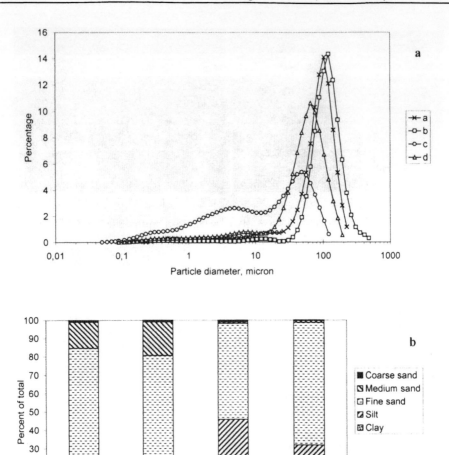

Fig. 2.a.,b. Salt-free particle size distribution of the crusts and soils. **a** as determined by Laser, *a* salt crust, site I; *b* soil below crust, site I; *c* Takyr crust, site II; *d* soil-plough layer, site III; **b** as determined by the sedimentation method

Mineral Composition

In the (salt-free) coarse sand of the solonchak crust (site I), quartz dominates, being accompanied by sizeable amounts of calcite and mica. Chlorite and kaolinite appear in minor amounts (Fig. 3.a). Only quartz is present in the medium and fine sand. In the silt, quartz is accompanied by kaolinite, mica, chlorite and feldspar. Below the crust, only quartz and calcite were identified in the coarse sand. In the finer fractions, calcite disappears, and feldspar, kaolinite, mica and chlorite appear in minor quantities. In the coarse sand of the takyr crust, quartz is accompanied by minor calcite. The fine sand has a similar composition, but in the medium sand, mica, chlorite and kaolinite are sizeable (Fig. 3.c). The silt has a similar composition. In the clay fraction also, quartz dominates, with minor kaolinite and mica. As in the crust of site I, calcite decreases with decreasing particle size.

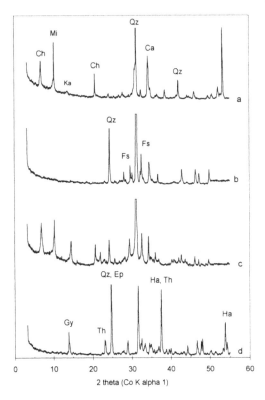

Fig. 3.a.-d. Mineral composition of crusts and soils as determined by XRD. a salt-free coarse sand of the salt crust at site I; b salt-free medium sand of the salt crust at site I; c medium sand of the takyr crust at site II; d bulk material including salts from the salt crust at site I; *Ch* chlorite; *Mi* mica; *Ka* kaolinite; *Qz* quartz; *Ca* calcite; *Fs* feldspar; *Gy* gypsum; *Ha* halite; *Th* thenardite; *Ep* epsomite

The mineral composition of the plough layer of site III is similar to that of the crust in site II, but calcite is preserved even in the fine-particle size fractions. The data indicate that there are no marked differences in the mineral (salt-free) composition of the solonchak and takyr crusts, and also that of the muynak soil.

Soluble Salts

Soluble salts were extracted from the soils (or crusts) by shaking in distilled water at a 1:10 ratio. Soluble cations were determined by ICP and soluble anions by ion gas chromatography. Soluble salt composition is given in Tables 1, 2 and 3. X-ray diffractometry was carried out using a Philips Model 1720 diffractometer.

Salts in the Crusts

From Table 1, it can be seen that the crust sampled from site I consisted of about 19% salt. Evidently, when the crust was sampled in the field much silicate material from below the salt crust proper, that adhered to it, was included in the sample. The salt content in the soil below the crust decreased to 3.1%. The crust on site II contained only 5.7% salt. Only 0.3% salt were determined in the crust-free soil from site III. The electrical conductivity (EC) measured in the extracts reflected these salt contents.

Table 1. Weight concentrations of salts in the crust/soils.

Site	Depth (cm)	Cations (%)	Anions (%)	Salt
I	Crust	5.11	14.05	19.15
I	1–15	1.02	2.07	3.09
II	Crust	1.71	3.98	5.68
III	0–15	0.11	0.19	0.30

Chloride was the major anion, followed closely by sulfate in the salt crusts of sites I and II (Table 2, and Fig. 4a,b). In the soil below the crust of site I and in the soil of site III, the concentration of sulfates was higher than that of chlorides. Nitrates were not present, except for traces in the crust of site II. The dominant cation in the salt crust of site I is sodium. Concentrations of Ca, Mg and K are relatively low (Table 2, Fig. 4b). In the soil below this crust, however, calcium dominates over sodium. Such is the case also in the soil of site III. In the crust of site II, concentrations of sodium and calcium are balanced. Potassium concentrations are also relatively high in the soil below the crust at site I.

The equivalent concentrations of Ca^{2+} and SO_4^{2-} are roughly equal in the soil of site I and the crust of site II, both being in the range of 20–30 $mEql^{-1}$. In the crust of site I, Ca concentration is also within this range. This suggests saturation with respect to solid phase $CaSO_4$ during the extraction and may have caused underestimation of total salt content in these soil samples due to incomplete extraction of $CaSO_4$.

Table 2. Ionic composition of soluble salts in 1:10 soil:water extracts of some Amudarya River Delta soils. Table 2.a. Anions; Table 2.b. Cations

2.a. Anions (in $mEql^{-1}$)

Site	I	I	II	III
Depth (cm)	Crust	1–15	Crust	0–15
Cl^- Conc	159.3	17.7	54.1	0.2
(%)	54.9	43.9	62.9	6.7
SO_4^{2-} Conc.	131.1	22.6	31.7	2.7
(%)	45	56.1	36.8	90.0
NO_3^- Conc.	0	0	0.4	0.1
(%)	0	0	0.4	3.3
ΣA	290.4	40.3	86.2	3.0

2.b. Cations (in $mEql^{-1}$)

Site	I	I	II	III
Depth (cm)	Crust	1–15	Crust	0–15
Na^+ Conc.	189.0	11.4	38.3	0.8
(%)	81.8	24.3	47.5	15.4
K^+ Conc.	1.0	4.4	0.8	0.3
(%)	0.4	9.4	0.9	5.8
Mg^{2+} Conc.	13.5	4.4	4.8	0.5
(%)	4.3	9.4	5.9	9.6
Ca^{2+} Conc.	28.0	26.5	36.7	3.6
(%)	1.2	56.5	45.5	69.2
ΣC	231.4	40.3	80.6	5.2
RE	11.3	0	3.3	0

$$RE = \frac{\Sigma A - \Sigma C}{\Sigma A + \Sigma C}$$

It appears, even on the basis of these preliminary analyses, that Na and Cl are not the dominant cation and anion in this environment, and that it is dominated by Ca and SO_4.

The RE(%) is negative in three out of the four samples. This is probably due to the lack of data for carbonate/bicarbonate concentration in the extracts.

The soil from site III has low salinity and is practically well-leached. It is sandy-silty in texture and represents the eventual salinity state to which dried lake -bottom topsoils may evolve, following decades of downward leaching (by rain) and in the absence of capillary rise. Soils in sites I and II, on the other hand, represent conditions of active capillary-rise from shallow groundwater.

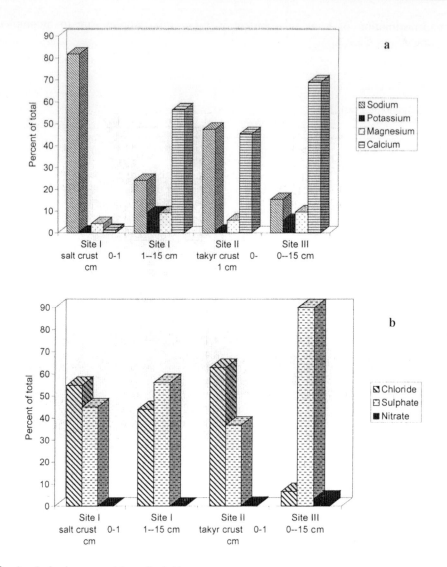

Fig. 4.a.,b. Ionic composition of soluble salts of the salt crust on site I. **a** ions; **b** cations

These results are corroborated by X-ray diffractograms of untreated (except crushing) bulk material. For the crust material on site I, a strong diffraction at 2.78 Å indicates the presence of halite (NaCl) and possibly of thenardite (Na$_2$SO$_4$) (Fig. 3d). A line at 7.50 Å indicates the presence of gypsum (CaSO$_4$·H$_2$O). The relatively high concentration of sulfates suggests that thenardite is one of the principal sulfate salts, though the presence of some bloedite (Na$_2$Mg SO$_4$·4H$_2$O) or konyaite [Na$_2$Mg(SO$_4$)·5H$_2$O] cannot be excluded. The line at 4.19 Å suggests also the presence of epsomite ($MgSO_4 \cdot 7H_2O$). Salt concentrations in the other materials were too low to be detectable by X-ray diffraction.

Micromorphological and Chemical Examinations

Micromorphology and chemical composition of preserved portions of the salt crust were examined using a scanning electron microscope (JEOL 5410 LV), with an EDAX attachment). Natural, undisturbed fragments of the crust were mounted on stubs, and coated under vacuum with either carbon or gold.

The salt crust was observed to consist of one layer, in which salt crystallites were arranged in one dense, interlocking matrix (Fig. 5a). Chemical scan showed the crystallites to be composed of sodium and magnesium sulfates and chlorides (Table 1). Spot analysis of individual crystallites suggested that the crystallites were either sulfates or chlorides, but not mixed salts. By their cubic habit, halite crystallites were easily recognizable (Fig. 5b). The chemical composition of these crystallites (by spot analysis) indicates only Na and Cl, with some small, additional amounts of Mg. Some of the crystallites were pitted by secondary solution channels. The size of the crystallites varied between 5 and 10 µm.

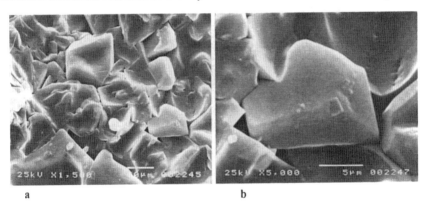

Fig. 5.a.,b. Micromorphology by SEM of portions from the salt crust at site I. **a** matrix of salt crystallites; **b** halite crystallites

In the chemical composition of many other crystallites, sulfates could also be identified. The most common cations associated with the sulfates were Na and Mg. By morphology alone, it was impossible to determine whether these salts were orthorombic thenardite (Na_2SO_4), monoclinic mirabilite, or the combined salt bloedite ($Na_2Mg\ SO_4 \cdot 4H_2O$).

Frequently, the dense halite interlocking mosaic layer was overlain by discontinuous gypsum crystallites. The gypsum crystallite appeared to be in a state of decomposition.

In conclusion, SEM+EDAX examination of crystallites support the data obtained by the chemical analysis of the dissolved salts from the salt crust, namely, that the major salt is halite, followed by sulfate salts such as Nasulfate (thenardite) and possibly mixed Na, Mg sulfate salts, such as bloedite and mirabilite. Casulfate salts (gypsum) were observed to form as second-generation salts on top of the halite layer.

The upper surface of the takyr crust, of a thickness varying between 5 and 8 μm, was smooth (between the cracks) and shiny. The upper surface appeared to be dense, but under magnification was seen to contain many fine pores of a diameter varying between 5 and 10 μm (Fig. 6a). The bottom part of the crust was outright vesicular, with the diameter of the vesicles varying between 200 and 300 μm (Fig. 6b). Chemical analysis indicated particles in the surface part of the crust to consist of mica, quartz and Fe, Al silicates.

Fig. 6.a.,b. Micromorphology by SEM of portions from the takyr crust at site II. **a** upper surface of the crust; **b** bottom part of the crust

Proposed Pathways for the Development of Soils and Crusts from the Aral Sea Bottom Following its Exposure by Desiccation

Salt efflorescences on soils are well-known phenomena in arid and semiarid regions where actual annual evapotranspiration greatly exceeds precipitation. Driessen and Schoorl (1973), for example, described evaporites from Turkish soils subjected to an annual rainfall of 240 mm, and Guzmuzzio et al. (1982) reported the mineral composition of efflorescences in saline soils of Spain. Eghbal et al. (1989) described soils with evaporites from California that have a xeric to aridic (150 mm a^{-1}) moisture regime. The common sources of the salts are seepages of runoff, springs and dry river channels that drain salt-containing rock formations. In less common cases, the origin of the salts was shown to be atmospheric deposition (Singer et al. 1999). In the present case, it is evident that the salts in the crust had accumulated by the evaporation of groundwater, though a contribution by salt-containing dust is probable also.

Vast areas of the former Aral Sea floor have become desiccated and are in the process of developing into soils. The study of salt distribution in the ADRD may help in predicting into what type of soils the exposed Aral Sea floor will eventually evolve.

Bottom material that is fine-sandy to sandy in texture desiccates to form sand dunes. There will therefore be only limited capillary rise of groundwater and accumulation of salts on the surface. Crust formation will be minimal. The land surface will remain bare, and serve as a source area for short-range aeolian transport of coarse-grained particles. The shifting dunes will be mostly unconnected with groundwater but fed by the, albeit limited, local rainfall. In due time, vegetation can establish itself on these dunes.

If groundwater is close to the surface, with time, active capillary rise and evaporative processes will lead to salt accumulation and crust formation. The time required for the salt crusts to develop is estimated to be approximately 5–10 years. In the presence of the salt crusts, the surface will remain bare (no vegetation) because of the high salinity. This is the common pathway for solonchak soil formation.

Well-developed and cemented crusts, even if totally composed of salts, can be expected to protect the surface from wind erosion. Dust generation will be reduced, because dust entrainment from crusts will be reduced. While it can be expected that salt crusts will be more resistant towards deflation, electron microscopy has shown (see below) that the salt crust is frequently overlain by second generation powdery and fine-grained salt crystallites – not yet very consolidated.

Salt crusts will develop not only in the presence of groundwater close to the surface, but also when periodic flooding takes place. Salt crusts and salt pans develop upon evaporation from various small waterbodies in the Amudarya River Delta. The source of these water bodies is not clear. They possibly represent (1) entrapped sea water or (2) surface runoff of rainwater that had collected in depressions. Mostly likely they are (3) artesian seepage of shallow groundwater fed by drainage water from the irrigated areas of the delta lying to the south.

In the absence of groundwater (or surface water) close to the surface, salt crusts will not form, or will form only to a very minor degree. Clay crusting will take place and ultimately takyr soils will form. The rate of formation of these crusts is much slower than the salt crusts. Dust entrainment from these surfaces will be much more pronounced. After some time, vegetation of shrub-like xerophytes and halophytes (such as tamarisks and jangill) will take root in these areas. As vegetation covers the surface, deflation will decrease. Vegetation recovery apparently can take place relatively rapidly, within 15–20 years of the exposure of the sea bottom. The local rainfall (annual average 100 mm) appears to be sufficient to support this vegetation in non-cultivated areas, stabilizing their surface against further wind erosion. An example of this pathway are the extensive shrub-covered areas east and southeast of Muynak.

Acknowledgements. This research was supported under Grant No. CA16-013 US -Israel Cooperative Development Research Program, Economic Growth, US Agency for International Development.

References

Driessen PH, Schoorl R (1973) Mineralogy and morphology of salt efflorescences on saline soils in the Great Kenya Basin, Turkey. J Soil Sci 24:437–451

Eghbal MK, Southard RJ, Whittig LD (1989) Dynamics of evaporite distribution in soils on a fan-playa transect in the Carizzo Plain, California. Soil Sci Soc Am J 53:898–910

Guzmuzzio J, Battle J, Casas J (1982) Mineralogical composition of salt efflorescences in a typical salorthid, Spain. Geoderma 28:39–55

Singer A, Kirsten WF, Bühmann C (1999) A proposed fog deposition mechanism for the formation of salt efflorescences in the Mpumalanga Highveld, Republic of South Africa. Water Air Soil Pollut 109:313–325

Skarie RL, Richardson JL, McCarthy GJ, Maianu A (1987). Evaporite mineralogy and groundwater chemistry associated with saline soils in eastern North Dakota. Soil Sci Soc Am J 51:1372–1378

Soil Survey Staff (Antizatorova et al.) (1969) Soil map of the Karakalpak Autonomous Region. Scale of 1:200 000. Ministry of Interior, Nukus

Irrigation and Land Degradation in the Aral Sea Basin

Nicolai Orlovsky, Michael Glantz, Leah Orlovsky

Keywords. Central Asia, desertification, drainage, pollution, salinization

Abstract. The Aral Sea basin has experienced the development of large-scale irrigation since the 1940s–1950s. This policy has resulted in an increased area of secondary salinized lands and other lands unsuitable for cultivation under conditions of a limited supply of irrigating water. The natural environment was seriously damaged, including a lowering of the Aral Sea level and deterioration in the living conditions of the local population. For a long time the population in the area and large segments of the scientific community were misinformed about or unaware of the real ecological state of the region. This, coupled with the Soviet Command and Control system of government, made it difficult, if not impossible, to take the necessary action toward environmental remediation.

At present the Aral basin hosts a very complicated system of canals, reservoirs, irrigated fields and hydrotechnical construction including 7.9 million ha of irrigated lands, 323 200 km of irrigation network and 161 800 km of collectors and drains. In spite of such a wide scale of water construction, irrigation techniques themselves are the same as in Middle Ages – border and furrow irrigation. Under such a system of land use, water loss in agriculture is up to one -fifth of the total water intake for irrigation.

The total volume of drainage water from the Aral Sea basin is more than 33 km^3 a^{-1}. Of this volume, more than 20 km^3 a^{-1} returns to the rivers and about 13 km^3 a^{-1} flows to large depressions or other chaotic areas in desert pastures.

Improper irrigation leads to secondary salinization, waterlogging, and, eventually, desertification. Diversion of drainage water to the rivers results in deterioration of river water quality; diversion to the marginal desert areas results in waterlogging of pastures and environmental degradation of the total area.

Introduction

In recent years, the development of arid lands has been taking place at an increasing rate all over the world, and irrigation has become the most important factor for the improvement of productivity in these potentially arable arid areas. However, such human activities as irrigation, water drainage, regulation of the surface runoff, modification of hydrographic networks, and the creation of new water reservoirs and collectors of return flow, affect the stability and rhythm of natural processes. They cause significant changes in the ecological situation at the scale of entire regions. Thus, they often bring about negative consequences not envisaged by planning, construction, and land development authorities. Such consequences include, for instance, secondary salinization, waterlogging, and increases in the mineralization of the return flow and changes in the groundwater level. These processes are characteristic for all areas subjected to intensive development of irrigation, especially those located in arid zones. An extreme example of this process of adverse change is one of the largest inland drainage basins, that of the Aral Sea.

The Aral Sea basin includes all the territories of Uzbekistan, Tajikistan, and Turkmenistan, parts of the territories of Kazakstan and Kyrgyzstan, and northern provinces of Afghanistan and Iran. The total basin area constitutes 156.7 million ha within the borders of the former USSR. Agricultural development in the Aral Sea basin during the past 40 years was based on large-scale irrigation, massive mineral fertilization and application of toxic agents and defoliants. Crop rotation was not practiced, and the necessary volumes of water for irrigation were exceeded twice to three times. Irrigation of a total of 8 million ha of the agricultural lands (7% of the total area of the region) resulted in sharp degradation of the natural environment. Regions around the Aral Sea have been declared an ecological disaster zone.

Our purpose is to show the wide scale character of anthropogenic impact on the natural conditions of the region, to consider the consequences of such anthropogenic interference, and to identify lessons for future reference from the incorrect and unreasonable agricultural and water use policies in the region.

Scales of Land Reclamation

Irrigation-based agriculture in the Aral Sea basin has a long history, dating back several centuries. Traditional forms of land use – oasis irrigation agriculture and nomadic stockbreeding, which were adapted to the local ecological conditions – made it possible to develop and use millions of hectares of arid lands. Thus, lands of ancient irrigation in the lower reaches of the Amudarya and Syrdarya Rivers exceeded the area covered by modern irrigation networks by more than three times.

In the period of the greatest development of ancient irrigation (reached between the 3rd century B.C. to the 3rd century A.D.), the total area of cultivated lands fed by irrigation channels was estimated as 3.5 to 3.8 million ha. This included areas of stable irrigation of about 1.7 million ha. It is necessary to note that the coefficient of land use (defined as the ratio of irrigated lands to all possible irrigated agricultural lands) did not exceed 10% in this area at that time (Andrianov 1991).

Methods and techniques of irrigation and system of land use – crop rotation and other methods of land cultivation – evolved in response to existing conditions over the course of a few thousand years and were passed on from generation to generation. These techniques were suited to small irrigated plots. The specific feature of irrigated agriculture in that period was the subdivision of irrigated lands into small plots (0.3–0.8 ha) with small permanent earth banks. There was a permanent irrigation ditch on a plot with trees along it. The irrigation was carried out without drainage: trees along irrigation ditches served as biological drainage, reduced the deficit of humidity in the lower layer of air and reduced evaporation from the soil. Water was economized and soil was protected from salinization, because small earth banks served as "dry" drainage.

From the middle of the 20th century, because of Soviet style of "collectivization", the area of irrigated plots increased by seven to ten times, with plot size on average 3.5 ha. It became easier to use machinery, but at the same time water use increased sharply since irrigation with drainage became the technique for leveling of humidification by furrow irrigation, which means allowing some of the irrigation water to overflow the ditches and channels. In the beginning of the 1950s, the area of individual irrigated plots increased again by more than five times, reaching up to 20–60 ha and as high as 100 ha. However, until 1960, the increase in water use in the region did not significantly affect either the water balance of the Aral Sea or the quality of river water. The capacity of the natural environment favored the self-purification process of water.

In order to achieve ambitious goals in the 1950s, such as the so-called cotton independence of the former USSR, intensive water works construction was started, and the area of irrigated lands in the Aral Sea basin increased very fast. In the period from 1913 to 1950 the irrigated area increased by 0.5 million ha. During 1950–1990, it increased by 3.7 million ha (Table 1). At present, the total area under irrigation is 7.9 million ha. Of this total, 4.6 million ha belto the Amudarya River basin and 3.3 million ha to the Syrdarya River basin.

Table 1. Development of irrigated agriculture in the Aral Sea basin (×1000 ha)

State	1913	1940	1950	1960	1970	1980	1990
Kyrgystan	758	912	737	889	883	995	1042
Tajikistan	347	377	300	398	468	617	694
Turkmenistan	381	412	352	434	463	927	1245
Uzbekistan	1485	2138	2053	2389	2751	3476	4155
Kazakstan				305	373	635	760
Total	2908	3839	3442	4411	5118	6610	7896

At present, the Aral basin hosts a very complicated system of canals, reservoirs, irrigated fields and hydrotechnical constructions. In 1987, there were 967 irrigation systems here, and 915 hydrotechnical installations in the sources of irrigation, among them 260 dams at water intake points. The total length of the irrigation network reaches 323 200 km; the length of the main and interfarm canals is 53 500 km; only 13 000 km of these were constructed with a leak-proof cover. In an attempt to maintain a normal productive state of the lands,

161 800 km of collectors and drains were functioning (40 000 km of them being main and interfarm ones); 7700 wells were used for irrigation and 6300 wells for vertical drainage. For regulation of the river flow rate, 450 water reservoirs were constructed with a total volume of 41.4 km^3, of which 39.1 km^3 was useable. To provide mechanized water pumping, 2320 pumping stations were operating with a power consumption rate of 41.4 million. kW. The volume of pumped water was about 55 km^3 per year (Ziadulaev 1990; Rakhimov 1990). The region's water system is one of the most sophisticated systems of water use in the world.

In spite of such wide-scale construction of water-use units, the irrigation techniques themselves were preserved from the Middle Ages – border (land surface flooding) and furrow irrigation. In 1985, 71.6% of the total irrigated area (668 000 ha) was irrigated by furrows. Of this, 4.6% was irrigated by mechanical means, and 2% by sprinklers. Under such a system of arid land use, the total water loss in agriculture is up to one fifth of the total of water intake for irrigation. For example, in 1987, the water intake for agricultural needs in the Aral Sea basin was estimated at 125.4 km^3, and water losses by transportation were 21.9%. These losses, especially those resulting from infiltration, can be explained by the fact that all the main canals, major parts of interfarming (94%) and intrafarming (95%) irrigation networks and all the collecting and drainage network were made directly above ground on the earth's surface.

The relative consumption of water (gross weight) for irrigation is 12 000 m^3ha^{-1} in Kyrgystan, about 16 000 m^3 ha^{-1} in Kazakstan and Uzbekistan, and about 20 000 m^3 ha^{-1} in Tajikistan and Turkmenistan. Each of these figures far exceeds the biologically required norms of irrigation. The total coefficient of efficiency of water use does not exceed 0.60, which indicates the low technical level of water use in Central Asian agriculture.

However, water use in the Aral Sea basin does not improve with time, but even deteriorates. If in 1970 in Turkmenistan 15 740 m^3 of water (gross weight) was used for the irrigation of 1 ha of cotton, in 1980 it had increased to 21 560 m^3, i.e. 5820 m^3 more. If the relative consumption of water volume ha^{-1} in 1980 was as it had been in 1970, it would have meant that 5.6 km^3 of water would not have been needed in 1980. This quantity exceeds the water resources of Turkmenistan by 1.5 times (Kolodin 1982).

The Present State of Irrigated Lands

The low technical level (and inefficiencies) of these irrigation systems, high level of water intake and use for irrigation have resulted in significant losses of water resources, subsequent groundwater level increase, and waterlogging of irrigated lands. The unsatisfactory condition of the land is characteristic of 60% of the total area of irrigated lands where groundwater lies at a depth from the surface of only 1–2 m. In areas of new irrigation the groundwater level increases on average by about a half a meter annually, and sometimes even by as much as 1 m. During the first decade after the construction of the Karakum Canal, the area with groundwater table at a depth of less than 2 m increased from 75 000 to 145 000 ha in the Murgab River Delta; and from 3000 to 67 000 ha in the Tedjen River delta. In 1990, in Turkmenistan groundwater levels of 1–2 m were observed on 47.7% of its irrigated territory (591 000 ha of 1 245 000 ha of irrigated lands) (Kolodin 1994). Depending on the specific natural situation, such a rise in the groundwater level leads to waterlogging and salinization of lands. The area of salinized irrigated lands in Turkmenistan is 89%, of which 12.7% are heavyly and very heavyly salinized; in Uzbekistan the figures are 51.4 and 5.7%, respectively (Table 2).

Table 2. Dynamics of salt-affected irrigated soils in the 0–100 cm layer, thousands of ha / % of total area

State	1982	1983	1984	1985	1989
Uzbekistan	1925.2/52.2	2087.6/55.9	1948.6/53.0	2044.3/52.0	2132.1/51.4
Turkmenistan	659.0/85.0	762.7/87.6	805.8/86.7	869.1/85.0	1085.0/89.0
Tajikistan	115.8/26.1	118.8/18.5	116.9/23.9	115.1/17.6	106.3/15.4
Kyrgystan	139.2'/13.8	94.6/9.2	127.0/12.7	142.7/14.1	119.1/11.5

Besides irrational water use, the salinization of soil is caused by the fact that large expanses of newly irrigated lands occupy, as a rule, either not drained or slightly drained delta plains, alluvial piedmont plains and the lower parts of alluvial fans. These lands are either salinized in their virgin condition or are susceptible to salinization by irrigation. Therefore, both new irrigation lands and areas of potential irrigation in the plain portion of desert areas became salinized and require human intervention for reclamation purposes. In addition, the strong orientation towards the extension of existing areas under irrigation as a result of new construction and the lack of attention to the condition of the older irrigated lands have resulted in increases in the demand for water. Newly irrigated lands used to be monitored for degradation, e.g. salinity increases, because the lands that had formerly been left fallow for long periods of time were no longer available for drainage.

In the Syrdarya River basin the area of salinized lands increased from 1 076 000 ha to 2 150 000 ha, and part of it (15%) had to be excluded from land use.

Consequences of Drainage Collection Network Activity

To enhance the fertility of the soils and to improve their general condition, the construction of a collector-drainage network started in the early 1960s. Over a three-decade period, 161 800 km of this network was built. Today, the project has both positive and negative aspects. One positive aspect is that the coefficients of land use on the old-irrigated land increased from 0.3 to 0.7. Consequently, intraoasis drainage solonchaks (salinized soils) were washed; marshes and small lakes, where carriers of malaria and other diseases could breed, were dried. Simultaneously, some negative aspects appeared. First of all, water consumption per unit area and production increased in the oases, and larger volumes of drainage water were generated. Drainage flow in the Aral Sea basin is estimated to be about 34 km^3 per year; 13 km^3 of this volume returns to the Syrdarya River and is utilized, about 3 km^3 enters the peripheral water reservoirs, and about 7.5–8 km^3 returns to the Amudarya River. Drainage water flow into desert depressions amounts to about 10 km^3 (Dukhovny et al. 1984). In 1968, 0.3 km^3 of mineralized groundwater was withdrawn from the Murgab, Tedjen and Prikopetdag oases. In 1981–1989 drainage flow varied from 1.55 to 1.89 km^3 per year and was withdrawn to the adjacent oases of the desert rangelands. As a result of irrigation, more than 4 million ha of solonchaks appeared (Zaletayev 1991). The return of 38% of the drainage water to rivers and sources of irrigation was the main cause of the deterioration in the quality of river and irrigation water. Collector-drainage water is characterized not only by high salt content, but also by its poisonous chemicals, defoliants, chemical fertilizers and heavy metals.

The diversion of drainage water to natural closed depressions of sandy deserts resulted in the formation of reservoir-like accumulators of collector, drainage and return flow water. At present, there are 2350 such water reservoirs, both small (in the interdune depressions) and large (such as Lake Sarykamysh and Lake Arnasai). The total area of these reservoirs combined with accumulators, is 7066 km^2. 330 000 ha of pastures are affected by these reservoirs, where valuable fodder species in vegetative communities are being changed into unpalatable species that are of little value.

After 30 years of wide-scale irrigation, 4.5 million ha of natural ecosystems in the region have been degraded (Babaev and Zaletaev 1990). At present in Turkmenistan 484 600 ha flooded, 5000 ha are dried up, 150 000 ha are flooded periodically, 450 000 ha are heavily waterlogged, and 2 300 000 ha are moderately waterlogged (Babaev and Babaev 1994).

The extensive introduction of drainage was not adequately prepared. Thus, in solving the local task of preventing the deterioration of irrigated lands, the drainage system damages the ecological situation of the whole region. It does this by causing a movement which brings about salt deposition and, when the drainage system was put to use, the accumulated salts rose to the surface.

Pollution of the Environment

Nowadays, the main trend in agricultural development for the intensification of production involves not only the introduction of irrigation and drainage, but also the application of fertilizers, herbicides, insecticides, fungicides and other chemicals. At present, cotton occupies up to 90% of all irrigated land in the Aral Sea basin. Each hectare of a cotton plantation receives on average 245 kg of nitrogen and about 125 kg of phosphorus. Nowadays, agriculture in the Aral Sea

Irrigation and Land Degradation in the Aral Sea Basin

basin annually consumes about 820 000 tons of mineral fertilizers: 472 000 tons of nitrogen, 213 000 tons of phosphates, 136 000 tons of potash fertilizers. Utilization of pesticides on each hectare of plowed land exceeds safe limits by several orders of magnitude. About 90% of utilized pesticides are not selective and are dangerous to insects, crops, animals, fishes, birds and people.

The total amount of pesticides removed by collector-and-drainage water is about 2.3% of the amount of pesticides used. About 5.5% of nitrate and 0.64% phosphates are removed from the cotton plantations during the growing season. Losses by transportation, storage and utilization (14–15%) also contaminate the air, soil and water sources. Thus, 40 900 tons of mineral fertilizers and 3650 tons of pesticides are washed out from irrigated fields and transported to open reservoirs (Rakhimov 1990). Sewage water also significantly contaminates water sources. In 1987, about 110 metric tons of oil products, 42 metric tons of iron compounds, 1540 metric tons of nitrogen, and 110 metric tons of phenols entered the rivers with 1.7 km^3 of sewage water.

More than 65% of the total amount of return water formed in the Amudarya basin returns to the river. The annual volume of drainage water returned to the river exceeds 5.5 km^3 with a mineralization level of 3.4–7.7 g l^{-1}. As a result, the mineralization of river water reaches 2–3 g l^{-1} in the lower reaches of the Amudarya and 1.8–2.2 g l^{-1} in the lower reaches of the Syrdarya. Besides salts, Amudarya water contains a large amount of heavy metals: manganese (1.3–2 times the maximum permissible concentration, MPC), iron (1.5–3.3 times the MPC), lead (5–10 times the MPC), cadmium (6–8 times the MPC); harmful organic compounds: oil products (36–46 times the MPC), phenols (400–1000 times the MPC); and other toxic compounds (Kolodin 1994). Therefore, Amudarya water does not meet the standard requirement for drinking water, and in the lower reaches, it is not even suitable for irrigation. The high level of mineralization of water leads to salinization of irrigated lands. Taking into consideration the content of minerals in water and the total irrigation quota for cotton annually, 10–33 metric tons of salts are brought to each hectare of cultivated land. The danger of chemical desertification is created. Its undesirable consequences are concentrated in the lower reaches of the Amudarya and Sirdarya Rivers.

Destabilization of the Environment

Most of the major changes in ecological conditions in many parts of the Aral Sea basin are the result of irreversible removal of water from the rivers for irrigation purposes and of the overregulation of river flow. First of all, the overregulation of river flow led to a decrease in the alluvium content of river water. About 1.3 million metric tons of humus, 0.12 million metric tons of nitrogen, 0.16 million metric tons of phosphorus and 2.4 million metric tons of potassium were deposited in the Amudarya River basin. About half of this amount remained on the arable land (Babaev and Alibekov 1996). At present, up to 70% of alluvium has been retained in the reservoirs, which has changed the soil formation processes. The overregulation of river flow brought about the elimination of high flood levels and changed the hydrological regime of the flooded lands, the deltas, and the river basin as a whole. Meadows, marshlands and landscapes of wetlands and delta-tugai complexes became affected by drying off and secondary salinization. Degradation of the soil and plant cover takes place. The drastic decrease of Amudarya and Sirdarya flow, which started from 1961, caused a disastrous drop in the Aral Sea level. According to data of 1994, the level of the Aral Sea dropped by 17 m. The width of the dried Aral Sea floor exceeded 120 km and its total area approached 3600 km^2. The volume of the Aral Sea decreased by 70% while its average salt concentration increased from 10 to 32 g l^{-1}.

A new and distinctive desert has appeared on the map of Central Asia – the Aral sandy-solonchak desert. After the final possible drying out of the Aral Sea, the natural borders of the new 565-million ha desert will be formed. It will be a typical anthropogenically created desert. Some have referred to the desertified region as the Aralkum.

The drop in the Aral Sea level affects all components of the region's environment – soils, water, fauna, flora, climate and humans. The degree of deterioration in the environment, the population's health, and agriculture in the Aral basin has territorial differentiation. A comparison of the natural conditions and farming dynamics in different parts of the Aral Sea basin shows that, in the upper parts of river basins, the ecological situations are better than in the lower parts. The pollution of the environment, the degradation of natural ecosystems and the decrease of flow are revealed mostly in the Aral Sea region where all negative processes of the basin – physical, chemical and biogenic aspects of desertification – are concentrated. The Aral Sea region includes the Karakalpakstan and Khorezm districts of Uzbekistan, the Dashkhovuz district of Turkmenistan, Kyzyl–Orda, and the southern part of the Aktubinsk district of Kazakstan.

The area of the Aral Sea region encompasses 472 900 km^2 and is inhabited by about 3 374 500 people. About 1 190 800 ha of the land are irrigated. Four zones can be distinguished according to differences in the characteristics of the desertification processes. They are the aquatic zone of the Aral Sea, the dried out coastal area, the river deltas, and the adjacent territories.

Development of Sand–Salt Storms

An important feature of the desiccated zone's landscape is the increase in the formation of dust storms that originate there. Up to ten dust storms are observed annually in the Aral Sea region. Wind-blown dust can travel a distance of 200 –300 km, and in some extreme cases up to 500 km. The main directions of their movement are: towards the southwest, to the oasis of the Amudarya River Delta (60%); and towards the west, to the rangelands of the Usturt plateau (25%). Therefore, in 85% of the cases, dust is transported above the sea, which has a significant reducing effect because the main part of the sand aerosol is deposited on the water surface (Grigoryev 1991).

There are differing opinions about the volume of dust material transported from the dried exposed floor of the sea. In our opinion, scientists from Kazakstan obtained the most reliable data on the basis of measurements and the modeling of heavy admixture transfer. It is estimated as ranging from 8 to 30 million metric tons of earth. With such an amount of transported material, the rate of removal of the dried sea floor soil is about 2 mm per year. This is close to reality, because during the last three decades a 6- to 9-cm-thick layer of surface soils was removed by the wind from the newly dried surface. According to these observations, in the Amudarya Delta region, about 90 000–100 000 metric tons (90–100 kg ha^{-1}) of dust has fallen annually on an area of 10 000 km^2; and on the Usturt plateau about 40 000–50 000 metric tons (31–39 kg ha^{-1}) have fallen within an area of 13 000 km^2. From 0.2 to 5 metric tons of salts ha^{-1} are deposited annually in the Amudarya Delta during sandstorms (Tsitsasov 1990).

The process of salt transfer in the Aral Sea region has continued for more than two decades. Therefore, the responses of ecosystems affected by this process have already appeared. From the beginning of the 1970s, a constant reduction in land productivity, as well as increased salinization of water resources, has been observed.

Degradation of Delta Ecosystems

Processes of delta ecosystems degradation have developed rapidly during 20 years – since the beginning of the 1960s till the end of the 1970s. The subsequent ecological changes are: transformation of the surface and groundwater; replacement of hydrophilic soils by semihydrophilic and automorphic ones, disappearance of glycophilic vegetation, coastal water and tugai species and communities (Borovsky and Kusnetzov 1979; Novikova 1985,1999).

In the 1980s the relatively stable spatial landscape system of this area was formed. It reflects the process of desiccation of deltas connected with the Aral Sea level decrease and diminishing of the flow of the Amudarya and Syrdarya Rivers. The process of desertification which is occurring at present in the deltas area is only slightly visible, because of the absence of the sharp change in plant communities which characterized the previous stage (Novikolva 1996).

Conclusion

The ecological crisis in the Aral Sea basin, and especially in the region surrounding the Aral Sea, is the result of: (1) The introduction of industrial methods of irrigation by using low levels of agricultural technology; (2) The rapid increase of the irrigated area together with a sharp decrease in water and soil quality as a result of contamination by pesticides, out-of-control mineral fertilization, and excessive irrigation. This resulted in the salinization of fertile soil, increased waterlogging, and erosion.

Irrigation of about 8 million ha (slightly more than 7% of the total area of land that is involved in agricultural production) changed the water balance and deteriorated the ecological situation of the whole Aral Sea basin.

Drainage, on one hand supporting optimal groundwater level and productivity of soil in the irrigated area, became on the other hand one of the dangerous factors of desertification and environmental pollution. The widespread use of drainage without careful planning and preparation solves only the local problem of not aggravating the condition of irrigated lands. At the scale of the whole region, this measure worsens the ecological situation because of the redistribution of the salts into the desert pastures, both from deep soil horizons into the surface and from oases areas. Consequently, overflooding, waterlogging, and salinization occur in territories of desert pastures.

At present, the environment of the Aral Sea basin is under stress and its self-regulation mechanisms are directed toward the establishment of a new equilibrium. This new equilibrium level is adequate for neither the region's environmental requirements nor for the living conditions of the local population. Considerable material and labor investments are required if there is any hope of recovering the degraded ecosystems.

References

Andrianov BV (1991) History of irrigation in the Aral Sea basin.. In: The Aral crisis: historic-geographical Retrospective. AK Nauk SSR, Moscow, pp 101–122 (in Russain)

Babaev AG Alibekov LA (1996) The basin of the Aral Sea in the "mountains-plains" system. Probl Desert Dev 1:3–8

Babaev AM Babaev AA (1994) Aerospace monitoring of dynamics of desert geosystems under watering. Probl Desert Dev 1:21–29

Babaev AG Zaletaev VS(1990) Ecological monitoring of key objects in arid zone. Probl Desert Dev 5:3–9

Borovsky VM Kuznetsov NT (1979) The Siberian rivers transfer as a solution of the Aral Sea. Probl Desert Dev 2:18–24

Dukhovny VA, Razakov RM, Ruziev IB, Kosnazarov KA (1984) The Aral Sea problem and nature conservation measures. Probl Desert Dev 6:3–15

Grigoryev Al A (1991). Ecological lessons of the past and the present. Nauka, Leningrad 251 pp

Kolodin MV (1994) Water resources in Central Asia and the problem of the Aral Sea. Probl Desert Dev 4–5:73–87

Kolodin MV (1982) Methods of effectiveness raising of water resources use in Turkmenistan. Probl Desert Dev 6:39–46

Novikova N (1985) Dynamics of delta plains vegetation of arid regions as a result of rivers flow transformation. In: Voronov AG, Vyshivkin DD (eds) Biogeographic aspects of desertification. Academy of Sciences, Moscow, pp 31–40

Novikova N (1996) Current changes in the vegetation of the Amudarya Delta. In: Micklin PD, Williams WD (eds) The Aral Sea Basin. Springer, Berlin Heidelberg New York pp 69–78

Novikova N (1999) Priaralye ecosystems and creeping environmental changes in the Aral Sea. In: Glantz MH (ed) Creeping environmental problems and sustainable development in the Aral Sea basin, chap. 6. Cambridge University Press, Cambridge

Rakhimov ED (ed) (1990) Socio-economic problems of the Aral Sea and the Aral Sea region., FAN, Tashkent 145 pp

Tsitsasov GN (ed) (1990) Hydrometeorological problems of the Aral region., Gidrometeoizdat., Leningrad 277 pp

Zaletaev VS (1991) Pecularities of the modern development and degradation of natural desert systems. Problems of Desert Development 3–4:56–64

Ziyadullaev SK (1990) Utilization of land and water resources in Central Asia and Southern Kazakhstan. Problems of Desert Development 2:3–7

Ecology of the Aqueous Medium in the Area Surrounding the Aral Sea

G. A. Lyubatinskaya

The geographical situation of Kazakstan in the heart of the Euroasian continent, the aridity of its climate and the significant drainless part of the territory determine the exclusive importance of water resources as the factor providing stability for the environment and stability of social economic development of the Republic.

The annually replenished resources of freshwaters in Kazakstan are defined by the volume of local outflow and inflow of the river waters from neighbouring territories, and over many years, they amount on an average to 126 kma^{-1}. With an average precipitation of 308 mm a^{-1} and a coefficient of outflow of 0.08 within the borders of the Republic, about 66.8 $km^3 a^{-1}$ of river water resources and temporary water courses (52%) are formed; 59.2 km^3 (48%) of the river runoff comes from the adjoining states, including China, the People's Republic (Ili, Irtysh): 35%, Uzbekistan (Syrdarya): 40%, Russia (Ural and Tobol): 19%, Kyrgyzstan (Chu, Talas, Assa): 6%.

Kazakstan, with a specific water provision with local outflow per unit of territory (24 600 $m^3a^{-1}km^{-2}$) follows Uzbekistan (27.3), West Siberia (198) and Kyrgyzstan (245). For the water supply per person (4170 km $a^{-1}person^{-1}$) Kazaktstan follows Kyrgyzstan (13.8) and West Siberia (37.1).

The resources of river runoff are characterized by significant changes in the course of time. With a maximum annual volume of outflow of 172 $km^{3,}$ the minimum value of outflow comprised 64.4 $km^3 a^{-1}$, i.e. nearly three times less than the maximum and two times less than the mean value.

The variation in the river runoff is connected with global and regional climate changes. It is known that over the past 100 years air temperature at the surface of the continents and oceans has increased by 0.53 °C.

The distribution of resources of river runoff is very irregular. There is practically no local river runoff in three regions of the Republic: the Kzylorda, Atyrau and Mangyshlak regions. The largest volumes of river runoff are formed in the West Kazakstan region (24.7 km^3a^{-1}), the Taldy–Kurgan region (10.5 km^3a^{-1}), the Semipalatinsk region (5.58 km^3a^{-1}) and the Almaty region (5.55 km^3a^{-1}).

The space-time distribution of river runoff unfavorable from the point of view of economic use, is compensated for by existing regulating reservoirs and channels of interbasin transfer. The largest reservoirs in Kazakstan are the following: Buchtarma reservoir (multiyear) and Shulba reservoir (seasonal), regulating entirely the Irtysh outflow and its affluents for the purposes of energy, navigation and the standard of productivity on the overflow land; Kapchagai reservoir regulating the outflow of the Ili river in the interests of energy, irrigation and the natural complex of the lower reaches of the river and Balkhash Lake; Chardarya reservoir (seasonal) providing outflow supply from the Syrdarya to the irrigation systems of the lower reaches, as well as supplying the lake systems of the delta and the Aral Sea. The Irtysh–Karaganda channel is the largest passage of interbasin redistribution of outflow in Kazakstan. It supplies Irtysh water to the industrial regions of Central Kazakstan which are not provided with water.

In total, freshwater consumption in the Republic for the past years has comprised 33 to 35 km^3a^{-1}, i.e. approximately 28% of the replenished water resources. The volume of water disposal amounts to 5 to 7 km^3a^{-1} with a water intake of about 40 km^3a^{-1}.

Agriculture is the largest water consumer in Kazakstan, annually using up to 28 km^3 of the water resources (80%). About 80% of the water is used for regular irrigation and approximately 20% for catchwork irrigation.

As a result of extensive water resource exploitation, the scale of natural water consumption and pollution in many regions of Kazakstan has increased. This influenced first of all the territories of drainless basins. These end reservoirs (Aral, Balkhash, Kaspiy) bear the integral load of climate fluctuations and economic activity at the water catchment area. They become high-priority real or potential epicentra of zones of ecological instability.

The Aral ecological crisis connected with shrinkage and salinization of Aral Sea has thus become very famous. Analysis shows that the falling level of the Aral Sea beyond the limits of natural fluctuations is the main factor for ecological destabilization. In this context the level is reduced to 11.4 m (about 80%) due to anthropogenic factors such as using outflows of the Amudarya and Syrdarya Rivers for irrigation and filling reservoirs. Deep and irreversible changes in natural environment, impermissible impairment of the living conditions of the population and severe economic damage served as the basis for Priaralye to be officially proclaimed an economically distressed zone in 1989.

The deficiency of water resources and the bad quality of the water have led to the degradation of the ecological system of Priaralye, impoverishment of the animal and vegetal life of the region, desertification of the land and severe impairment of the sea biology. The dry sea floor became the source of salts escaping and silting within a radius of about 200 km. Calculations by specialists show that every year up to 75 million tons of salted dust is carried into the air. More than 500 kg ha^{-1} salts fell on some areas of Priaralye, which served as one of the reasons for soil plant degradation.

As a result of the ecological crisis in the region, desiccation and reduction of total humidification of Priaralye can be observed; the aridity of climate has been intensified, air humidity has decreased to 10 to 18%, the frost-free period has been reduced to 30–35 days.

Table 1 A. Data on water quality of the Syrdarya River. 2 km downstreem from the Chardarin dam; years 1985–1989; (depth: more than 0.5 m) (Hydrometeorology Institute, Almaty); Cr, Hg, Cd, Co, Ag, V, Sn were not detectable

Indicator	1985	1986	1987	1988	1989
Transparency (cm)	14.7	15.9	15.3	13.3	19
Temperature ($^\circ$C)	11.7	8.2	10.9	15.2	15.8
Weighed Particles (mg l^{-1})	-	-	295	101	127.9
Oxigen (mg l^{-1})	11.44	11.11	12.22	11.35	11.4
Oxidation (mg l^{-1})	15.8	12.8	24.12	14.87	21.68
Biological oxigen (mg l^{-1})	2.65	2.14	2.3	2.37	2.56
Ammonia (mg l^{-1})	0.07	0.06	0.056	0.037	0.178
Nitrite (mg l^{-1})	0.02	0.04	0.026	0.024	0.014
Nitrate (mg l^{-1})	2.48	1.59	2.041	1.66	1.238
Phosphate (mg l^{-1})	0.023	0.005	0.007	0.019	0.031
Silicate (mg l^{-1})	5.3	4.3	6.8	4.8	3.5
Resin, Asphalt (mg l^{-1})	0.55	0.2	0.2	0.030	0
Oil products (mg l^{-1})	0.09	0.16	0.083	0.05	0.06
Phenols (μg l^{-1})	0	0	1	0.3	2
Fats (mg l^{-1})	-	0.05	0.33	0.03	-
Detergents (mg l^{-1})	0.01	0	0	0.007	0.003
pH	7.62	7.57	7.65	7.57	7.60
Hydrocarbonates (mg l^{-1})	192	178	172	471.7	159.5
Chloride (mg l^{-1})	155.8	164	145.6	110.7	123
Sulfate (mg l^{-1})	524	405	595	595.3	683
Calcium (mg l^{-1})	123	104	111	119	154.5
Magnesium (mg l^{-1})	80.5	82.6	71.8	82.1	78.6
Sodium (mg l^{-1})	140	-	-	-	-
Potassium (mg l^{-1})	-	-	-	-	-
Hardness (Ca+Mg) (mg l^{-1})	12.68	12.34	11.44	13.01	14.18
Mineralization (mg l^{-1})	1255	1038	1297	1233	1410
Iron (total) (mg l^{-1})	0.05	0.03	0.127	0.12	0.035
Copper (μg l^{-1})	0	0	0.43	4	0
Zinc (μg l^{-1})	1.5	0	4.71	5.0	0
Nickel (μg l^{-1})	-	0	1.75	2.67	2.3
Lead (μg l^{-1})	-	0	8	0	0
Molybdenum (μg l^{-1})	-	3.9	8.62	6.37	7.9
Aluminium (μg l^{-1})	-	5	10.45	6.5	5.65
Manganese (μg l^{-1})	-	1.35	3.03	0	3.0
Titanium (μg l^{-1})	-	0	4	0	0
Bismuth (μg l^{-1})	-	0	0.83	0	0
Fluorine (mg l^{-1})	-	1.2	0.71	0.57	0.66
Dichlordiphenyldichlorethane (DDD) (μg l^{-1})	0.004	0.011	0.006	0.008	0
Dichlordiphenyldichlorethane (DDE) (μg l^{-1})	0.008	0.021	0	0.004	0
Dichlordiphenyltrichlorethane (DDT) (μg l^{-1})	0.327	0.086	0.027	0.025	0
Hexachlorcyclohexane (HCCH) (μg l^{-1})	0.02	0.03	0.033	0.025	0.009
Lindane (μg l^{-1})	0.02	0.02	0.019	0.017	0.007

Ecology of the Aqueous Medium in the Area Surrounding the Aral Sea 129

Table 1 B. Data on water quality of the Syrdarya River. 2 km downstreem from the Chardarin dam; years 1990–1994; (depth: more than 0.5 m) (Hydrometeorology Institute, Almaty); Zn, Cr, Hg, Cd, Co, Ag, V, Sn, Bi were not detectable

Indicator	1990	1991	1992	1993	1994
Transparency (cm)	16.5	-	-	-	-
Temperature ($^\circ$C)	17.1	-	-	-	-
Weighed Particles (mg l^{-1})	41.8	30.6	38.6	24.4	28.1
Oxigen (mg l^{-1})	10.2	10.99	10.6	10.4	10.4
Oxidation (mg l^{-1})	20.22	22	24.9	22.6	24.8
Biological oxigen (mg l^{-1})	2.59	2.126	2.261	2.33	2.489
Ammonia (mg l^{-1})	0.06	0.052	0.087	0.066	0.064
Nitrite (mg l^{-1})	0.02	0.02	0.03	0.02	0.04
Nitrate (mg l^{-1})	2.51	3.384	2.59	2.44	2.58
Phosphate (mg l^{-1})	0.027	0.039	0.03	0.04	0.056
Silicate (mg l^{-1})	12.4	-	-	-	-
Oil products (mg l^{-1})	0.13	0.063	0.12	0.18	0.11
Phenols (μg l^{-1})	3	1	0	0	0
Fats (mg l^{-1})	-	-	-	-	-
Detergents (mg l^{-1})	0.006	-	-	-	-
pH	7.7	-	-	-	-
Hydrocarbonates (mg l^{-1})	173.5	-	-	-	-
Chloride (mg l^{-1})	128.5	107.4	69.2	58.5	63.5
Sulfate (mg l^{-1})	512	515.4	566	588	541
Calcium (mg l^{-1})	173	111.6	111	107	99.3
Magnesium (mg l^{-1})	62.2	41.27	71.7	67.6	72.5
Sodium (mg l^{-1})	162.0	-	-	-	-
Potassium (mg l^{-1})	5	-	-	-	-
Mineralization (mg l^{-1})	1245	1005	1051	1127	1013
Iron (total) (mg l^{-1})	0.44	0.272	0.52	0.281	0.511
Copper (μg l^{-1})	2.33	0.001	0.001	0	0
Nickel (μg l^{-1})	3.4	-	-	-	-
Lead (μg l^{-1})	0	0	0.001	0	0.001
Molybdenum (μg l^{-1})	1.25	-	-	-	-
Aluminium (μg l^{-1})	31	-	-	-	-
Manganese (μg l^{-1})	2	-	-	-	-
Titanium (μg l^{-1})	5.5	-	-	-	-
Fluorine (mg l^{-1})	0.62	0.531	0.604	0.6	0.58
Dichlordiphenyldichlorethane (DDD) (μg l^{-1})	0	0.002	0.002	0	0
Dichlordiphenyldichlorethane (DDE) (μg l^{-1})	0	0.001	0.005	0	0
Dichlordiphenyltrichlorethane (DDT) (μg l^{-1})	0	0.009	0.02	0	0
Hexachlorcyclohexane (HCCH) (μg l^{-1})	0.01	0.034	0.02	0.015	0
Lindane (μg l^{-1})	0.01	0.023	0.005	0	0

At the present moment, the lack of drinking water has become the main problem of Priaralye. The Syrdarya River has become unsuitable for supplying the population with water of the required standard.

The hydrochemical regime of the Syrdarya River is determined basically by the wastewaters from irrigated massifs, and inhabited localities and towns. Water from the Syrdarya River flowing to the territory of Kazakstan is very polluted and does not meet sanitary requirements. At the zones located near the borders the river runoff has increased the mineral content to more than 1.0 g l^{-1} and is polluted with mineral and organic nitrogen substances. In the Kzyl–Orda region the contents of harmful substances has increased the MAC value three to four times, the mineral content reaches about 1.7 g l^{-1} (Table 1a, b). Quality of water in the Syrdarya River does not correspond to the standard requirements for sanitary households and fish farms. In the Syrdarya River there are pesticides of the dichloro-diphenyl-trichloro-ethane group and of chlororganic concentration (to 0.244 mg l^{-1}) which is impermissible in fishery sites; nitrite and oil products exceed the content of MAC value in water (nitrite: 3–11 MAC value, petrolium products: 3–13 MAC value). Phenols and mineral salts may be found periodically in high concentrations.

Wastewaters are the main source of water pollution in the Syrdarya River. The volume of wastewaters from irrigation massifs of the Kazakstan part of the Syrdarya basin amounts to 2842 million m^3 a^{-1}. Mineral content of drainage wastewaters has increased from 2510 to 7200 mg l^{-1}. The volume of return waters exceeds project indicators by 1.5–2 times, caused by significant amortization of the irrigation and drainage wastewater network. For 20 to 25 years of exploitation the amortization of irrigation systems of water-regulating constructions has comprised 14 to 25%, of drainage and wastewaters system 20 to 30% correspond and vertical drainage 61 to 81%.

Wastewaters everywhere contain residual concentrations of nitrogen-bearing and mineral fertilizations and toxic chemicals; concentrations of nitrates, nitrites, propanide, saturn and other pesticides exceed the permitted limits. In separate samples their content shows about 10 MAC (Table 2; chemical analysis of the samples of drainage and waste water was made under the methods of Lurs et al. 1974).

Conclusions

Considerable withdrawal of water from the Syrdarya river for irrigation (> 80% of the run-off) and significant water losses at irrigation systems (up to 37% of the water intake) form a large volume of drainage and discharge effluent up to 20% of the run-off of the river. With the return of the water, a high amount of pollutants flows into the Syrdarya river. Water resources shortage and high pollution level of the run-off of the Syrdrya have led to an ecological tension in the Prearalye.

Managing water, which is an economic value, is one of the most important ways to achieve effective and stable water use in the Prearalye. Application of the principle "one who pollutes must pay", which has been justified in many countries, under the conditions of reasonable and economically grounded payments for water, would promote to harmonious use and protection of water resources and preservation and rehabilitation of natural water objects (Myrzakhmetov 1996; Zaurbekov 1998; Development of methodological principles of ecological and irrigation requirements for operation of irrigation systems in the Priaralye 1996).

Table 2. Properties of soils and soil water

Soil layer (cm)	0–20	20–40	40–60	60–100	100–160
Volume weight (g cm^{-3})	1.40	1.48	1.50	1.52	1.47
Specific weight (g cm^{-3})	2.59	2.69	2.69	2.74	2.61
Porosity (%)	46.0	45.0	44.0	45.0	44.0
Hygroscopic water (%)	4.6	4.7	4.6	4.4	4.0
Volume humidity before ponding (%)	32.2	28.5	30.4	30.7	31.3
Volume lowest water capacity (%)	32.4	34.8	39.7	40.3	37.6
Volume of soil water drainage (%)	9.8	6.7	1.2	1.7	3.5
Volume of salt-dissolved water (1 m^{-3})	74.5	72.4	70.9	146.2	216.0
Volume of water leached through the soil (1 m^{-3})	591.8	477.5	309.5	-	-
Water exchange in soil horizon	7.9	6.6	4.4	-	-

References

Development of methodological principles of ecological and irrigation requirements for operation of irrigation systems in the Priaralye (1996) Scientific work report, Kazakh Research Institute of Water Management, Zhambyl

Lurs Y, Rybnikova A (1974) Chemical analysis of industrial waste water. Chimiha, Moscow, pp 334

Myrzakhmetov M (1996) Kazakhstan s water resources and complex approach to their protection and harmonious exploitation. Problems and ways of scientific technological development in the Republic of Kazakhstan, materials of Republican conferences, Almaty, pp 283–287

Zaurbekov AK (1998) Scientific basis for harmonious exploitation and protection of water resources of river basin water resources. Doctoral Thesis, Taraz

Potable Water: Research into Seasonal Changes and Conditions in the Aral Sea Region

S. Sokolov

One of the key issues and problems of the living standard of the inhabitants in the Aral Sea region is that of a safe drinking water supply system. This problem is especially urgent for the population of the remote villages, as people living in these villages have to use water for drinking purposes from shallow wells and boreholes. As the physical conditions of the groundwater are characterized by high salinity, the problem of a safe potable water supply has become very important.

In accordance with data from the Giller Institute, which estimated the social problem of Kazakhstan Prearal, water supply is a factor which can actually decrease the life quality of the population and prevent economic development. The cost of water is enormously high, and water from the Syrdarya River must be used to irrigate channels and prepare ice in winter. In many villages during the Soviet time (1963-1965), wells were built and equipped with a pump set to utilize underground water. However, preliminary investigation shows that this water, with its high level of mineralization, is very harmful to health.

This project plans to observe the conditions of drinking water sources in all inhabited places in the Kazalinsk and Aralsk regions, and discusses these problems with the local authority, to suggest practical decisions for drinking water supply systems for each village, and also to interview local people, and to attract the attention of regional sanitary epidemiological stations and health departments.

According to the plan in operation in the former Soviet Union for the development on agricultural water supply system in the Kyzylorda oblast, meeting requirements for drinking water in inhabited places were planned to build mainly water pump systems; but the realization of this plan was far from satisfactory. Establishing a sustainable system of quality drinking water supply from Aral-Sorbulak underground water over a period of 10 years remains still the unsolved key to the problem of social rehabilitation of the region.

According to the locations previously mentioned during hydrochemical research in the region of the Syrdarya River Delta, it seemed interesting for us to observe the general character of changes in water composition depending on place and time of testing, as the changes of quantities and water levels in reservoirs at either flooding, or at low levels are accompanied by dilution or concentration, respectively, of the substances dissolved in the water, i.e. a more or less significant change in conditions of balance and hydrochemistry.

Chemical analyses of samples included volumetric titration methods (analysis of main components) and spectrometric analysis using the Merck SQ–118 standardized systems specially designed for water analysis. The method of AAS spectrometry was also used for analysis of metals (Cu, Zn, Pb, Cd, Cr); pesticides were analyzed by chromatography.

- Water, both river and underground, is rich in sulfate and mainly sodium-magnesium. The concentrations of sodium and magnesium are comparably equal.
- The row of cations Na > Mg > Ca varies more with increase in total salinity, and is a normal phenomenon in superficial water hydrochemistry.
- The influence of this is more apparent from the data on water composition at some localities, connected with level and resource fluctuations of superficial water at different seasons of the year.

It is apparent from these data that in the period of flood and low water level, the total mineral content of water can sometimes double. The concentration of hydrocarbonate ion remains almost constant, the concentration of calcium hardly changes, and the main changes take place in water sulfates, and chlorides of magnesium and sodium contents. Concentrations of chloride in localities near the middle of the Aral Sea change considerably.

Seasonal tables show the degree of pollution by the chemical elements: nitrates, nitrites, metals and particularly heavy metals.

The content of heavy metals (copper, zinc, lead and cadmium) is similar in all waters; but it is rather high on average and (except for zinc) sometimes exceeds the water standard. Arsenic, molybdenum and uranium were analyzed in waters, but these metals were either not found, or their concentration was below standard.

The high nickel concentration found in all tests of river water attracts particular attention. The concentration of this metal is not usually taken into account in ecological tests, but the abnormally high content was unexpected and puzzling, especially since widespread regional hydrochemical measurements of nickel are as a rule, much below those of manganese and zinc and lower than for copper. Further hydrochemical research located a source of nickel pollution in the Syrdarya River. Divergent results require additional zinc analysis by spectrophotometric (Merck SQ-118) and AAS methods. This may be explained by the fact that zinc in sulfate waters (pH 7 –8.5), present in the forms Zn^{2+}, $ZnOH^+$ and $Zn(OH)_2$, $ZnSO_4$ and some others (as well as the other metals – copper, lead etc.), can be bound by organic substances which are rather abundant in the waters, judging from the chemical consumption of oxygen.

Thus, along with the organic substances present in waters, as a result of natural metabolic process of microflora and microfauna in reservoirs, it is necessary to take anthropogenic pollution into account.

Total mineralization of underground water in the Kazalinsk and Aralsk regions is very high and parts of the Syrdarya Delta have a dangerous pollution concentration in autumn.

One example of the results of the chemical analyses of drinking water conditions is shown in Table 1. Recommendations were given to local administrations, UNDP and UNESCO for the Prearalian region.

134 S. Sokolov

The water samples were taken in 22 villages in the surroundings of Aralsk and Kazalinsk.

Table 1. Chemical analysis of drinking water in the Aralsk and Kazalinsk region (october 1998; mg l^{-1})

Points of water sampling		pH	EC ms/cm	DO mgl^{-1}	Temp C°	Na	K	NH$_4$	Ca	Mg
1	Zhalandos	8,1	1,6	11,6	13,1	123	6,1	0,1	108	77,4
2	Zhalandos (water tank)	7,6	1,6	12,8	11,7	179	6,4	0,0	111	70,5
3	Baskara	8,2	4,3	10,3	14,7	840	7,3	1,5	16	5,8
4	Syrdaria (river)	8,2	1,6	13,5	11,0	247	5,4	0,0	117	70,8
5	Syrdaria (water pump station)	8,2	1,6	11,4	12,0	925	5,6	0,1	106	77,7
7	Primov (water tank)	8,0	1,7	11,0	13,8	1211	7,3	0,1	105	91,6
8	Primov (underground)	8,3	3,7	11,2	15,0	171	5,1	0,1	7	8,7
9	Muratbai (well)	8,0	1,6	11,2	14,3	385	7,2	0,2	92	87,9
10	Abai (underground 1)	8,3	4,0	11,4	14,8	918	4,9	1,7	8	9,9
12	Aktan Batur (und. 1)	8,3	4,4	11,0	13,1	771	15,4	2,0	10	2,6
14	Zhankosha Batur (und.)	8,3	4,2	9,1	17,7	1112	6,8	1,2	12	10,2
15	Zhankosha Batur (well)	7,7	2,2	10,7	14,2	80	10,3	0,8	162	99,8
16	Koshebahy (und.)	8,3	3,8	9,5	19,0	411	2,1	1,9	8	5,8
17	Koshebahy (chenal)	8,2	1,7	14,2	10,3	137	5,7	0,0	114	81,8
18	Boskol (und.)	8,1	5,0	11,0	16,2	1018	4,5	1,1	12	7,3
19	Kaukei (und. 1)	8,4	2,9	9,4	21,2	77	3,7	1,0	13	4,4
21	Kaukei (und. 3)	8,4	2,9	8,2	20,5	338	4,4	0,7	10	4,8
22	Tasaryk (und. 1)	8,3	3,3	11,1	16,5	748	7,0	1,0	8	11,0
24	Tasaryk (und. 3)	8,4	3,3	11,2	16,0	505	5,5	0,9	9	7,7
26	Lahaly (und. 2)	8,3	4,0	9,8	14,8	913	9,7	1,3	12	3,5
27	Urkendey (und.)	8,5	3,9	10,8	15,5	841	7,9	0,9	13	5,5
28	Urkendey (well)	7,4	3,8	11,7	13,9	513	13,6	0,0	244	244,8
29	Novokazalinsk (home)	8,1	1,6	13,7	12,2	33	9,8	0,0	105	72,2
30	Tastubek (well 1)	6,8	3,8	10,3	14,1	63	13,8	0,0	94	60,9
32	Zhalanash (well)	7,3	1,9	11,2	13,1	154	15,3	0,0	109	58,6
33	Zhalanash (water tank)	8,4	0,8	14,1	8,0	122	15,5	0,0	11	3,8
34	Aralsk (home)	8,5	1,0	11,8	12,7	156	11,5	0,0	11	6,4
35	Koshar (und.)	7,3	21,4	8,0	17,9	3989	13,1	1,2	205	131,7
36	Koshar (home)	7,9	1,5	9,2	8,7	212	6,0	0,0	82	76,0
37	Old Kazalinsk (home)	7,6	1,7	11,3	15,7	98	5,1	0,0	118	70,8

n.d. not detectable

Table 1 continued.

	Cl	SO$_4$	NO$_3$	NO$_2$	CO$_3$	Fe	Cu	Zn	Cd	Pb	Mn	Ni	Cr
1	42,4	138,5	3,1	0,19	171	0,00	0,01	0,003	0,001	n.d.	0,02	0,01	0,000
2	24,0	140,1	3,1	0,01	358	0,01	0,02	0,010	0,001	n.d.	0,00	n.d.	0,001
3	28,0	140,1	1,3	0,01	140	0,34	0,01	0,003	0,001	n.d.	0,00	n.d.	0,001
4	171,5	506,1	1,5	0,14	159	0,55	0,17	0,034	0,004	0,005	0,04	0,03	0,004
5	149,3	207,6	3,5	0,04	140	0,60	0,07	0,012	0,002	0,002	0,01	0,02	0,002
7	411,0	141,4	2,2	0,03	134	0,82	0,09	0,021	0,001	0,001	0,07	0,01	0,000
8	82,0	137,9	0,4	0,02	201	0,08	0,02	0,003	0,001	0,002	0,01	0,00	n.d.
9	24,0	145,5	1,1	0,01	207	0,13	0,11	0,017	0,002	0,004	0,03	0,00	n.d.
10	38,5	202,2	0,3	0,02	183	0,08	0,07	0,011	0,003	0,001	0,01	0,00	0,001
12	37,0	220,5	0,4	0,02	110	0,03	0,04	0,003	0,001	0,000	0,04	0,00	n.d.
13	31,5	224,6	0,2	0,01	119	0,04	0,02	0,002	0,001	0,000	0,01	0,00	n.d.
14	125,0	139,9	2,2	0,07	141	0,13	0,01	0,002	0,000	n.d.	0,00	0,00	n.d.
15	48,5	146,6	17,3	0,14	158	0,09	0,07	0,009	0,003	0,006	0,02	0,02	n.d.
16	212,0	191,1	0,2	0,02	140	0,00	0,11	0,004	0,001	0,001	0,01	n.d.	0,001
17	67,0	140,4	2,7	0,15	157	0,01	0,21	0,011	0,007	0,002	0,06	0,03	0,001
18	817,0	141,8	0,9	0,02	354	0,87	0,02	0,004	0,001	0,001	0,02	n.d.	0,000
19	438,0	136,7	1,8	0,11	224	0,25	0,03	0,002	0,002	0,000	0,01	n.d.	0,000
21	418,0	152,5	0,7	0,02	212	0,26	0,03	0,003	0,002	n.d.	0,00	n.d.	0,000
22	423,0	134,9	2,2	0,05	154	0,10	0,01	0,002	0,000	n.d.	0,00	n.d.	0,000
24	565,0	150,8	2,1	0,02	175	0,02	0,02	0,002	0,000	n.d.	0,00	n.d.	0,000
26	769,0	138,5	0,9	0,02	118	0,10	0,01	0,003	0,001	0,001	0,02	n.d.	0,000
27	312,0	137,5	2,2	0,02	370	0,26	0,02	0,004	0,001	0,000	0,01	n.d.	0,000
28	618,0	155,8	4,4	0,02	145	0,09	0,10	0,012	0,009	0,002	0,07	0,02	0,002
29	440,0	122,6	0,0	0,04	148	0,38	0,07	0,022	0,002	0,001	0,04	0,02	0,001
30	415,0	146,9	2,0	0,09	221	0,00	0,00	0,000	0,000	n.d.	0,01	0,00	0,000
32	100,0	134,5	48,7	1,57	218	0,07	0,11	0,009	0,004	0,002	0,04	0,03	0,000
33	100,0	98,3	0,3	0,02	134	0,84	0,01	0,000	0,000	n.d.	0,00	n.d.	0,000
34	140,0	100,1	0,4	0,03	150	0,42	0,00	0,000	0,000	n.d.	0,00	n.d.	0,000
35	415,0	156,6	0,4	0,03	362	0,76	0,00	0,011	0,002	0,002	0,01	0,01	0,001
36	377,0	441,1	4,0	0,02	131	0,90	0,14	0,035	0,004	0,005	0,03	0,02	0,002
37	143,0	108,4	2,7	0,02	152	0,02	0,07	0,010	0,004	0,001	0,02	0,01	0,001

Part II:

Salt Stress

Halophytes on the Dry Sea Floor of the Aral Sea

Siegmar–W. Breckle, Anja Scheffer and Walter Wucherer

Keywords. Halophyte types, ion pattern, salt desertification, dry sea floor

Abstract. Within a relatively short period, less than a half century, the area of salt deserts in Central Asia has grown by about 60 000 km². The process of salt desertification is tremendously active in the region of the former Aral Sea. The coastal plain and the dry sea floor of the Aral Sea are an evident model for studying salt desertification. The vast occurrence of salinization processes is the main reason for a very diverse halophytic flora on the dry sea floor. In comparison to other ecophysiological life forms, halophytes thrive on saline soils and are able to grow even on rather strongly salinized substrates. Investigation of the adaptive mechanisms of the various halophyte types is essential for an adequate species composition for phytomelioration of these saline soils. Phytomelioration by artificial planting on the dry sea floor for more rapid closure of the vegetation cover is a great need to minimize the widespread negative effects of salt desertification.

Introduction

Saline soils and halophytic vegetation are a common feature and part of the habitat pattern in deserts and steppes. The percentage of saline ecosystems in a distinct area is increasing with aridity. In Central Asia about 90% of the area are arid, semiarid or at least subhumid. Solonchak soils and halophyte plant communties are most typical for the Caspian Basin, the Aral Sea and the Balkhash Basin. In the Aral Sea Basin, within the short time of about 40 years, the development of a huge, new salt desert has taken and is still taking place, caused by the drying of the Aral Sea. This process is comparable in size with the Great Iranian Salt desert and is even larger than the Great Salt Desert in Utah. The area of saline soils on the dry sea floor amounts to about 38 000 km². Additional areas with saline soils are found within the irrigation zones. These secondary salinized soils have an area of about 22 000 km² (Glazovskii and Orlovskii 1996). This means that in a relatively short time period, within less than a half century, the area of salt deserts in Central Asia has grown by about 60 000 km². The process of salt desertification is tremendously active in the region of the former Aral Sea. The coastal plain and the dry sea floor of the Aral Sea are an evident model for studying salt desertification. The dry sea floor is covered with a mixture of shells, clay, loam, sand and salt. Only the older sandy soils are almost free of salt.

Since the middle of the 1980s, open salt deserts have formed on a large scale. The fast increase in the salt desert area and the rapid change have caused an absolute dominance of halophytes. Other life forms are more and more lacking. In comparison to other ecophysiological life forms, halophytes thrive on saline soils and are able to grow even on rather strongly salinized substrates. The formation of vegetation types on the solonchak soils of the newly formed sea floor stands can only be understood by studying the ecological characteristics of the relevant halophyte species. During the 1998–1999 expeditions on the coast of the Aral Sea, hundreds of plant and soil samples were collected and analyzed.

Halophyte Flora

The vast occurrence of salinization processes is the main reason for a very diverse halophytic flora on the dry sea floor. From the 266 species hitherto known, 200 species (75.2%) occur on solonchaks or other saline soil types of the dry sea floor. The remaining species may be influenced episodically or periodically by salt after germination or during other life phases. This results in a rich halophytic flora of the dry sea floor which, on the one hand, is affected by salinity to various degrees and, on the other has evolved adaptations for survival on those saline stands. From the halophytes 45 species (22.5%) belong to the group of euhalophytes (Fig. 1).

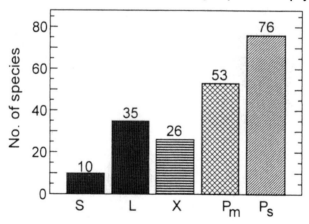

Fig.1. Spectrum of halophyte types on the dry sea floor of the Aral Sea. *S* stem succulent halophytes; *L* leaf succulent halophytes; *X* recretohalophytes; *Pm* pseudohalophytes with moderate salt tolerance; *Ps* pseudohalophytes with slight salt tolerance

Within the euhalophytes the group of the stem-succulent euhalophytes (S) with *Salicornia europaea* (s.l.), *Halocnemum strobilaceum*, *Halostachys caspica* and *Ofaiston monandrum* comprise annuals and perennials as well as the group of leaf-succulent euhalophytes (L) with several *Suaeda* species (*S. crassifolia*, Fig. 2; *S. acuminata, S. microphylla, S. physophora*), *Climacoptera* species (*C. aralensis,* Fig. 3; *C. ferganica, C. lanata*) and *Petrosimonia* species (*P. triandra, P. squarrosa, P. brachiata*) as typical examples.

Fig. 2. The leaf-succulent halophyte *Suaeda crassifolia* on the marshy solonchak of the transect Karabulak on the north coast of the Aral Sea

Fig. 3. The leaf-succulent halophyte *Climacoptera aralensis* on the coastal solonchak of the transect Bayan on the east coast of the Aral sea

Fig. 4. The accumulation form (chokolak) of *Halocnemum strobilaceum* on the coastal solonchak on the east coast of the Great Aral sea, forming nepkha hummocks

Fig. 5. The recretohalophyte *Tamarix hispida* on the degraded coastal solonchak of the dry sea floor of the Small Aral Sea

Halocnemum strobilaceum is a typical semishrub and forms a widespread vegetation cover along most parts of the east coast of the Aral Sea. Such vegetation types with *Halocnemum strobilaceum* and other perennial halophytes, with a cover percentage of 20–30% have a low salt dust output. On the dry sea floor from the 1980s and 1990s the perennial vegetation is lacking and the salt content of the substrate is much higher. Often *Halocnemum* exhibits the typical growth-form of a sand-accumulating nepkha shrub, the so called chokolaks (Fig. 4). The plant communities with *Salicornia europaea* and *Suaeda* species dominate on the marsh solonchaks close to the present coastline. The recretohalophytes (salt-excreting species, X) are represented by *Tamarix hispida* (Fig. 5), but also by

Frankenia hirsuta, Limonium gmelinii, Aeluropus littoralis, all with salt glands, and *Atriplex* species with bladder hair .

The *Atriplex* species are grouped to the recretohalophytes, since their bladder hair recrete salts (Schirmer and Breckle 1982). Two of the Atriplex species are especially interesting. *Atriplex fominii* grows preferably on sand-covered coastal solonchaks, but also on many other localities, and exhibits a considerable morphological plasticity. *Atriplex pungens* is quite common on the dry sea floor of the northern coasts of the Aral Sea, but apparently prefers soils enriched with gypsum. Often small-scale strictly monodominant vegetation units can be found.

Five species belong to the pseudohalophytes with a slight recreto function (several Poaceae species). These are part of the large group of pseudohalophytes with moderate salt tolerance (55 species or 27.5%): examples are *Bassia hyssopifolia, Artemisia scopaeformis, Euclidium syriacum, Kochia stellaris* etc. This group is very heterogenous, concerning their ecomorphology as well as their ecophysiology. They are predominantly restricted in their distribution to the degraded coastal solonchaks of the dry sea floor. They can withstand moderate salinity in soil (moderate pseudohalophytes, P_m). The largest group (35.5%) are the slight pseudohalophytes (P_s), which are intermediate to the non-halophytes (N), these are not treated here. The latter may be glycophytic species, which can for some time tolerate a slight salinity of the topsoil. Examples of this group are, e.g. some Brassicaceae species, *Strigosella* species, and Asteraceae such as, e.g. *Tragopogon marginifolius, Acroptilon repens*. Table 1 gives a list of typical plant families with a high proportion of halophytes. Most species of Chenopodiaceae are leaf- and/or stem succulents (Table 1).

Table 1. Halophyte types within some plant families of the dry sea floor

Family	S	L	X	P_m	P_s
Chenopodiaceae	10	33	10	12	10
Brassicaceae				13	11
Asteraceae		2		8	9
Poaceae			2	9	6
Fabaceae				4	9
Tamaricaceae			7		
Limoniaceae			4		

S stem succulent halophytes; *L* leaf succulent halophytes; *X* recretohalophytes; P_m pseudohalophytes with moderate salt tolerance; P_s pseudohalophytes with slight salt tolerance

The Brassicaceae and Poaceae comprise mainly pseudohalophytes. All species of Limoniaceae are recretohalophytes.

Ion Patterns of the Several Halophyte Groups

In studies on the ecology of halophytes along salt gradients in Iran und Afghanistan (Breckle 1986) several Irano–Turanian species were discussed. A similar study is done here for the halophytes of the Aral Basin. Some preliminary results are shown in Table 2, giving an impression of the main cation and anion content in stems and leaves (Cl, Na, K, Ca, Mg) of selected plant species.

Table 2. Ion pattern of several halophytes (mmol kg^{-1} dw)

Species	Locality	Cl	Na	K	Mg
Stem succulent halophytes					
Halocnemum strobilaceum	B	3042	3748	506	133
Salicornia europaea	B	4402	4094	522	152
Ofaiston monandrum	Ka	1604–2789	1272–3096	323–525	490–574
Leaf succulent halophytes					
Sueda acuminata	Ka	2659–5133	3399–5018	589–860	185–356
Sueda crassifolia	Ka	4076–4895	3254–3676	426–427	471–619
Climacoptera aralensis	Ka	2136–3216	1774–4142	517–910	130–197
Petrosimonia triandra	B	510	966	550	320
Pseudohalophytes					
Euclidium syriacum	Ka	314	127	497	95
Non-Halophytes					
Eremosparton aphyllum	B	155	33	324	78
Stipagrostis pennata	B	78	45	327	43

B Bayan; *Ka* Karabulak

It is obvious that leaf- and stem succulents, like species from the genera *Suaeda, Salicornia* and *Halocnemum,* accumulate considerably more Na and Cl (3000–5000 mmol kg^{-1}) in comparison with other species. The ionic content (Na and Cl) of *Climacoptera* species and of *Ofaiston monandrum* are lower (2000 –3500 mmol kg^{-1}) in comparison with species from *Salicornia* and *Suaeda.* Even lower are the values from *Petrosimonia triandra.* On the other hand, the Na and Cl accumulation of pseudohalophytes like *Euclidium syriacum* and *Stripagrostis pennata* is very low.

It is always an open question to what extent the edaphic conditions influence the ionic pattern and content in plants. The Pontic–Irano–Turanian *Suaeda acuminata* (Fig. 6) is also very common in Central Asia (Wucherer 1986). This species exhibits a wide ecological amplitude and thus can be found on very contrasting saline stands. Within the transect Karabulak on the northern coast of the Aral Sea, four localities were chosen for investigation (Fig. 7).

It is obvious that the Na and Cl content of the above ground plant organs of *Suaeda acuminata* on degraded coastal solonchaks and puffy coastal solonchaks is significantly lower. These soils contain significantly less salt in the top soil. On these stands also the Na content is higher than the Cl content in comparison with the marshy solonchaks and crusty coastal solonchaks. This example with *Suaeda* demonstrates that the ion content in halophytes growing on real solonchaks with high salinity is distinctly influenced by the edaphic conditions. Balnokin et al. (1991) studied the Na, Cl, and S content in *Salicornia europaea, Climacoptera aralensis* und *Petrosimonia triandra* along the transect Bayan on the east coast of the Aral Sea.

Fig. 6. The leaf-succulent halophyte *Sueda accuminata* on the coastal solonchak of the transect Bayan on the east coast of the Aral Sea

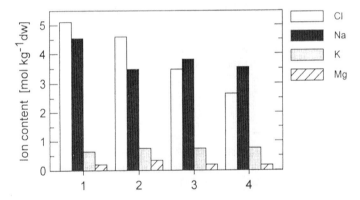

Fig. 7. The content of the main cations in the above-ground parts of *Sueda accuminata* localities from the Aralkum desert. *1* Crusty coastal solonchak; *2* marshy solonchak; *3* degraded coastal solonchak; *4* puffy coastal solonchak

It was shown that the Na and the Cl content in *Salicornia europaea* and *Climacoptera aralensis* was more or less constant in the roots, but increased in the stems and leaves with increasing soil salinity. In *Petrosimonia triandra*, the ion content remained rather constant also in the above ground organs even with varying soil salinity. This is an indication that also in the stem and leaf succulent halophytes, in the recreto- and pseudohalophytes from the dry Aral Sea floor, different mechanisms and strategies for the adjustment and regulation of the salt concentration in the plant tissues are operating (Breckle 1995) and thus a differing salt tolerance in the various species leads to a specific pattern of species and halophyte types along salt gradients.

Conclusion

Investigation of the adaptive mechanisms of the various halophyte types is essential for an adequate species composition for phytomelioration of these saline soils. The salinization of the substrate on the dry sea floor varies to a great extent, causing a wide variety of saline soil types; various solonchaks have developed: marshy solonchaks, crusty and puffy solonchaks, solonchaks slightly covered by sand, degraded coastal solonchaks, takyr solonchaks etc. Phytomelioration by artificial plantings on the dry sea floor for a more rapid closure of the vegetation cover is a great need to minimize the widespread negative effects of salt desertification in the whole area. Studying natural halophytes is thus very important for all those regions where salinity has reached such a level that desalinization techniques are much too costly. Investigations of halophytic ecosystems, of the salinity process in agrarial systems and of plant strategies for salt regulation are an urgent need in the Aral Sea region, where salt desertification has become dominant.

Acknowledgements. This research is funded by the German Federal Ministry for Education and Research (BMBF grant 0339714).

References

Balnokin YuV, Myasoedov NA, Baburina OK, Wucherer W (1991) Ion content of Na^+, Cl^-, S and prolin in the halophytes by the different salinity level on the dry sea floor of the Aral Sea. Probl Osv Pustyn' 2:70–78 (in Russian)

Breckle SW (1986) Studies on halophytes from Iran and Afganistan. II. Ecology of halophytes along salt gradients. Proc R Soc Edinb 89B:203–215

Breckle SW (1995) How do halophytes overcome salinity? In: Khan MA, Ungar IA (eds) Biology of salt tolerant plants. Department of Botany, University of Karachi, Pakistan. Book Graffers, Chelsea, Michigan, USA, pp 199–213

Glazovskii NF, Orlovskii NS (1996) The Problems of the desertification and droughts in GUS and the ways of their solution. Izv Akad Nauk Ser Geogr 4:7–23 (in Russian)

Schirmer U, Breckle S–W (1982) The role of bladders for salt removal in some Chenopodiaceae (mainly *Atriplex* species). In: Sen DN, Rajpurohit KS (eds) Tasks for vegetation science. Vol 2, Junk, The Hague

Wucherer W (1986) The spreading of *Suaeda acuminata* (C.A. Mey.) Moq. in Kazakhstan. In: Baitenov MS, Vasyagina MP (eds) Botanical materials of the herbarium of the Institute of Botany. Nauka, Alma–Ata, pp 38–39 (in Russian)

Halophytes: Structure and Adaptation

A.A. Butnik, U.N. Japakova and G.F. Begbaeva

Keywords. Anatomy, fruit, leaf, morphology, root, stem

Abstract. Halophytes are an ancient ecological type the origin of which is connected with the zone littoralis.A Saline environment of their habitat has a positive effect on them, as it reduces competition, prevents diseases and vermin and creates humidity at the expense of accumulation of moisture by salt. The negative qualities of this habitat are the high osmotic pressure of the soil solution and the toxic effect of salt. The morphogenesis and structure of vegetative and generative organs of hyperhalophytes (*Halocnemum strobilaceum, Salicornia europaea*) and euhalophytes from *Suaeda* (*S. arcuata, S. microsperma, S. prostrata*) and from ephemers (*Hymenolobus procumbens, Spergularia microspermoides*) were studied in the Kyzylkum desert with chloride-sulphate salting of soil, showing the characteristics of plant development caused by the specific habitat. The fruits of halophytes are not sclerified. The pericarp is parenchymatous. The testa have two layers of cells. Their adaptations to extreme desert conditions, including salinization, are implemented at the expense of the submerged position of fruits in fleshy cortex, the presence of tannins and melanins in the testa and fat in the embryo, which makes it difficult for salt to penetrate. The assimilating organs of hyperhalophytes are the shoots with reduced leaves and chlorenchymatic non-Kranz cortex of stems. Euhalophytes have small cylindrical leaves with Kranz (*Suaeda arcuata, S. microsperma*) and non-Kranz structure (*S. prostrata, Hymenolobus, Spergularia*). However, the main line of adaptation of halophytes in both group is the succulent strategy whereby moisture is preserved at the expense of abundant water-bearing cells with thin walls. The structure of stems and roots of the species studied is anomalous, polycambialous and sclerenchymatous, which guarantees protection of lateral meristems and their substitution when they are damaged.

Introduction

Salinization of soil in an arid zone is a typical phenomenon.The area of saline land in Uzbekistan is 70% of its total territory (Akzhigitova 1982). The drying of the Aral Sea is increasing this process, prompting the spread of salts over large areas.

Plants which grow and end their life cycle on soils with a high salt content are named halophytes (Waisel 1972). Many researchers have paid particular attention to this plant group under different aspects (Keller 1927; Joshi 1937; Ungar 1962; Medvedeva 1962; Joshi and Rao 1986; Jefferies and Rudmik 1991; Butnik et al. 1997). There are many classifications of halophytes, based on various principles (Stocker 1928; Iversen 1936; Henkel and Shakhov 1945). According to Akzhigitova (1982), halophytes are divided into four groups: hyperhalophytes, euhalophytes, hemihalophytes and haloglykophytes. Accepting this classification we propose another type – haloxeromesophytes. Basically, ephemers, which grow on saline soils of desert land in moist periods, belong to this group. Details of biology and structure of halophilous ephemers are absent in publications.

This chapter attempts a more detailed study into the ontogeny of some halophytes, to reveal correlations and process of formation of all organs and tissues, the adaptive strategy of plants and the specialization of their structures in the process of evolution.

Material and Methods

The structure of vegetative (leaf, stem, root) and generative (fruit, seed) organs of species from the Caryophyllaceae (*Spergularia microspermoides*), Chenopodiaceae (*Halocnemum strobilaceum, Salicornia europaea, Suaeda arcuata, S. microsperma, S. prostrata*) and Brassicaceae (*Hymenolobus procumbens*) has been studied.

The materials were collected in southwest Kyzylkum, which is situated in the Iranian-Turan phytogeographical region. Hyperhalophytes (semidwarf shrub *H. strobilaceum* and an annual plant with long vegetation, *Salicornia europaea*) and haloxeromesophytes (*Hymenolobus, Spergularia*) grow on puffy, moist salt marshes with chloride-sulphate salinization, but euhalophytes (annual species and *Suaeda* genus with long vegetation) on dry salt marshes. The micropreparations were made by accepted methods of anatomy.

Results and Discussion

Based on an analysis of the morphogeny and structure of all organs of seven halophyte species, the following characteristic features were revealed. The fruits of *H. strobilaceum, S. europaea* and *Suaeda* species belong to one group like most of Chenopodiaceae species (Butnik 1981). In this group, ripe fruits have modified perianth petals; but in comparision with psammophytes and petrophytes (*Salsola* and *Nanophyton* species etc.); the perianth and pericarp of halophytes are thin and parenchymatous. The spermoderm of *H. strobilaceum, S. europaea* and *Suaeda* species has two layers (Fig. 1a, 2a). The testa of all the studied halophyllous species from Chenopodiaceae contain tannins. Heterocarpy in *S. europaea* is conditioned by a different location of the fruit on the plant and is evident from the difference in sizes and absolute weight of the seeds. Each type of fruit of *S. prostrata* has its ripening period and they differ from each other in morphological and anatomical features. Most probably, there is correlation between morphological types of fruits and degree of tolerance of seeds to salts. The different salt tolerance of large and small seeds of *S. europaea* was discovered by Ungar (1979). Our investigations have shown a difference in the structures of seeds, i.e. the spermoderm of large seeds is thin and parenchymatous, while small ones are thicker with hooked trichomes.

The trivalve capsule of *Spergularia* has a parenchymatous perianth and a pericarp with glandular trichomes. The species with salt-producing trichomes, according to Nagolevsky and Tilba (1989), are on a lower stage of development compared to salt-accumulating plants (*H. strobilaceum, S. europaea*). Nevertheless, *Spergularia* becomes similar to hyperhalophytes because of its thin parenchymatous perianth and pericarp and also its few-layered testa with tannins. The silicle of *Hymenolobus* is dehiscent with slimy seeds. Unlike other species of Brassicaceae (which have five or six layers), the testa of *H. procumbens* has two layers: the exotesta and the pigment layer.

According to the shape of the embryo and the abundance of endosperm and perisperm, *H. strobilaceum, S. prostrata* and *Spergularia* are more primitive (Fig. 1b) and *S. arcuata, S. europaea, Hymenolobus* more progressive. The type of mesophyll of cotyledons in halophytes are homogenous iso-spongy (*H. strobilaceum, Spergularia*), dorsiventral (*S. europaea, Hymenolobus*, Fig. 2b,c), Kranz-suaedoid (*S. arcuata*), Kranz-atriplicoid (*S. microsperma*). The macro- and microstructure of fruits of halophilous species are simplified, probably connected with the conserving effect of salts.

Slight linear growth of vegetative organs, considerable atrophy of annotinous shoots, short metameres, aphylly, presence of a few buds (three to five) in the same node, partition and growth of adventitious roots are natural for the typical semidwarf shrub *H. strobilaceum*.

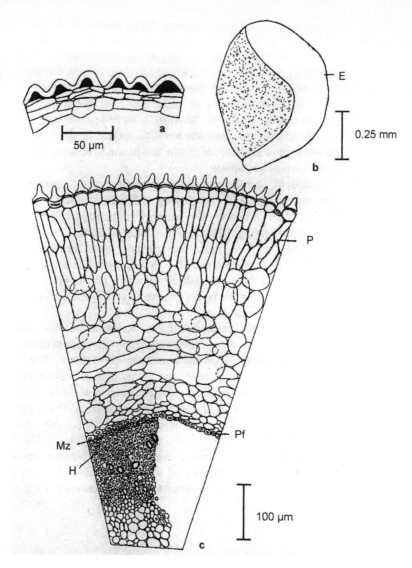

Fig. 1. a.-c. *Halocnemum strobilaceum,* portion of seed coat with tannins in exotesta **(a)**, schematic drawing, showing location of embryo in seed **(b)**, cross section of generative shoot **(c)**. *E* embryo; *H* Halodermis; *Mz* meristematic zone; *P* palisade cells; *Pf* pericyclic fibres.

Halophytes: Structure and Adaptation

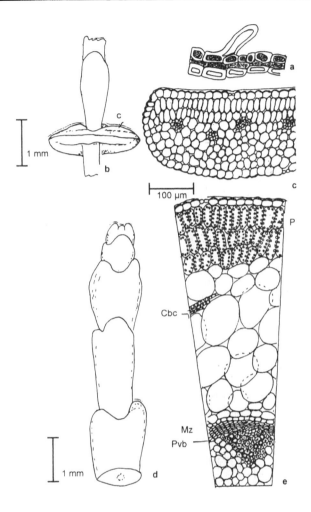

Fig. 2. a.-e. *Salicornia europaea*, cross-section of seed-coat (**a**), schematic drawing and cross-section of cotyledon (**b, c**), articulate axis (**d, e**). *C* cotyledon; *Cbc* conducting bundle of cortex; *Mz* meristematic zone; *Pvb* primary vascular bundle.

The lifetime depends on ecological conditions and varies from 8–12 to 20–25 years. The high location of gyps (at a depth of 30–50 cm) depresses the growth and fructification of plant. Rosette growth form in the juvenile stage of ontogeny, basitonic branching and numerous lateral shoots with short metameres are natural for *S. europaea* and *Suaeda* species (*S. arcuata, S. microsperma*), which ensures the formation of numerous generative organs and high seed productivity.

S. europaea is an aphyllous plant but *Suaeda* species have small (1–2 sm) cylindrical leaves.

Ephemers (*Hymenolobus, Spergularia*) are described as being short-term (1.5 and 2 months) and small-growth plants. Lability of the growth form (from monopodial to hemisimpodial, basitonic branching) is a peculiar feature of the ephemers studied, *Salicornia* and *Suaeda* species.

The fleshy shoots with non-Kranz structure of bark are assimilatory organs of hyperhalophytes (Fig. 1c, 2d,e). In this manner of keeping moisture, aphyllous and succulent plants are similar to each other but their adaptive strategies are different. They occur due to change of cell divisions in the growth cone, expansion of function of primary bark (substitution of function), and formation of special tissue, the haloderm (multilayer, small-celled parenchyma, located between the bark and the central cylinder of stems in *Halocnemum strobilaceum;* Medvedeva 1962). The presence of papillas on the epidermal cells, their thick outer walls and cuticle and submerged stomata protect assimilatory shoots against surplus transpiration. According to their position in the evolutional system, the hyperhalophytes are the culmination of evolution for the non-Kranz group of plants. It is natural for the leaves of *Suaeda* species to have abundant water-bearing tissue. Water- and salt-accumulating functions are carried out by different tissues: swollen epidermal cells (in all species), large-celled hypodermis (*S. microsperma*), water-bearing parenchyma in the middle of the leaf (*S. microsperma, S. arcuata*). Kranz species have larger cellular epidermis, small and few stomata of anomocytic and hemiparacytic types. The leaf type of *Suaeda* species correlates with its area. Species with wide area (*S. prostrata*) have non-kranz, less specialized structure of leaf; but endemic species from Central Asia (*S. arcuata, S. microsperma*) have a Kranz structure leaf because of its formation in arid conditions.

The axial organs of *Halocnemum, Salicornia* and *Suaeda* species are sclerotized. Secondary thickening of polycambial type occurs due to formation of numerous layers (up to 15) of successive cambium which are localized inside the roots in a spiral. A similar structure guarantees good protection for meristems and their mutual substitution in the case of damage. Considerable sclerification due to formation of intermediate fibres is found, too. In epidermis and bark in *Suaeda* species fat and slime have been discovered. Mechanical elements (sclerenchyma) contain pectine substances which are capable of swelling.

Adaptation of ephemers (*Hymenolobus, Spergularia*) corresponds to the above -mentioned general regularities. Their leaves are reduced to small sizes; they have well-developed aerenchyma and water-bearing parenchyma. For axial organs it is natural to have a wide core with a cavity in the basis of stem, wide phloem and less-developed xylem. The structural elements of the stem of *Spergularia* contain salt-storage cells.

The adaptation of euhalophytes and haloxeromesophytes to halo- and xero factors is implemented by succulent strategy based on the preservation of moisture in the leaf because of salts (up to 30%). It is correlated with thin walls of cells and an abundance of water-bearing tissue.

Halophytes: Structure and Adaptation

The total tendency of adaptation, particularly expressed in hyperhalophytes, is reduction of leaf organs, tissue (cambium) as a result of suspension of effect by salts, switching over to the functions of assimilation on the bark of shoots (substitution of organs) and adaptation on the subcellular level due to the high salt-containing capacity of the cell plasm.

References

Akzhigitova NI (1982) Galofilnaja rastitelnost Srednej Azii i eje indikacionnye svojstva. FAN Uz SSR,Tashkent, 192 pp

Butnik AA (1981) Karpologicheskaya kharakteristika predstaviteley sem. Chenopodiaceae Bot J 10. 66:1433–1443

Butnik AA, Japakova UN, Begbaeva GF (1997). Adaptivnye svojstva galofitov v ontogeneze ras- tenij. Probl Osvo Pustyn. 4:61–65

Henkel PA, Shakhov AA (1945) Ekologicheskoe znachenie vodnogo rezhima nekotorykh galo
-phitov. Bot Zh SSSR 4. 30:154–166

Iversen I (1936) Biologische pflanzentypen als Hilfmittel in der Vegetationsforschung. Dissertation Middl fra Skalling lab, Copenhagen

Jefferies RL, Rudmik T (1991) Growth, reproduction and resource allocation in halophytes. Aquat Bot 1–2. 39:3–16

Joshi AC (1937) Some salient points in the evolution Amarantaceae and Chenopodiaceae. Am J Bot 1. 24:3–9

Joshi AC, Rao V (1986) Stomatal types in some halophytes from the Mediterranean coast. Indian Bot Soc 65:525–526

Keller BA (1927) O solerose (*Salicornia herbacea*) i ego otnoshenii k zasoleniyu. Priroda i selskoe khozyaistvo zasushlivykh pustynnykh oblastei SSSR. 1–2:71–76

Medvedeva RG (1962) Materialy po anatomii triby Salicornieae Dum.Soobsch.1. Osobennosti stroeniya sarsazana *Halocnemum strobilaceum*.Izv. AN KazSSR: 81–91

Nagolevskiy VYa, Tilba AI (1989) Ekologo-fiziologicheskiy i geograficheskiy analiz galofitov Sev. Kavkaza. Ekologiya. 4:3–8

Stocker O (1928) Das Halophytenproblem. Ergeb Biol 3:265–353

Ungar JA (1962) Influence of salinity of seed germination in succulent halophytes.Ecology 4 43:763–764

Ungar JA (1979) Seed dimorphism in *Salicornia europaea* L.Bot Gaz 1. 140:102–108

Waisel Y (1972) Biology of halophytes. Academic Press, New York, 396 pp

Environmental State and an Analysis of Phytogenetic Resources of Halophytic Plants for Rehabilitation and Livestock Feeding in Arid and Sandy Deserts of Uzbekistan

Ch.N. Toderich, R.I. Goldshtein, W.B. Aparin, K. Idzikowska and G.Sh. Rashidova

Keywords. Human impact, soil salinity, halophytic biodiversity, biosaline agriculture, Kyzylkum deserts

Abstract. The extent of salinity and pollution of soils and cover vegetation by various contents of salts, organic pollutants, traces of heavy metals is presented for different regions of the South-Central Kyzylkum deserts (Uzbekistan). The Buchara oasis was chosen as the object for biomonitoring and as model site for afforestation and establishment of halophytic pastures for undertaking cattle/goat/sheep farming in arid and sandy deserts of Kyzylkum. The ecological differentiation of phytogenetic halophytic resources, the botanical and economic characteristics of plants and the allocation of edaphic resources to maintenance growth, reproduction and survival were examined for environmental heterogeneity within salt-affected and degraded sandy lands of South-Central Kyzylkum. A computer database of 120 native wide-ranging and narrowly distributed taxa, belonging to 80 genera and 25 families is proposed. Four large ecological halophytic groups in respect to their growth and life strategies in saline habitats were analyzed. For Kyzylkum plant vegetation it was found that C_3 species dropped to low levels in both numbers and biomass, whereas C_4 plants have become the dominant vegetation. This finding throws new light on the C_3/C_4 leaf anatomy of plant diversity of Kyzylkum as a cool temperate desert. More than 25 annual and perennial shrubs/semishrubs, herbs, especially species of genera *Salicornia*, *Halostachys* and *Halocnemum*, *Ephedra*, *Haloxylon*, *Suaeda* and several perennial species of *Salsola*, *Tamarix*, *Artemisia*, *Ceratoides*, *Climacoptera* etc. were tested and recommended both for undertaking halophytic agroforesty and biosaline agriculture in arid and sandy Kyzylkum deserts.

Introduction

Uzbekistan, with a population of 24.4 million and an area of 447 000 km^2, where arid and sandy deserts occupy around 80%, is one of the leading countries of Central Asia. In the presence of salt-affected lands and degraded sandy deserts, insufficiency of good-quality of water for agriculture, inability to meet the demand for food for an increasing population, shortage in agricultural production, handicaps in cattle/goat/sheep farming (due to overgrazing and destruction of pastures), there appears a strong case for developing a systematic strategy for rational usage, rehabilitation and conservation of desert agriculture lands for economic development and reclamation. Kyzylkum as a cool temperate desert could become an area of sustainable development in Uzbekistan. However, intensive use of the natural resources of the desert without taking into consideration the stability of the basis of ecological systems, as well as a large -scale development of the gold mining and gas industries, especially in Central and southwest Kyzylkum for 15–20 years have aggravated the processes of desertification and land degradation on these territories. Large areas of desert (35 million ha) of Uzbekistan are subject to underground or irrigation water salinization, aeolian erosion and worsening of their previous state because of human activities, and perhaps trends to aridity of the climate. The consequences are moving sand, sand encroachments and salinization; in lower reaches of the Syrdarya and Amudarya Rivers about 60% of the land suffers from various aspects of salinization. The appearance of large irrigation drainage lakes in the Kyzylkum desert in past two decades, especially in the Buchara oasis, changes fundamentally the water-salt balance of the environments of the adjoining territories. Poor natural draining capacity, floating soils, absence of antiseepage, coating of canals, old irrigation and drainage systems and poor efficiency of washing irrigation lead to a raise in groundwater level and soil salinity in these territories. Much of the salt-affected land occurs in areas where there is a severe food deficit during seasons when halophytic biodiversity can be utilized and the environment is favourable for its growth. We suggest that utilization and conservation of the native halophytic vegetation for further salinity control, economic development and reclamation through genetic improvement and natural selection, employing proper agromanagement practices, are crucial for Uzbekistan.

Few studies have been made on the effects of planting and grazing of halophytic communities in various ecological desert zones of Uzbekistan: Sheveluha et al. (1990), Kamalov (1995), Kamalov and Ashurmetov (1998). Several adaptive strategies, based on geobotanical, morphological and structural feature diversity in different groups of halophytes of the Kyzylkum desert were distinguished (Akdjigitova 1982; Butnik 1984, 1995; P'yankov et al. 1997, 1999; Toderich 1996). However, screening trials of a wide variety of halophytes for agroforestry and biosaline agriculture in arid saline sands of the Kyzylkum desert have not been yet investigated. This chapter focuses on the contemporary environmental state and analyzes the phytogenetic resources of halophytic plant diversity of the southern regions of the Kyzylkum desert that have been collected during many international field expeditions (1996–1998) in Uzbekistan.

The Effects of Environmental Factors and Human Impact in the Southern Kyzylkum Desert

The data analyzed during field work throughout the Kyzylkum deserts to the Aral region and back along the Amudarya River show that a wide salt-affected area is located in the Aral region, Central and southwest Kyzylkum, especially in the Aygakitma and Minbulak depressions, the Buchara oasis, the Navoi and Horesm districts, the Karshi steppe and the Shahrisabs tracts. The Buchara oasis, situated in the lower part of the Zerafshan River Valley, was chosen as the object of biomonitoring and as model site for afforestation and establishment of halophytic pastures for undertaking cattle/goat/sheep farming in the arid and sandy Kyzylkum desert. Soils are characterized by low productivity and high salinity (rarely more than 3% with a predominance of carbonates and sulfate chloride type of salinization: the chemical content of the water is closely related to the collecting point and varies widely [Vol. Eq.%]: HCO_3 8–18; Cl 17–27; SO_4 55–67.6; Ca 27 –35; Mg 24–45; Na+K 28–37), heterogeneous particle-size composition, unfavourable physical and physicomechanical properties of the water and a high degree of compaction. The humus content ranges from 0.5 in sandy desert and grey-brown sites to 0.7–1.2% in the virgin and newly irrigated takyrs. The mineralization of superficial water changes from 0.7 to 4.6 g l^{-1}; and subsoil water from 1.5 to 5 g l^{-1}, while the mineralization and hardness of subsoil water in the Kashkadarya valley has 1.5–2 times more. High human impact (urban, industrial, and agricultural activities, handicrafts and traffic) in the southern part of the Kyzylkum desert leads to pollution of sands and irrigated lands with pesticides, nitrates, organic pollutants and various heavy metals, which results in poor-quality surface waters. The appearance of various pollutants and/or high concentrations of heavy metals in certain parts of the Zerafshan River reveals that a serious effort must be made in order to avoid further deterioration of the water quality.

Traces of some metals were found also in the vegetative organs of different arid plants. The contents of metals such as: (mg kg^{-1}) V (0.5–14); Cu (1.4–5.4); Zn (8.1–72); Mn (11.8–91); Cd (0.01–0.5); Ni (0.22–12.4); Cr (0.4–44); Sb (0.01 –0.48); Fe (110–5020); Se (0.1–3.2); Co (0.12–2.7); Th (0.06–2.46); Sr (21–980); Pb (0.01–5.7); Mo (0.1–9.3) were closely related to edaphic factors, types of species and their habitats, reaching a maximum among the genera *Salsola, Tamarix, Climacoptera, Artemisia, Alhagi, Kalidium, Lycium* and *Aeluropus*. The most polluted areas occur in Uchkuduk, Zerafshan, Buchara, Navoi and the Karshi steppe. Thus, the appearance of a maximum concentration of various salts, organic pollutants and trace metals in various parts in soils, water and plant vegetation indicates environmental problems in the whole area of the Buchara oasis. In view of these alarming conditions halophytic agroforesty and biosaline agriculture based on ecological farming appear to be the only solution for restoring salt

Analysis of Phytogenetic Resources of Halophytic Plants of the Kyzylkum desert

-affected and/or degraded lands and providing alternative fodder for cattle/sheep/goat farming in the arid and sandy Kyzylkum deserts.

In the territories of Middle Asia more than 710 species of halophytes are described, belonging to 34 families of flowering plants and 219 genera. According to data of Granitov (1964), and Akdjigitova (1982) about 30% (representatives of 19 families) of them are endemic to Central Asian flora. A comparative study of the high diversity of halophytic life forms was undertaken by us to investigate interactions between genetic diversity, photosynthetic pathway, reproductive specificity, seed ecology and economic potential in respect to growth and survival in saline habitats. As a result, screening trials and a computer database for estimating the botanical and economic potential of halophyte species of Kyzylkum were examined. This computer database contains more than 120 native wide -ranging polymorphic annual and perennial species as well as narrowly distributed taxa belonging to 80 genera and 25 families indexed alphabetically. Species included are known or strongly suspected of being able to grow and produce/reproduce under saline conditions in at least part of their natural range and/or under irrigation with saline brackish water. The matrix of species, based on detailed information of life form, botanical and cytoembryological characteristics, pastoral and nutritional values, economic interest and maximum reported salinity tolerance, with bibliographical references are analyzed for each species. Great attention is also given to the anatomy of the basic C_3/C_4 type of Kyzylkum vegetation, fruit and seed morphology (scanning microscopic analysis), the ecology of seeds and the creation of a seed bank. For the cover vegetation of Kyzylkum we found that C_3 plants dropped to low levels in both numbers and biomass, whereas C_4 types became the dominant vegetation. This observation was quite unusual for Kyzylkum as a cool temperate desert. Many halophyte species have also evolved a number of different C_4 leaf anatomies, ranging from a typical bundle sheath arrangement in the genera *Salsola*, *Suaeda*, *Atriplex* to cylindrical chlorenchyma sheaths surrounding internal water-storage tissue in the leaf or vegetative tissue of flower organs of various succulent species of *Salsola*.

We propose more than 25 species that can be used to rehabilitate and clean up contaminated and/or salt-affected lands of the arid and sandy Kyzylkum desert (Table 1). As seen from Table 1, useful halophytic plants could be selected for soil conservation and halophytic pastures.

Representatives of the Chenopodiaceae have a major place in the composition of vegetation cover of salt-affected desert lands. Based on the specificity of distribution of underground water and its level of mineralization, we completely agree with the data of Akdjigitova (1982) and divide halophytic plant diversity into four large and several intermediate ecological groups. The hyperhalophyte

Table 1. A Biological characteristics of some halophytes for rehabilitation and livestock feeding in saline arid and sandy deserts of Uzbekistan

Species	Life form	Genome C_3/C_4 -anatomy	Botanical characteristics	Reproduction
Aeluropus litoralis (Poac.)	salts excreting H; perennial dry grass (20–60cm)	2n=14, 28 C_4	anemogam.; B: IV–VI; F: VII	vegetative, rarely sexual
Alhagi pseudalhagi (Fabaceae)	accumulative, H, Ps; thorn-tipped herb (50–100 cm)	2n=16 -	entomogam.; B: V–VII; F: VIII–IX	sexual, seedling / planting
Anabasis salsa (Chenopod.)	salts accumulative X; perennial aphyllous semishrublet (10–60 cm)	2n=18,36 C_4	autogam. to cross entomogam.; B: V–VI; F: VII–IX	sexual; vegetative
Artemisia halophylla (Asteraceae)	salts and toxic ions excluders Mx; perennial semishrub (40–100 cm)	2n=24,48 -	entomogam.; B: IX–X F: IX–XI	sexual; parthenocarpic, apomictic
Atriplex heterosperma (Chenopod.)	M (facultative halophyte); annual, dry herbs (50–150 cm)	2n=18,36,72 C_4	anemogam. B/F: IV–IX	sexual ; self-sowing
Camphorosma lessingii (Chenopod.)	X, P; semishrublet (25–80cm)	2n=12 C_4	anemogam.; B: VI–VIII; F: IX–X	sexual; seedling/ planting
Ceratoides ewersmanniana (Chenopod.)	accumulative of toxic ions; Ps; perennial, semishrub (7–120cm)	2n =18 C_4	anemogam.; B: VII–VIII; F: X–XI	sexual, seedling/ planting
Climacoptera lanata (Chenopod.)	salts and toxic ions hyperaccumulative H; annual succulent herbs (10–60 cm)	2n=18 C_4	anemogam.; B: VII–VIII; F: IX–X	sexual ; parthenocarpic and apomictic
Ephedra strobilacea (Ephedraceae)	X; perennial, dioecious, evergreen, rhizomes, aphyllous shrub (1.0–1.5 m)	2n=14,28 C_3	B: V; F: VI–VII	vegetative, rarely sexual
Gamanthus gamocarpus (Chenopod.)	salts excreting M; annual, ephemeric dry herbs (3–20 cm)	No data	entomogam., rarely anemogam.; B/F: IV–IX	sexual, parthe-nocarpic; self-sowing
Halocnemum strobilaceum (Chenopod.)	accumulative H; perennial succulent semishrublet (40–75 cm)	2n=18 -	autogam., often anemogam.; B/F: VIII–X	sexual; vegetative
Halostachys caspica (Chenopod.)	hyperaccumulative H; perennial succulent shrub (up to 3.5 m)	2n=18 -	anemogam.; B: VII–VIII; F: VIII–IX	sexual
Halothamnus subaphylla (Chenopod.)	X, Ps; perennial, multibranched shrub (120–160 cm)	No data	anemogam., rarely entomogam. and autogam.; B: V–VIII; F: X–XI	sexual, seedling/ planting

Environmental State and Analysis of Phytogenetic Resources of Halophytic Plants 159

Species	Life form	Genome C₃/C₄ -anatomy	Botanical characteristics	Reproduction
Haloxylon aphyllum (Chenopod.)	salt and ions toxic; accumulative X; perennial aphyllous, stem-succulent, tree-like (4–10m)	_-_ C_4	anemogam.; B: IV–V; F: X–XI	sexual; seedling/ planting
Kalidium caspicum (Chenopod.)	accumulative, Mx; perennial, semishrub (15–75 cm)	2n=18 C_4	anemogam.; B: VII; F: VIII–IX	sexual
Lycium ruthenicum (Solanaceae)	accumulative Mx; perennial thorn-tipped shrub (70–150 cm)	No data	entomogam.; B: IV; F: V–VI	sexual
Nanophyton erinaceum (Chenopod.)	X; perennial, microphyllous semishrublet (15–25cm)	_-_ C_4	anemogam., rarely entomogam.; B: VI–VIII; F: IX–XII	sexual; vegetative
Nitraria retusa (Nitrariaceae)	accumulative Ps; perennial multibranched shrub (up to 2m)	_-_ C_4	No data	prevails vegetative
Phragmites communis (Poaceae)	H, X; perennial rootstocks grass (1.5–4m)	No data	anemogam.; B: VI– VIII; F: VII–IX	prevails vegetative; sexual
Salicornia europaea (Chenopod.)	hyperaccumulative H; annual, succulent herbs (40–50cm)	2n=18 C_3	anemogam.; B: VI–VII; F: VII–VIII	sexual
Salsola arbuscula (Chenopod.)	Ps, X; perennial, leaf succulent shrub (50–120cm)	_-_ C_4	anemogam., rarely entomogam.; B: V–IX; F: IX–X	sexual, seedling/ planting
Salsola gemmascens (Chenopod.)	X, Ps; semishrublet, slow -growing, succulent (15–55cm)	2n=18,36 C_4	anemogam., rarely entomogam.; B: VIII–X; F: IX–X	sexual
Salsola orientalis (Chenopod.)	H; perennial polymorphic semishrub (15–70cm)	2n=18,36 C_4	anemogam., occa -sionally autogam. or entomogam.; B: V–VIII; F: X–XI	sexual parthenocarpic (30–65%) and apomictic
Salsola sclerantha (Chenopod.)	M; annual, leaf succulent herbs (10–45 cm)	2n=18,36 C_4	anemogam., rarely entomogam.; B: VI –VIII; F: IX–X	sexual
Suaeda microphylla (Chenopod.)	salts accumulative Mx; perennial succulent, poly -morphic semishrub (30–75 cm)	2n=18,36,54 C_3/C_4	anemogam., rarely entomogam.; B: VII–VIII; F:VIII–X	sexual parthenocarpic
Suaeda salsa (Chenop.)	H; annual, leaf succulent herbs (25–100 cm)	2n=18,36 -	anemogam.; B/F: VIII–IX	sexual

Species	Life form	Genome C_3/C_4 -anatomy	Botanical characteristics	Reproduction
Tamarix hispida (Tamaricaceae)	salts excreting Mx; perennial polymorphic tree -like (2–6 m)	No data	entomogam., rarely anemogam.; B: V– IX; F: VIII–X	sexual; seedling/ planting
Zygophyllum eichwaldii (Zygophyllac.)	salts accumulative Ps, X; perennial leaf succulent semishrublet (30–75 cm)	2n=8–13	entomogam.; B: IV–VI; F: VII–VIII	sexual

H hyperhalophytic species, able to grow on habitats with superficial (0.5–1.5m in depth) and high mineralized underground water; *X* xerohalophytes; *Ps* psammoxerohalophytes (underground water > 4.0 m in depth); *M* mesohalophytes (underground water 1.5–2.5 m); *Mx* halomesoxerophytes (underground water 2.5–4.0 m); *B* anthesis; *F* fruiting period

Table 1. B Ecological and economical potentials of some halophytes for rehabilitation and livestock feeding in saline arid and sandy deserts of Uzbekistan

Species	Seeds ecology	Economic interest	Habitat
Aeluropus litoralis (Poaceae)	D: A$_1$; Gs: 38–74%; Ls: 8–9 years; sST: 14 days 5/10 /28 °C	fodder (around a year), Hr: 0.1–0.20 t ha^{-1}, rare 0.4 t ha^{-1}	wet-crust-plump solonchak and salty sands and clay soils
Alhagi pseudoalhagi (Fabaceae)	D: A2; Gs: 12–38%; Ls: > 3 years; mechanical or chemical treatment	fodder; medicinal; industrial; commercial products; sand-fixing	in dry old valleys of the rivers; low wet salted lands; weak saline semi -arid and sandy deserts
Anabasis salsa (Chenopod.)	D: B1; Gs: 12–36 % ; Ls: 7 to 12 mo; lSt: 2 mo/ 5 / 25 °C or removal of fruiting covers	rehabilitation of desert lands; fodder (around year), commercial products, medicinal	solonchak alkaline and sodic, grey-brown soils, loam sands; salt-marshes and takyrs
Artemisia halophylla (Asteraceae)	D: B1,B2; Gs: 20–94%; Ls: 2 years; sequences of 13/31 °C/14 mo	improvement and/or creation of autumn-winter pastures; Hr: 0.6–2.38 t ha^{-1}; fuel; medicinal	dump to plump salt -marshes; alkaline and sodic soil, grey-brown and heavy clay desert and semidesert
Atriplex heterosperma (Chenopod.)	D: B1; Gs: 26–58%; Ls: 7–9 mo; lSt: 15 /25 °C night/day	desalinization of sites; fodder, Hr: 0.6–0.8 t ha^{-1}	salt-marshes, solonchak rarely on clay loam on semidesert areas
Camphorosma lessingii (Chenopod.)	D: B1–B3; Gs: 79–84; Ls: 8–10 mo; light-sensitive /28 °C	improvement/creation of autumn/winter pastures; fodder; medicinal	grey brown-sandy; clay loam weak saline soils in arid and semiarid lands
Ceratoides ewersmanniana (Chenopod.)	D: B1; Gs: 20–80%; Ls: 3 years; sSt: 3 °C /0.5 mo/28 °C day/night cycles	improvement of sandy/saline in arid foothills areas; fodder: Hr: 0.03–1.0 t ha^{-1}	salt-marshes sands, clay slopes gravely stony soils, endemic of Middle Asia

Environmental State and Analysis of Phytogenetic Resources of Halophytic Plants   161

Species	Seeds ecology	Economic interest	Habitat
Climacoptera lanata (Chenopod.)	D: A1; Gs: 35–80%; Ls: 8–10 mo; lSt: 1.5–2.0 mo / 28 °C or action of organic acids	desalinisation of sites, autumn-winter-spring halophytic pastures Hr: 1.0–1.5t ha^{-1}	clay-sandy /saline and margins of salt-marshes soils, coast of salty lakes
Ephedra strobilacea (Ephedraceae)	Gs: 20–40%	rehabilitation of desert lands; fodder Hr: 0.08–0.5 t ha^{-1}; medicinal and sand-fixing	compacted weakly salted sands, grey-brown desert soils
Gamanthus gamocarpus (Chenopod.)	D: B1,B3; Gs: 12–26%; Ls: 8–9 mo; lSt: 1–12 mo 5 °C /30 °C light /dark cycles	rehabilitation of takyrs and takyr-like lands	slightly saline sands and silt clay loam soils, takyrs
Halocnemum strobilaceum (Chenopod.)	No data	Rehabilitation and soil conservation; fodder, Hr: up to 50 kg ha^{-1}	swampy, crust-plump higher salted sands, clay loam silty soils, solonchak
Halostachys caspica (Chenopod.)	D: B$_1$; Gs: 10–21%; Ls: 8–10 mo; lSt: 0.5–1.0 mo/ 5 °C light/dark cycles	Improvement of saline affected lands fuel; medicinal; decorative, industrial	pioneer in the colonization of hyper-saline, wet-compacted and solonchak-alkaline and sodic soils
Halothamnus subaphylla (Chenopod.)	D: A2–B1; Gs: 9–35%; Ls: 7–9 mo; lSt: 1–2.5 mo /2–4 °C/28 °C night/day cycles	fodder: summer-autumn pastures, Hr: 0.3–1 t ha^{-1}; medicinal, sand-fixing; industrial	skeletal salted slopes, salted marsh margins, grey-brown sandy soils
Haloxylon aphyllum (Chenopod.)	D: A2–B1; Gs: 40–88%; Ls: 8–10 mo; sSt: 2 /4 °C 0.5 mo) or removal of fruiting covers	creation of pastures and improvement of sandy/saline lands; Hr: 0.05–1.4 t/ ha^{-1}; fuel and sand-fixing	salty sand ridges and heavy clay soils forming there a rarefied forest-like areas
Kalidium caspicum (Chenopod.)	D: B1,B3; Gs: 14–29%; Ls: 8 mo	poisonous; rehabilitation of sandy/ saline lands	wet and plump salt-marshes on coast of salty lakes
Lycium ruthenicum (Solanaceae)	Ls: > 1 year	rehabilitation of desert lands; fuel and sandy-fixing	irrigated stripes, salt-marshes, takyrs; weakly saline sand
Nanophyton erinaceum (Chenopod.)	D: A1; Bs: 12–26%; Ls: 8–10mo	improvement of sceletal, gravelly gypsonous sites; fodder; medicinal	gravely grey-brown sandy soils; solonets and solonchaks
Nitraria retusa (Nitrariaceae)	No data	desalinization of salt/affected arid and sandy lands ; sand-fixing; decorative	salted sands, solonchak and takyr-like soils; sceletal gravelly soils

Species	Seeds ecology	Economic interest	Habitat
Salicornia europaea (Chenopod.)	D: B_1; Gs: 12–18 %; Ls: 10–12 mo; lSt: 1.0–2.5 mo/3 °C	desalinization of sites; fodder: autumn-winter halophytic pastures; Hr: 0.1–0.2 t ha^{-1}	damp and wet-salt marshes, coast of rivers, lakes and channels
Salsola sclerantha (Chenopod.)	No data	creation of halophytic pastures on condensed sandy salted soils	compacted and grey -brown sandy weakly salted soils; rarely on loam clay
Salsola arbuscula (Chenopod.)	D: A2–B1; Gs: 12–20%; Ls: 10–12 mo; lSt: 0.5 –2.0 mo /3 °C/28 °C	improvement of salt affected lands; fodder: summer-autumn pastures Hr: 0.2–0.25 t ha^{-1}	fixed saline sands; stony gravely or gypsum contents sites
Salsola orientalis (Chenopod.)	D: B1; Gs: 18–52%; Ls: 7–9 mo; sSt: 3 °C/26 °C /0.5–1.5 mo)	improvement of saline arid and semiarid lands; fodder: summer-winter pastures, Hr: 0.5–2.1 t ha^{-1}	grey-brown, takyr-like soils and sands with high gypsum contents
Salsola gemmascens (Chenopod.)	Gs: 32–46%; Ls: 8–10 mo; lSt or remove of fruiting covers and action with organic acids	creation of pastures on salted and takyr-like soils; rehabilitation of salt -affected sandy sites	edificator on takyr-like; clay loam gypsiferous soils; compacted saline sands
Suaeda microphylla (Chenopod.)	D: B1–B3; Gs: 12–38%; Ls: 7–10 mo; lSt: 1–10 mo 5 °C /15 °C and dry storage at 24 °C	desalinization of arid lands; fodder: autumn -winter pastures; pharmaceutical	saline-alkaline and sodic soils; wetlands, takyrs and secondary salted sites
Suaeda salsa (Chenopod.)	No data	fodder; halophytic pastures Hr: 0.5–1.1 t ha^{-1}	salt-marshes with weakly mineralised superficial underground waters
Tamarix hispida (Tamaricaceae)	No data	fodder; ornamental fuel and sand-fixing; improvement of lands	salty sands on valleys of rivers and lakes; loam salted slopes
Zygophyllum eichwaldii (Zygophyllac.)	D: B1; Gs 24–32% 28 °C dark cycles	rehabilitation and soils conservation; poisonous; pharmaceuticals	salted loam clay and sand deserts; along irrigated channels and margins of lakes

Gs germination of seeds; *Ls* longevity of seeds germiability; *Hr* crop capacity; *lSt* long term stratification; *sSt* short term stratification of seed; *D* type of dormancy of seeds; *A1* weak exogenic; *A2* strong exogenic; *B1* deep endogenic; *B2* deep endogenic; *B3* intermediate endogenic type of seed dormancy; *mo* month(s)

group consists of 4,3% (about 31 species) and is dominated by *Salicornia*, *Halostachys*, *Halocnemis*, *Aeluropus* and *Climacoptera*, which inhabit all salt marshes and wet salt-affected sites with superficial and highly mineralized underground waters (0.5–1.5 m in depth).

Environmental State and Analysis of Phytogenetic Resources of Halophytic Plants 163

In most cases they have a high species cover and low species diversity. Their life strategies are characterized by prominent aphylous and stem succulence; deep immersed reproductive organs in a fleshy stem; presence of thinner, highly pigmented and non-sclerified fruiting covers; significant delay in reproductive organ development (in late summer); slow growth and differentiation of the embryo and accumulation of various substances, e.g. oils, which prevent the penetration of toxic salt ions into the embryo; low rate of seed germination. In addition to their high reclamative potential the representatives of this group have good forage value and can be used for biomass production in the arid and sandy deserts of Kyzylkum.

Mesohalophytes, represented by some graminous and annual *Salsola*, *Suaeda*, as well as by species of genera *Bassia*, *Halogeton*, *Halocharis* and *Atriplex* can be used for further growing mineralized underground water of various contents of salts setting down 1.5–2.5 m in depth. Halomesoxerophytes characterized by *Tamarix*, *Halolachne songarica*, *Atriplex*, *Seidlitzia rosmarinus*, *Limonium gmelinii*, *Kalidium caspica*, *Phragmites communis*, several *Salsola*, *Artemisia halophylla* etc. posses a wide range of salt-tolerant properties to salinity with a groundwater level of 1.5–4 m. Most frequently distributed in the sandy deserts of Uzbekistan is the haloxerophyte group, which is able to grow and reproduce on different salt-affected and destroyed sandy lands with very deep groundwater. These three groups are dominated fully or partially by species that often inhabit disturbed salty sand areas and possess a wide range of salt-tolerant properties to salinity with a groundwater level of 2.5–4 or more in depth. Many of these ecological groups have low species cover and relatively high species diversity. Perennial plants usually show a burst of vegetative and reproductive growth early in the season before hypersaline conditions develop. Significant plasticity in relation to flowering and fruit formation also occurs in these groups. It is likely that there is competition for resources between maintenance of the individual, growth and sexual reproduction. Most halophytic plants are perennial, alternatively vegetative reproduction leads to clonal perpetuation of the genotype. Phenotypic plasticity involving both morphological and physiological changes in response to episodic events is an important characteristic associated with the survival of these long-lived halophytes. In this case, the low frequency of sexual reproduction in perennial halophytes species is a limitation to the availability of seed supplies on a commercial scale. We find that the reproductive stages appear to be more sensitive to saline conditions. Obviously, this generative tissue could be a most important source of germplasm for future breeding.

Most halophytic forage shrubs-semishrubs are salt sensitive during germination or contain various inhibitors and, as shown in Table 1, special temperature requirements and/or chemical treatment, destruction of intact seed coat or removal of fruiting covers are effective for increasing germination of seeds. Thus, study of seed morphology and ecology of germination, as well as creation and conservation of a unique seed bank of native halophytic plant diversity for further economic development and reclamation of arid lands of Uzbekistan will be needed.

Conclusion

A suitable combination of saline strata, saline water of reasonable depths and salt concentration for genetic diversity and phenotypic plasticity of halophytes plants holds promise for future rehabilitation and conservation of salt-affected lands and providing fodder for cattle/sheep/goat farming in the arid and sandy desert of Kyzylkum. Proper prospects and agromanagement practices in respect to environmental heterogeneity and specificity of plant growth, reproduction and survival should be undertaken to develop biosaline agriculture and agroforesty research in the desert areas of Uzbekistan. Great attention should also be given to the development of processes of phytoremediation using native halophytes and salt-tolerant plants. Many woody/semiwoody and herbaceous species of the genera *Haloxylon, Salsola, Halothamnus, Anabasis, Nanophyton, Kalidium, Tamarix, Alhagi, Peganum, Artemisia* and *Halostachys* have shown an ability to break down or convert toxic ions and traces of heavy metals, and can be used to clean up contaminated lands in the Kyzylkum deserts.

The problems of utilization and conservation of the botanical diversity of the wild halophytes gene base could be solved by organizing a landscape structure with hydro- and phytomelioration within natural saline/sandy and irrigated lands, as well as by a system of conserving areas of the natural halophytic and tugai vegetation and by land use control.

The wide inter- and intraspecific variation and phenotypic plasticity of halophyte species provides a good model system for better understanding of mechanisms of plant adaptation and competitive ability of plants to harsh desert environments. Our results should enable more rigorous selection of halophytic plants for rehabilitation of saline/sodic lands of Kyzylkum and for the intensive evaluation of sources as fuel wood, timber, commercial products, forage and medicine.

References

Akdjigitova NI (1982) Halophytic vegetation of Central Asia, FAN, Tashkent,189pp

Butnik AA (1984) The adaptation of anatomical structure of the family Chenopodiaceae Vent. to arid conditions. Summary of biological science doctor degree thesis. Academy of Science Uzbekistan, Tashkent, 18pp (in Russian)

Butnik AA (1995) Adapting strategies of wood and semiwoody plants in arid environment (xerophylization problems). J Arid Stud, 5S: 73–76

Granitov II (1964) Vegetation cover of southwestern Kyzylkum, vol I, Nauka, Tashkent, 334pp

Kamalov SK (1995) Working out of technology of clayey saline phytomelioration in the southern part of Aral Sea. J Arid Lands Stud, 5S: 311–315

Kamalov SK, Ashurmetov AA (1998) Phytomelioration of the dry sea floor of the Aral Sea and Amudarya Delta. Proc Int Conf on Desert Technology, W Australia

P'yankov VI, Voznesenskaya EV, Kondratschuk AV, Black CC Jr (1997) A comparative anatomical and biochemical analysis in *Salsola* (Chenopodiaceae) species with Kranz -type leaf anatomy: a possible reversion of C_4 to C_3 photosynthesis. Am J Bot 84 (5): 597–606

P'yankov VI, Black CC, Artyusheva EG, Vosnesenskay EV, Ku MSB, Edwards GE (1999) Features of photosynthesis in *Haloxylon* species of Chenopodiaceae that are dominant in Central Asian deserts. Plant Cell Physiol 40 (2): 125–134

Sheveluha VS, Shamsutdinov ZS, Nazariuk LA (1990) Halophytii prirodnoi florii ispolzovanya ih dlya virashivanya v pustinyah Srednei Azii na osnove oroshenya solenoi vodoi. Sb.: Ecologo-populatsionui analiz kormovich rastenii estestvenoi florii, introduktsiya i ispolzovanie. Siktivkar, pp 35–42

Toderich CN (1996) Structural analysis in *Salsola orientalis* S.Gmel. with a Kranz-type sepal anatomy. Proc Int Conf on Morphology and Anatomy of plants, St Petersburg, pp 87–91

Salinity: A Major Enemy of Sustainable Agriculture

Yoav Waisel

Keywords. Salinity, salt accumulation, irrigation

What Is the Problem?

Sustainability is the commonly used term for "development which meets the needs of the present without compromising the ability of future generations to meet their own needs" (Pereira 1996). The adoption of such an approach means that the current handling of agricultural development, and especially of its expansion policy, should be continuously checked and evaluated. This would assure that development meets the above stated requirements and will sustain long-term stability.

It is interesting to note that most articles that deal with sustainable agriculture refer to the effects of changing agricultural structure, the dependence on fossil fuel, use of chemical fertilizers and pesticides or to effects on the people involved. Very few publications are concerned with the use of irrigation water or with its quality (WCED 1987; Thomas and Middleton 1993; Roeling and Wagenmakers 1998).

What is the present situation? The food supply of the world is gradually approaching the verge of insufficiency. In part this is a problem of wealth distribution, but to a greater extent it is the outcome of the shortage in irrigation water of high quality. The available sources of water that can be used for the expansion of crop production by irrigation are the low-quality, marginal sources of water, that in arid regions are mostly saline (Waisel 1972).

The use of such water has caused salinization in almost 10% of all arable soils of Africa, and of 50% of the irrigated land there (Thomas and Middleton 1993). It affects some 30% of the irrigated land in Pakistan (Barrow 1991).

Except for rare and limited sites, salinity does not constitute a serious problem in humid regions. Even when farmlands of such regions become salt-affected, there is enough freshwater to wash the salts off. Saline irrigation water is a unique but doubtful privilege that was solely granted to the people of the arid and semiarid regions. With the growing population of the world and the increased pressure on resources for food production, the proper use of marginal water for sustaining agriculture becomes a critical matter. The growing population of the world increases the pressure on resources for food production, a process that will worsen in the near future. Thus, understanding the basic principles of salt fluxes, and of the salt and water balances of fields that are irrigated with water of low quality is of supreme importance.

Salinization is an old problem. Analyses of ancient records of irrigation agriculture in the Tigris–Euphrates basin (pre-Sargonic texts) imply that already at that time agriculture decline had occurred. Probably this happened because of salt accumulation (ki-mun) in the irrigated farmlands. The postulation that salinization played a role in the decline of agriculture is based on several aspects. For example, it is based on hints that the ancient farmers shifted, at certain times, from cultivation of the salt-sensitive wheat to that of the salt-tolerant barley. Moreover, the fact that salinization became a serious factor in such fields is evident from the introduction of a soil-leaching procedure (ki-duru). Such methods would have not been developed unless the fields had suffered from excessive accumulation of salt. The Sumerians understood the nature of salinization and, in the relatively small scale of their operations, tried to cope with it as much as they could. Depending on climatic changes, it seems that salinization had repeated itself in the Diyala river region of Iraq a few times (Jacobsen and Adams 1958).

Apparently, sustainability was not successful in the past, when population pressure was low. Would it be successful now under the enormous pressure of population growth?

The Facts

Arid zone agriculture has three sources of salinity: rainwater, capillary rise of saline underground water and irrigation water.

Rainwater (Cyclic Salts)

Depending on the distance from the sea and on the regional topographic contour of the coastline, salt concentrations of rainwater vary between 10 and 50 mg l^{-1}. A rough estimate of rainwater in the Middle East yields an average value of some 30 mg NaCl l^{-1}. This means that even under dry farming, i.e. in areas with up to 300 mm mean annual rainfall, salt is added to the fields at an annual rate of 10 g $l^{-1}m^{-2}$. Upscaling of these figures to agricultural dimensions means an annual addition of 100 kg of NaCl ha^{-1}. In drainage basins of highly arid areas, where surface runoff is an important factor, such additions of salts are even higher.

Capillary Rise of Saline Underground Water

Such sources of water are rather spatial affairs, that usually develop in and around terminal drainage basins. Nevertheless, in some regions, because of an improper use of irrigation practices, the rise of saline water into the root zone of cultivated fields may reach a disastrous scale. Such situations can be prevented if the groundwater level is kept low. Actually, because of the increasing exploitation of underground water, e.g. for several date palm groves in the Tunisian Sahara (Cloudsley–Thompson 1977), the underground water levels of some basins have dropped considerably.

However, under the prevailing agricultural procedures, e.g. in Central Asia, the ample use of saline water for field irrigation in such sites has caused a rise in the ground water level, stopped the leaching out of the excess salts and caused salinization of the upper soil layers (Khabilbullaev et al. 1998)

Irrigation Water

All irrigation waters contain salts, with some sources containing more and others less. The average salt concentration of irrigation water in Israel is approximately 500 mg NaCl l^{-1}. If we assume an annual use of such irrigation water, at a rate of 1 m^3 of water 1 m^{-2} of irrigated land, the salt balance would show an annual addition of at least 500 g NaCl to each 1 m^2 of arable land. On a field scale this means the annual addition of 5 tons of NaCl ha^{-1}. In other regions of the world this might be even worse. Some of the irrigation water that was used in the Western USA (Texas, New Mexico or Arizona) contained even higher concentrations of salt, e.g. 2000 mg l^{-1} of dissolved salts or higher.

What is the fate of such salts, and how much of it can be removed? Most of the major crop plants accumulate in their tissues only insignificant quantities of NaCl. Thus, the balance between the input of salt by the irrigation water and the output of it by crop harvesting is annually added to the soil. Apparently, irrigation with saline water means a massive and constant addition of NaCl to any irrigated field. As in arid regions there is not enough rainwater to leach such salts out of the soil, the use of saline water causes salt accumulation that develops into a crucial problem.

Only few groups of crop plants can tolerate irrigation with brackish or saline water. These include date palms, coconut palms, barley, sugar and fodder beets, cotton, rape, kale, Bermuda grass, Rhodes grass etc. Indeed, the problem of finding varieties with a reasonably high salt resistance does not seem to be the major problem. Such cultivars are available and reasonable yields of several crops can be obtained even under irrigation with water whose salt concentration is as high as that of the Baltic Sea. Moreover, using modern genetic manipulations such as induced mutagenesis or bioengineering may certainly help to develop even better genotypes. Thus, the direct physiological effects of salinity can be solved. However, the crucial point is not the temporary irrigation of a crop with saline water but the long-term accumulation of salts in the arable horizons of the soil.

Because of the shortage of good irrigation water, the use of marginal water and of recycled water seems nowadays to be rather essential. However, such water contains not only elevated concentrations of NaCl, but also rather high concentrations of borates. Boron becomes toxic to sensitive crop plants, once the concentrations rise above 1 mg l^{-1}. When B concentrations increase to above 4 mg l^{-1} water, growth and productivity of even the most tolerant crop plants are reduced (Wild 1988). Plants may be affected even more by the synergistic influences of B and NaCl (Shani and Hanks 1993).

How long is it feasible to use saline irrigation water without an irreversible poisoning of the fields? This would depend on the local climatic conditions as well as on the soil type and drainage conditions. In any case, the bottom line for such a situation is: the time is very short.

Are there any solutions?

Various solutions were proposed.

Stop Irrigation and Move Back to Dry Farming

Agriculture under such conditions would be sustainable even in most of the arid regions. However, the capability to grow plentiful food under such conditions is limited. Therefore, the return to dry farming is not a solution for the rapidly increasing human population of the arid zones.

Save Water: Use of the Minimal Possible Quantities of Water

Some scientists have treated the problem of reclamation of salt affected soils as "simple, because it is only necessary to supply sufficient water over and above the consumption by the crop" (Wild 1988). This is called by irrigation scientists the leaching requirement, a factor that includes the salt content of the soil solution, the salt concentration of the irrigation water and the salt tolerance of the irrigated crop.

This is another example where the irrigation literature contains only few references to the environmental aftereffects of leaching programs. Water is regarded as an inexhaustible resource that can be freely used. Salinization of underground water did not bother anyone. However, this is wrong. Groundwater is a major source on which arid-zone agriculture depends, and protection of its quality should be of prime concern to the people of such regions. Prevention of salinization of the scarce water sources should outweigh even temporary gains that can be made by reclamation of certain areas of land.

Thus, a good irrigation scheme must consider reasonable ways not only to remove the accumulated salts from the arable topsoil layers but also to solve the complex situation that has produced them.

The wisest way to fight salinization of arable land is by avoidance: by reducing the use of low-quality water for irrigation and by introducing modern saving techniques. No doubt, this will spatially reduce the capability to produce food.

Therefore, additional measures have to be sought to enable a minimal use of saline water. Improvement of the traditional irrigation systems and employment of advanced ones such as trickle irrigation may help a lot. However, even by such irrigation practices, the salts of the irrigated water still accumulate at the periphery of the wetted soil volume (Fig. 1).

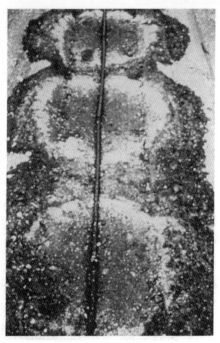

Fig. 1. Salt accumulation around the wetting zone of a trickle irrigation system

What happens then? The use of trickle irrigation segregates between two types of roots: roots that grow near the trickler, and are supplied with freshwater, and roots that grow in the periphery of the wet soil "pear" and suffer there from a highly salinized environment. Thus, not all roots have an equal supply of water, or an equal exposure to salinity, even under the sophisticated modern irrigation techniques.

How does a root system function when some of its parts are exposed to a saline medium? The functional capability of the affected roots is usually lost and therefore the plants' water supply is reduced.

This was demonstrated experimentally by measurements of the water status of split-root grape plants. The plants were strongly affected when one part of the root system was exposed to freshwater and another part to a saline medium. The result of such a situation was a significant reduction in yield (Shani et al. 1993).

Saving irrigation water can be achieved also by wise selections of suitable water-saving cultivars. Attempts along these lines were made in Kazakhstan by substitution of the commonly used "wet" cultivars of rice by "dry" ones.

Saving of water, and subsequently lowering salinization, can be achieved also by other simple means, i.e. by using crop residues to mulch the ground and reduce evaporation. Mulching of maize fields with plant residues significantly reduced the soil surface temperature. Indeed, soil temperature under mulch was 8 °C lower than that of an unmulched field (cf. Buerkert et al. 1996). Insulation with organic mulch has additional advantages; it reduces wind erosion and lowers fertilizer losses, in addition to the saving of water.

Nevertheless, by the employment of even the most modern methods only slowed the advancement of salinization but did not prevent it..

Reduction of Salinity by Intensive Leaching

Application of excessive quantities of water can leach much of the accumulated salts out of the soil, or at least reduce the salt content to a bearable level.

Such an approach was employed in Israel in the early 1940s in order to transform the salt flats between Jericho and the Dead Sea into arable land. Each 1 m^2 of land had to be washed with some 6 m^3 of fresh Jordan water. However, at that time the area was not limited in water availability, and leaching of the salts into the Dead Sea did not affect its salinity.

Leaching can remove most of the chloride out of the soil. However, depending on the calcium content of the soil and on its texture, leaching may cause excessive adsorption of the sodium ions on the fine soil particles. Consequently, this will result in sodification of the soil. Thus, the use of marginal water for such purposes not only would add new salts into the field but may also intensify sodification.

In any case, water is a valuable commodity and wasting large quantities of water just to delay salinization for a while would be a serious mistake.

Following intensive leaching, salts percolate into the lower soil horizons and either precipitate there or reach the groundwater. Such a solution can be applicable for coastal sand dunes, where the percolating water with the leached salts are drained directly into the sea. Similarly, leaching can be used in some terrestrial habitats but only near other terminal drainage basins.

However, this leads to a serious dilemma: on the one hand, the drained water is usually too precious to be dumped into the sea. On the other hand, recycling such water for irrigation will only worsen salinization (cf. also Alsaeedi and Elprince 1999).

In sites where leaching of soils may still be practiced, the use of huge quantities of water reduces the salt content of the soil but induces conditions of hypoxia or even of anoxia. Not only does this replace one restrictive factor by another, but combines the two factors to act synergistically. The bottom line is that in either case there are no winners.

The Use of Salt Accumulating Crops to Combat Soil Salinization

Some crop plants such as beet and spinach have the capability to accumulate rather large quantities of NaCl in their shoot tissues. By harvesting the tops of these plants, some of the salt that was added to the field with the irrigation water is removed.

This has triggered the idea of using salt-accumulating plants (e.g. *Suaeda monoica*) for reclamation of salinized citrus orchards (Waisel 1972). *Suaeda* plants accumulate NaCl up to some 50% of the dry weight of their leaves. So it seemed to be a promising solution. However, there is a catch to such an approach. Reasonable growth of *Suaeda* plants requires irrigation. The salt that is added to the orchard with the additional irrigation water equals more or less the maximum amount of salt that otherwise would have been removed from the site by the very same *Suaeda* plants.

Similar attempts to remove salts from the soil by salt-tolerant crops such as sorghum, sugarbeet or *Glycerrhiza glabra* were reported from Uzbekistan. However, because the irrigation water there was only mildly saline, the removal of the extra salt from the soil by the biomass of salt-tolerant cultivars of such crops was possible (Khabilbullaev et al. 1998).

Some halophytes, e.g. *Salicornia herbacea* L.; *Aeluropus* sp.; *Suaeda salsa*; *Atriplex* sp.; *Suaeda microphylla; Salsola arbuscula* etc., can be used simultaneously for fodder and for removal of some salt. However, even such an approach does not provide a real solution.

A Search for a Genetic Breakthrough

Development of highly salt-tolerant transgenic plants, using the currently available molecular methods of bioengineering and mutagenesis, can be foreseen. As the amounts of seawater are not limited, the use of such water for food production may replace some of the freshwater that is currently being used for irrigation and save it for other essential uses.

Summary

Application of saline irrigation water to agricultural land results in a constant addition of NaCl. The water is transpired by the plants but the salt accumulates in the soil. The salt content of the top soil of arable land increases with every irrigation cycle, and eventually may reach unacceptable levels. Thus, at our present state of knowledge it seems that the utmost outcome of all recommended measures would yield only a delay in salinization, but cannot really prevent it.

Actually, such facts should have been enough to convince the policy-makers that the chances of sustaining crop growth under an ever-increasing salinity are next to nil, and other solutions should be sought.

The take-home message is that, with the present agricultural practices, the use of saline water for sustaining reasonable agricultural production is doomed to fail. Saline water and sustainability of agriculture are two incompatible concepts.

References

Alsaeedi AH, Elprince AM (1999) Leaching requirement conceptual models for reactive salt. Plant Soil 208:73–86

Barrow CJ (1991) Land degradation – development and breakdown of terrestrial environments. Cambridge University Press, Cambridge, UK, 295 pp

Buerkert A, Lamers JPA, Marschner H, Bationo A (1996) Input of mineral nutrients and crop residue mulch reduce wind erosion effects on millet in the Sahel. In: Burkert B, Allison BE, von Oppen M (eds) Wind Erosion in West Africa: The problem and its control. Margraf Verlag, Weikersheim, Germany, pp 145–160

Cloudsley–Thompson JL (1977) Man and the biology of arid zones. Edward Arnold, London

Jacobsen Th, Adams RM (1958) Salt and silt in ancient Mesopotamian agriculture. Science 128:1251–1258

Khabilbullaev A, Razakov P and Kosnazarov K (1998) Combined hydroponic and drip irrigation on saline and sandy soils of the lower reaches of the Amudarya. In: Ecological research and monitoring of the Aral Sea Deltas. A basis for restoration. UNESCO Aral Sea Project, Final Scientific Reports 1992–1996, Paris, pp 343–353

Pereira LS (1996) Education, research and training for sustainable use of water resources in agriculture. Medit 7:10–15

Powell MA (1985) Salt, seed and yields in Sumerian agriculture. A critique of the theory of progressive salinization. Z Assyriol 75:7–38

Roeling NG, Wagenmakers MAE (1998) Facilitating sustainable agriculture: participatory, learning and adaptive management in times of environmental uncertainty. Cambridge University Press, Cambridge, UK

Shani U, Hanks RJ (1993) Model of integrated effects of boron, inert salt and water flow on crop yield. Agron J 85:713–717

Shani U, Waisel Y, Eshel A, Xue S, Ziv G (1993) Responses to salinity of grapevine plants with split root systems. New Phytol 124:695–701

Thomas DSG, Middleton NJ (1993) Salinization: new prospectives on a major desertification issue. J Arid Environ 24:95–105

Waisel Y (1972) Biology of halophytes. Academic Press, New York

WCED (1987) Our common future. Oxford University Press, New York

Wild A (ed) (1988) Russell's soil conditions and plant growth. Longman, Burnt Mill, Harlow, Essex, UK

Part III:

Impact of Grazing

Remarkable Differences in Desertification Processes in the Northern and Southern Richtersveld (Northern Namaqualand, Republic of South Africa)

Norbert Jürgens

Key words. Desertification, Namib Desert, Richtersveld, communal grazing, biodiversity

Abstract. A seminomadic traditional land use system, based on the ecological properties of a unique environmental constellation at the boundary between winter and summer rainfall climate, might have been instrumental in conserving natural resources in the Northern Richtersveld (Northwestern Namaqualand, Northern Cape Region, RSA). In the arid to semiarid northern part of the Richtersveld, desertification processes are of relatively low intensity, if compared with the southern part of the Richtersveld, in spite of the generally higher level of aridity on the Northern Richtersveld. This observation is of wider interest because the two regions of the Richtersveld have experienced a different history of land-tenure and land use practices, which might, in part, have caused the different level of resource degradation:

- In the Northern Richtersveld until today traditional seminomadic pastoralism in a communal rangeland tenure system has been maintained, although locally replaced by mining areas, permanent settlements and, since 1991, a National Park.
- In the Southern Richtersveld, the communal rangeland was subdivided by fences into economic units. These have been managed by farmers for a number of years. In spite of the short survival of the economic units concept, the fences still subdivide the landscape.

This chapter describes patterns of environmental factors, biodiversity, land use and desertification indicators. Several possible functional interactions between desertification processes and land use are discussed and focal areas for further research are indicated.

Introduction: the Richtersveld Scenario

The Richtersveld is a semiarid to hyperarid landscape, situated in northern Namaqualand in the northwesternmost corner of the Northern Cape Region of the Republic of South Africa. The region is of special interest with respect to the factors controlling arid ecosystems, as it allows comparison of several contrasting scenarios. The contrasts are defined by differences in rainfall seasonality, degree of aridity, phytogeographical characteristics and by differences in land-tenure and land use practices.

Two Different Climatic Zones

Climatically, the Richtersveld is divided into a temperate winter rainfall zone in the west and a tropical summer rainfall zone in the east.

Two Different Floral Kingdoms

Controlled by the above mentioned rainfall seasonalities and by temperatures and air humidity, the southwestern portion of the Richtersveld forms part of the Greater Cape Flora with its Succulent Karoo Region, while the northeastern portion is covered by plants of the Nama Karoo Region of the Palaeotropical Kingdom (Jürgens 1991; Jürgens et al. 1997)

Several Different Intensities of Aridity

While the moister parts of the Richtersveld receive as much as 300 mm annual precipitation, the most arid parts are exposed to hyperarid climate with ca. 30 mm mean annual rainfall.

Two Different Land Use Histories

In contrast to the ecological zonation of the Richtersveld in longitudinal direction, land-tenure and land use practices divide the Richtersveld in latitudinal orientation: while the Northern Richtersveld has always been communal land, used by seminomadic small stock pastoralism, in 1982 the Southern Richtersveld was subdivided by fences into economic units (Boonzaier et al. 1990; Kröhne and Steyn 1990; cf. Fig. 1). These have been managed by farmers for a number of years. In spite of the short survival of the economic units concept, the fences still subdivide the landscape today.

These factors form the environmental background for the peculiar traditional land use system of the Richtersveld: seasonal migrations between the winter rainfall zone and the summer rainfall zone (and the gallery forests of the Orange

River) and between the drier lowlands and the moister mountain areas allow most efficient pastoral land use linked to lowest degradation of vegetation, because the latter is not damaged during the dry seasons.

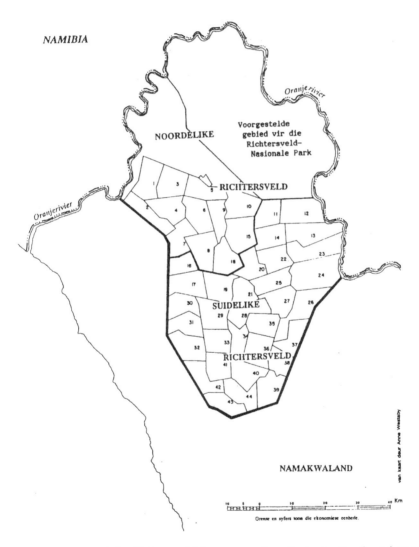

Fig. 1. Subdivision of the Richtersveld into economic units in the South and communal ground in the North

Indicators of Desertification Processes

In the course of a major mapping of the vegetation of the Richterveld, several obvious effects of land use related human activities on soils and vegetation could be observed. Some of these observations are listed here, including biodiversity (Fig. 2).

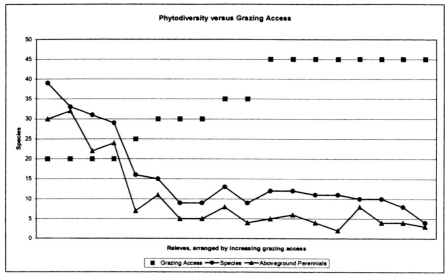

Fig. 2 Decrease in phytodiversity with increasing grazing access in the *Ruschia senaria* community (cf. Table 1)

Erosion

Intensity of erosion can be estimated by the presence and growth of gullies (gully erosion) or by the increase in the depth of exposure of roots (sheet erosion). Both types of erosion are very obvious in parts of the Richtersveld

Southern Richtersveld: Large gullies are present in parts of the Southern Richtersveld, e.g. in the foothills of the region NNE of Eksteenfontein. Here, strong grazing pressure has led to strong reduction of the more stable life forms of vegetation (for details see below) and the characteristic sandy loam / loamy sand of the region is rapidly eroded once the biocrusts at the soil surface are destroyed. Even single car tracks can be an initial reason which later results in gully systems (Colour Plate 7), while deeper roads produce very large gully systems in many places.

Besides gully erosion, sheet erosion has reached strong intensity in the 5-km radius around Eksteenfontein and in the wider mountain region in W, NW and SW of Eksteenfontein (Colour Plates 7–9). In extreme cases, exposed granite (Colour Plate 8), silcrete or calcrete indicates the complete loss of fine material (and therefore water storage capacity).

An obvious correlation exists between the position of stockposts and sheet erosion. Good examples can be observed along the road from Eksteenfontein to Kuboes, where larger parts of the landscape have lost their fine material cover and are now invaded by annually expanding exotic plant species (Colour Plates 11, 12). Populations of very few species (mainly *Ruschia senaria*) have been able to resist the process and still protect islands of deeper fine material soils (Colour Plates 9, 10).

Northern Richtersveld: All the above mentioned structures and processes are much less present in the Northern Richtersveld. Gully erosion can be observed in places on roads in the steeper parts of the Hells Mountains. In the open landscape, stronger gully formation is present around the Goariep Mountain (Ploughberg), especially at its foothills (Colour Plates 7–10) and in the plain close to the permanent settlement of Kuboes. However, these gullies are small and young in age. The pastoralists from Kuboes are aware of this damage, and explained during a workshop with the author that this area has been overutilized. In comparison with the Southern Richtersveld, the erosion in the Northern Richtersveld is much less important and limited to specific regions and situations.

Invasion of Neophytes

The invasion of neophytes is obvious along river courses (*Nicotiana glauca, Prosopis glandulifera, Datura* spp., *Ricinus communis*, more rarely: *Argemone* spp.). However, during the past 20 years, the author could observe no strong expansion of these taxa. In contrast, in the Southern Richtersveld, a very strong expansion of the regional neophytes *Fingerhuthia africana, Galenia africana* and of the continental neophyte *Atriplex lindleyi* can be observed, and is directly linked to overgrazed and/or eroded surfaces (Colour Plates 11, 12).

Increase in Indicators of Disturbance

There are numerous possibilities to identify indicators of disturbance: destruction of permanent vegetation can be studied along rivers after floods, at abandoned stockposts, in the context of roadworks and mining activites and by monitoring clearing experiments. Based on all this information, a list of indicators of disturbance can be established, including taxa with different environmental preferences: *Atriplex semibaccata, Didelta carnosa, Drosanthemum hispidum, Gorteria diffusa, Lebeckia multiflora, Mesembryanthemum pellitum, Mesembryanthemum squamulosum, Opophytum aquosum, Plantago caffra, Psilocaulon subnodosum, Tetragonia echinata.*

The general increase in these taxa in the landscape can be interpreted as an indicator of disturbance. Their presence in the general landscape shows a general

182 — N. Jürgens

increase from the Northern Richtersveld to the Southern Richtersveld (Colour Plate 11).

Decrease of Phytodiversity

While the previous section shows that after disturbance a number of weedy species contribute to general phytodiversity, strong grazing pressure nevertheless results in decrease of overall species richness. This can be shown with the example of a plant community which is occurring over the whole range of the Richtersveld and which can be analyzed with respect to a gradient of grazing impact. Table 1 shows 20 releves (10 x 10m) of the *Ruschia senaria* community, arranged in a sequence of increasing accessability for grazing and trampling stock (grazing access). The scale of the parameter grazing access is an estimation between 1 (no stock grazing = protected wilderness area) to 10 (very strong grazing impact due to close vicinity, i.e. less than 1 km to the next permanent settlement or permanent waterhole). Fig. 2 shows a decrease in species richness with increasing access to grazing. Again, the degraded poorer variants show a strong concentration in the Southern Richtersveld.

Table 1. Species composition of vegetation of a number of releves of the Ruschia senaria community, sorted from left to right by increasing access to grazing

Releve	1	2	3	4	5	6	7	8	9	10	12	13	14	15	16	17	18	19	20	21
Region (N=North, S=South)	N	N	S	N	N	S	S	S	S	S	S	S	N	S	S	S	S	S	S	S
Locality	K	N	D	N	N	Kl	E	E	E	E	E	E	K	E	E	E	E	E	E	E
Grazing access	3	4	4	4	4	4	5	6	6	6	7	7	9	9	9	9	9	9	9	9
Species	32	39	38	33	31	29	16	15	9	9	13	9	12	12	11	11	10	10	8	4
Above ground perennials	18	30	29	32	22	24	7	11	5	5	8	4	5	6	4	2	8	4	4	3
Perennials	18	29	31	32	24	24	8	11	6	5	9	4	5	8	4	5	8	4	4	3
Therophytes	14	9	7	1	7	5	8	4	3	4	4	5	7	3	7	6	2	6	4	1
Geophytes	0	0	2	0	2	0	1	0	1	0	1	0	0	3	0	3	0	0	0	0
Disturbance+invaders	0	0	0	0	0	0	2	1	0	0	0	0	3	2	1	2	1	0	1	0
Ruschia senaria	5	2	9	1	18	20	7	1	20	2	5	20	22	11	0	0	7	7	10	15
Euphorbia mauritanica	2		1	r		3	1	1	1	0	2	1	+	1	1		0	+	+	1
Heliophila variabilis	+		+		+		0	0	0	0	1	1	+		+	r		+	+	
Didelta carnosa	+	2	+	r	3	r				0		0							+	
Pentzia pilulifera	+		+					0	1	0	1	2		1	r		+			
Oncosiphon suffruticosum	+	1			+	+							+							
Ruschia sarmentosa		4		+	2	2														
Crassula muscosa muscosa	r			r	r	+														
Euphorbia chersina		1		r	+	1														

Remarkable Differences in Desertification Processes in the Richtersveld

Releve	1	2	3	4	5	6	7	8	9	10	12	13	14	15	16	17	18	19	20	21
Senecio corymbiferus		1		+				1												
Solanum namaquanum	+	+				+														
Sphalmanthus decurvatus		2		r	3															
Forsskaolea candida	+	1												0						
Sphalmanthus deciduus		+	1			1														
Galenia dregeana		4		+	8															
Antimima clavipes		1		1	1															
Astridia speciosa		1		+	1															
Cotyledon orbiculatum		+	+	r																
Crassula deceptor		r		r		+														
Crassula expansa			+	r			+													
Euphorbia dregeana		1		1	r															
Euphorbia ephedroides		1		r	+	+		1	0										+	
Euphorbia gummifera	3			1																
Berkheya canescens			1				0													
Blepharis furcata					r			0												
Crassula elegans		+		r																
Ceraria fruticulosa				+	+															
Conophytum gratum		+		r																
Euphorbia hamata				+	+								+							
Galenia crystallina		1	1			1									+		+			
Galenia namaensis							0		0	0										
Hebenstreitia parviflora		r			r															
Diascia orange							0	0												
Senecio arenarius	+		r																	
Stoeberia beetzii		3		+																
Didelta spinosa			+			3												+		
Drosanthemum hispidum		1	+																	
Helichrysum obtusum		+					0						+							
Mesembryanthemum squamulosum		10				r														
Microloma calycinum			+															+		
Microloma sagittata							0										0			
Mitrophyllum clivorum						2		0												
Nemesia ligulata		r					0													
Nymania capensis	+			+																

184 N. Jürgens

Releve	1	2	3	4	5	6	7	8	9	10	12	13	14	15	16	17	18	19	20	21
Othonna opima		+		r	+															
Oxalis pes-caprae									0				0							
Peliostomum leucorrhizum			+																+	
Prenia sladeniana	+			r																
Tylecodon paniculatus		2					0													
Aloe pearsonii		2		r																
Tylecodon wallichii		+		r																
Zygophyllum cordifolium		+	1																	
Arenifera stylosa				r	r															
Aridaria ovalis			+		+															
Tetragonia verrucosa			+		1															
Trachyandra bulbinifolia			r	+												+				
Ornithoglossum viride				r		0				0			0							
Pteronia ciliata							1			1						+				
Ruschia glauca							1	1		1										
Tetragonia fruticosa		1			2															
Tetragonia reduplicata						1	0			1							1		+	
Osteospermum microcarpum	r	r	1			+	0			0	0									
Osteospermum sinuatum										0	1									
Cheiridopsis robusta									0	1	0									
Mesembryanthemum pellitum	r				+			1	4	0	1									
Aristida adscensionis	+														10					
Sutera fruticosa	+												0		0					
Gorteria diffusa	r	r	+		5							9			6	+		6	1	
Amellus nanus	r				2										+	r		+		
Tetragonia echinata															+			+		
Atriplex semibaccata						0							1		+	+		+		
Lasiospermum brachyglossum													0				1			1
Lycium kleinblä.													1	1	1					
Plantago caffra					2	0							5		1	1			5	
Zygophyllum retrofractum													0	+			0	1		1

Locality Abbr.: *N* Numees; *Kl* Kliphoogte; *K* Kuboes; *E* Eksteenfontein; *D* Doringrivier; perennial species = bold; for further explanation compare text

Feedback Loops

In the Richtersveld, disturbance often leads to increase in therophytes, which show halophytic properties. Especially *Mesembryanthemum pellitum*, *M. squamulosum*, *Opophytum aquosum*, *Psilocaulon subnodosum and Psilocaulon dinteri* result in toxic amounts of salinity in the topsoils, causing plants to die at the onset of the summer drought. This effect is removed after a number of years, when strong winter rains cause leaching of the salts in the topsoils.

Role of Stockposts

This process can be studied in detail at stockposts, where the dung of the stock is adding to the soil salinity and the leaching of soil salinity takes much longer.

Role of Biocrusts

Closely related to the above mentioned processes, the formation and destruction of biocrusts (composed of cyanobacteria and lichens) is also affected by the changes in land use, especially trampling, and soil salinity.

Discussion

Several processes of desertification have been mentioned which are related to pastoral land use. (Other, very important processes, related to mining etc., are outside the scope of this chapter). The geographical distribution of indicators for these processes shows clear differences between the Northern and Southern Richtersveld.

In the Northern Richtersveld, degradation in the overall landscape is less frequent than in the Southern Richtersveld, in spite of the higher aridity of the north. Here, strong degradation can be observed locally, close to preferred grazing areas, e.g. in the Hells Mountains and the Kourougab Plain. Very strong local degradation is present around the sole permanent settlement, Khubus, with the adjacent lower parts of the Ploughberg (Goarieb). During the past 15 years, it has become increasingly obvious that stockposts are concentrated at the areas which can be reached by car. During interviews with stock farmers, organized by F. Archer and the author over a number of years, farmers mentioned that the margins of the Ploughberg had been overexploited in the past.

In contrast to the situation in the north, the Southern Richtersveld shows a more widespread degradation, defined by frequent sheet and gully erosion, stronger invasion of neophytes, stronger replacement of natural vegetation by therophytes and other indicators of disturbance. Local gradients around present and historical stockposts support the hypothesis of a strong influence of pastoral land use.

Based on these observations, two hypotheses can be proposed, which are able to explain the stronger level of degradation in the more humid Southern Richtersveld.

1. The seasonal movement of stock activities (in the traditional seminomadic land use system of the Northern Richtersveld) into those areas which very recently have received rain (or which are not dependent on rain, in the case of the gallery forests of the Orange river) resulted in low degradation of vegetation and soils. The biological background is the fact that during the rainy season the stock is able to crop annuals and herbaceous parts of perennials and does not have to damage important parts of the shoot system. In contrast, farmers of the economic units had to overexploit the vegetation even during the dry season, as they are not able to move with their stock and follow the rains. The concentration on the limited area of the economic unit increases the intensity of grazing activities during the drought season and forces the stock to increased damage of those parts of perennial plants which are essential for survival during the drought season. Over the years, this might have caused the overall difference between Northern and Southern Richtersveld. Such a system would present a very interesting example for the non-equilibrium model proposed by Behnke et al. (1993), that in such a system stock density has less impact on the long-term composition of vegetation. If this hypothesis should be true, a communal seminomadic land use system would be of advantage in comparison to sessile economic units, for ecological reasons.
2. Due to the more arid nature of the Northern Richtersveld, stock movements in the whole Richtersveld resulted in lower stock numbers and, therefore, lower impact per surface area from grazing than in the more humid Southern Richtersveld. If this hypothesis should be true, the differences between the Northern and Southern Richtersveld are simply caused by intensity of exploitation and allow no interpretation with respect to communal-seminomadic versus sessile farming.

Discrimination between these two hypotheses is problematic due to lack of historical data on land use. In spite of the fact that estimations of stocknumbers are reported in Boonzaier et al. (1990); Hendricks (1998) and Kröhne and Steyn (1990), there is no reliable information on the total stock numbers in relation to the rainfall seasonality areas.

Therefore, a more thorough reconstruction of monthly stock numbers in the rainfall seasonality areas or the different economic units (Fig. 1) would be of interest. However, it is questionable whether adequate reliable data can be gathered.

Alternatively, a more thorough study on the medium-term vegetation dynamics (cf. Jürgens et al. 1999) under different grazing pressures over the rainfall seasons could be established as an experiment. Such an experiment should include the wilderness in the Richtersveld National Park areas as well as test sites under different grazing regimes during different seasons (or in direct response to rainfall events, cf. Behnke et al. 1993).

Acknowledgements. I would first like to thank Annelise Le Roux (on flora and vegetation) and Fiona Archer (on peoples' rights and land use) for the initial and strong motivation to document land use and vegetation changes in the Richtersveld. The people at Kuboes and Sanddrift, and especially many stock farmers, contributed information on flora and land use practices. My thanks go to Robbi Robinson, Harold Braack, Paddy Gordon, Hugo Bezuidenhout, Peter Novellie, Howard Hendricks and Johan Taljard for supporting and enabling permanent plots and exclosures in the Richtersveld. For cooperation and discussion in the field I would like to thank Dieter von Willert, Volker Bittrich, Annelies Le Roux, Zelda Wahl, Mike Struck, Dave MacDonald, Loretta Van Zyl, Graham Williamson, Heidi Hartmann, Michael Dehn, Angela Niebel, Stephan Rust, Berit Hachfeld, Alex Gröngröft, Ursula Jähnig, Judy Beaumont, Elke Erb, Inge Gotzmann, Stefanie Nußbaum, Tanja Osterloh, Andrea Gimborn, Gitta Schüttler, Iris Oguz, Britta Stöcker, Jens Boenigk and Pascale Chesselet. On the administrational level, the South African National Parks, the Administration of Nature Conservation of the Northern Cape, the National Botanical Institute, the German Science Foundation, the Schimper Stiftung, the Merensky Foundation and the GTZ are thanked for their support.

References

Behnke RH, Scoones I, Kerven C (eds) (1993) Range ecology at disequilibrium. New models of natural variability and pastoral adaptation in African savannas. Overseas Development Inst Regent's College, London

Boonzaier EA, Hoffman MT, Archer FM, Smith AB (1990) Communal land use and the "tragedy of the commons": some problems and development perspectives with specific reference to semi-arid regions of southern Africa. J Grassl Soc S Afr 7/2:77–80

Hendricks H (1998) Traditional stock farming in the Richtersveld. J Bot Soc S Afr 84(3): 86–87

Jürgens N (1991) A new approach to the Namib Region. I: Phytogeographic subvdivision. Vegetatio 97: 21–38

Jürgens N, Burke A, Seely MK, Jacobsen KM (1997) The Namib Desert. In: Cowling RM, Richardson D (eds) Vegetation of Southern Africa. Cambridge University Press, Cambridge, pp 189–214

Jürgens N, Gotzmann I, Cowling RM (1999) Remarkable medium-term dynamics of leaf-succulent Mesembryanthemaceae shrubs in the winter-rainfall desert of northwestern Namaqualand, South Africa. Plant Ecol 142: 87–96

Kröhne H, Steyn L (1990) Grondgebruik in Namakwaland. Suplus People Project, Cape Town, 98 pp

Colour Plates 1-2: Wucherer, Flora

Colour Plate 1. *Calligonum aphyllum*, an example of an abundant psammophytic leafless shrub

Colour Plate 2. *Matthiola stoddartii*, an example of a typical ephemeral plant

Plates 3-6: Wucherer, Vegetation dynamics

Colour Plate 3. The chinks are the main sources for the seed bank at the north coast of the Aral Sea

Colour Plate 4. *Salicornia europaea* community on the marshy solonchaks (succession stage I)

Colour Plate 5. Pioneer plants of the second generation of *Climacoptera* and *Petrosimonia* species on the coastal solonchaks (stage II)

Colour Plate 6. The intermediate plant community of *Stipagrostis pennata* and *Phragmites australis* on the southeast coast of the Aral Sea

Plates 7-12: Jürgens

Colour Plate 7. Initial gully formation caused by a car track. NNE of Eksteenfontein

Colour Plate 8. Gully erosion, running uphill on a granitic slope at the western margin of the Goariep (Ploughberg)

Colour Plate 9. Soil protection by the *Ruschia senaria* community in a gully system at the western margin of the Goariep

Colour Plate 10. Closer look at the *Ruschia senaria* community, with its soil-protecting property. Note the dark colour of soil due to persistance of biocrusts and the soil accumulation under the bushes of *Ruschia senaria*

Colour Plate 11. Sheet erosion and degraded vegetation in the wider landscape around a stock post in the Southern Richtersveld

Colour Plate 12. Sheet erosion and invasion of exotic flora and indicators of disturbance (here *Fingerhuthia africana, Galenia africana, Euphorbia ephedroides*) in the Southern Richtersveld

Colour Plates

Plates 13-16: Yair, Sedimentary Environments

Colour Plate 13. Tamarix trees along a low linear sandy recessional ridge

Colour Plate 14. A Saxaul forest in an area of deep sand

Colour Plate 15. View of a thick and saline topsoil silty-clayey layer

Colour Plate 16. View of a typical saline Solonchak surface

How Grazing Turns Rare Seedling Recruitment Events to Non-Events in Arid Environments

Suzanne J. Milton and Thorsten Wiegand

Keywords. Florivory, Karoo, recruitment event, seed production, vegetation composition

Abstract. When livestock numbers are high in relation to the percentage of the vegetation comprising preferred forage plant species, and where no provision is made for occasional livestock withdrawal during periods of flowering and seed set, grazing will deterministically lead to the near-eradication of certain forage plant guilds from rangelands. Recruitment events for long-lived plant species, and even for ephemeral plants, are uncommon in arid (<200 mm/a^{-1}) environments and are closely tied to weather sequences that favour the particular plant species by promoting first seedset and then germination and seedling survival. Preferred forage plant species, whether ephemeral or perennial, with or without soil-stored seed banks, decrease or disappear from areas where the grazing regime, by preventing seed-set, turns potential recruitment events to non-events. This chapter provides evidence for the rarity of recruitment events in an important forage plant *Osteospermum sinuatum* (DC) T. Norl. (Asteraceae) in South African Karoo shrubland and analyzes population structure and seedling densities in protected (9.4 SD 7.7 seedlings m^{-2}) and grazed areas (0.02 SD 0.3 seedlings m^{-2}) 2 years after a sequence of years that favoured seedling recruitment. Seedling to adult ratios were 5:1 in protected areas and 0.7:1 in grazed areas on the same ranch. Although refuges have potential to increase seed availability in heavily grazed rangelands, seed input to sink populations will be constrained by the dispersal mechanism. Seeds of *O. sinuatum* are tumbled over the ground by wind, and an exponential decrease in seedling density with distance from the seed source ($n = 25$, $r = -0.699$, $p < 0.001$), indicated that most seeds are dispersed within 5 m of the source. When grazing control by fencing is not economically feasible, many small refuges (such as provided by spinescent and toxic plants) may be more effective than seed plantations in maintaining key forage species in rangelands.

Introduction

On intensively used parts of arid and semiarid rangeland landscapes worldwide, perennial palatable forage plant populations tend to decrease and to be replaced by spinescent or toxic plants that have less value to the grazier (Bucher 1987; Skarpe

1990; Milton et al. 1994; Fusco et al. 1995), or by seed-bank ephemerals that disappear during dry periods (O'Connor 1991; Grice and Barchia 1995). Vegetation changes of both types occur in the semidesert Karoo shrublands of South Africa (Milton and Hoffman 1994; O'Connor and Roux 1995; Steinschen et al. 1996; Todd 1997; Palmer et al. 1999) and are a matter of concern for communal and commercial ranchers (Dean and Macdonald 1994; Hoffman et al. 1999), who depend on domestic livestock for food or income. The indigenous shrubs that replace perennial grasses and palatable shrubs and succulents in the Karoo generally contain high concentrations of herbivore deterrents (alkaloids, cardiac glycosides or salts), and live for decades, possibly delaying further shifts in rangeland composition (Milton et al. 1994; Todd 1997).

The problem of declining forage plant populations has stimulated research aimed at understanding the processes leading to such changes. First, some experimental evaluation of the effectiveness of reseeding (Barnard 1987; Milton 1994a), bush clearing and livestock manipulations (O'Connor and Roux 1995) in reversing unwanted changes has been attempted, but the scarcity of weather conditions suitable for germination and establishment necessitates long-term funding and dedication to achieve meaningful results from field trials. Second, a modelling approach based on available information on plant life-history attributes and demographic responses to weather, competition and grazing (Wiegand and Milton 1996) gives some indication of likely timescales for grazing-induced vegetation change and recovery potential. A third approach to acquiring information on demographic processes in arid rangeland plants, that is difficult to obtain experimentally, is through monitoring of rare recruitment events at sites differing in grazing management. This approach is of special importance since there is growing evidence that arid zone plant populations are structured by infrequent large recruitment events (Eldridge and Westoby 1991, Pierson and Turner 1998). The notion that the vegetation composition in arid areas is event -driven rather than determined by interactions among species (succession and grazing), is currently favoured by many range scientists (Ellis and Swift 1988, Westoby et al. 1989, Behnke and Scooner 1993), as well as by South African policy makers (Hanekom 1996). In order to evaluate this thinking for management of arid rangelands, it is therefore important to understand whether infrequent recruitment events are influenced by grazing management.

Here we follow the third approach to understanding vegetation change in Karoo shrubland vegetation. Focusing on *Osteospermum sinuatum* (DC) T. Norl. (Asteraceae), a widespread and important forage plant in Karoo shrublands, we use data from an unusual recruitment event to show how grazing, in combination with rare recruitment events and short-distance seed dispersal, leads to population decreases. We also consider the potential value of refuges for the persistence under grazing of local populations of this and other species with comparable life -history attributes.

Study Sites and Management History

All observations were made at the Tierberg Karoo Research Centre (TKRC) and on the adjacent ranch Argentina in the southern Great Karoo of South Africa (33°10S, 22°16E, 780 m asl). The study site (details in Milton et al. 1992) lies in a valley between mudstone ridges, and the fine-textured soils, although deep, are stony and saline. The climate is arid (annual mean 170 mm, range 50–400 mm) and rainfall shows no predictable seasonality. Succulents (Mesembryanthemaceae, Zygophyllaceae) and non-succulent (Asteraceae, Aizoaceae) dwarf shrubs (<0.5 m high) dominate the vegetation, and annuals, grasses and geophytes contribute little to the cover or biomass. A peculiarity of the vegetation is that most species have short-lived seeds that are either stored on the plant in hygroscopic capsules or are produced opportunistically after major rain events and are non-dormant (Esler 1993).

Livestock have been excluded from the TKRC since 1987 when it was fenced off from the moderately grazed sheep ranch, Tierberg. The adjacent ranch Argentina was stocked twice as heavily as Tierberg between 1913 and 1960, but livestock densities were reduced to the recommended level in the 1960s. Vegetation projected canopy cover (as measured by the line intercept method) did not differ significantly between at the TKRC (23.2%) and Argentina (19.3%) in 1991 (Milton and Dean 1993), but palatable shrubs constituted 72% of the total cover at Tierberg and only 20% at Argentina. Forage plant density averaged 19 000–25 000 plants ha^{-1} at TKRC and 6 000–7 000 plants ha^{-1} on Argentina (Milton 1992). Although these data suggest a two-thirds decrease in carrying capacity for sheep on Argentina compared to TKRC, livestock densities are only 15% below the average stocking rate for the area, resulting in intensified browsing effects on the remaining forage plants.

Air photographs (scale 1:8300) show no differences in the appearance of the vegetation across the fence separating Tierberg and Argentina ranches in 1939, but by 1974 a fence-line contrast had developed and this was more marked in 1991. This boundary fence runs north-south across the broad, flat valley of the Sand River, so the fence-line contrast provides opportunities for comparative studies of the effects of weather patterns and grazing on vegetation dynamics within an otherwise homogeneous landscape unit. It can also be used for quantifying seed dispersal distances from protected (source) to grazed (sink) forage plant populations. Five livestock exclosures, circular, 12 m in diameter, 200 m apart and 15 m from the boundary fence separating this ranch from TKRC, were constructed on Argentina in 1989 in order to document the effects of protection from grazing on plant population and vegetation dynamics.

Study Species and Methods

The study species *O. sinuatum* is a broad-leaved, deciduous shrub that stores carbohydrate in its roots and stems, giving it resilience to grazing and drought (van der Heyden and Stock 1996). It flowers and grows opportunistically after rain and can set seed 0–4 times annually depending on rainfall distribution, but does not maintain a soil-stored seed bank (Esler 1993). High population densities occur on some ranches, but on other ranches densities are low, with little evidence of recruitment (Milton 1992). Seed production per plant of *O. sinuatum,* and of other species palatable to domestic livestock, was reduced by 80–99% on the Argentina ranch in comparison with the TKRC exclosure, whereas there were no differences in seed crops of toxic plants (such as *Pteronia pallens* L.f. Asteraceae) between grazing treatments (Milton and Dean 1990).

Germination of available fresh seeds occurs after heavy rain events in all but the hottest months of the year (Milton 1992). An unusual weather sequence led to abundant recruitment of the shrub on long-term ecological research sites and surrounding ranches. An above-average rainfall of 150 mm in March–May 1995 (autumn mean 60 SD 33 mm) stimulated flowering. This was followed 18 months later by an exceptional 190-mm spring rainfall (mean 36 SD 26 mm) that resulted in abundant germination and recruitment of *O. sinuatum* on TKRC and surrounding ranches.

From 1989 until 1996 during a study of temporal and spatial patterns in seedling germination and establishment (Milton 1995), seedlings of all higher plant species were monitored in 16-cm diameter rings ($n = 100$, total area 2 m^2) fixed to the soil surface. Eighty of these monitoring rings were at TKRC, and 20 at Argentina. *Osteospermum sinuatum* seedling density data from this study provide a background against which to compare the magnitude of the 1996 recruitment event. In February 1997, towards the end of the hot and fairly dry summer that followed this spring germination event, the density of surviving *O. sinuatum* seedlings was assessed in 32 randomly placed 1-m^2 plots, of which 25 were at TKRC and 7 on Argentina.

In order to find out whether *O. sinuatum* recruitment differed with grazing treatment and whether the grazed area functioned as a sink for seed, seedlings (< 5 mm basal diameter) and established plants (B.D. > 5 mm) were counted in 700 quadrats in November 1998. These 1-m^2 quadrats were arranged in 20-m-long transects ($n = 35$) running parallel to the TKRC-Argentina boundary fence. Five sample transects (each comprising 20 quadrats), spaced 200 m apart over a distance of 800 m, were placed 5 m from the boundary fence in the protected TKRC, and five parallel set of transects were set out on the sheep ranch Argentina at distances of 2, 5, 10, 30 and 60 m from the fence separating this farm from TKRC. In addition to transects exposed to sheep grazing, 20 quadrats were sampled in each of the five livestock exclosures on Argentina, by placing two 10 -m transects at right angles to one another within the exclosures. Analysis of variance followed by a Tukey test was used to compare seedling densities on transects placed at various distances from the boundary fence. Seedling adult ratios were assumed to indicate the reproductive success of established plants under grazing and protection, and were compared using Chi-square tests with Yates correction.

Results

Magnitude of the Recruitment Event

Osteospermum sinuatum germination events significant enough to be recorded in the miniquadrats occurred in 6 of 9 years, but the November 1996 event was 16 times greater than other years (Table 1). None of the plants recorded in the first three events survived for more than 24 months. Assuming that miniplots gave realistic estimates of the densities of seedlings that emerged in November 1996, then densities of 8–9 seedlings m^{-2} recorded 3 and 24 months later indicate a survival rate of 13–15% (Table 1).

Table 1. Densities and frequencies of Osteospermum sinuatum seedlings (< 5 cm high) at Tierberg Karoo Research Centre (TKRC) and on an adjoining sheep farm between 1989 and 1998 provide evidence that the abundant emergence in November 1996 followed by recruitment was an unusual event. Data for 1989–1996 collected 1–6 times yearly from fixed 0.02 m2 circular plots (total area 1.6 m2 at TKRC, 0.4 m2 on grazed farm), data for 1997 obtained from randomly placed 1-m2 plots, and data for 1998 is for 1-m2 plots arranged in contiguous linear sets of 20 plots

Year	Plot size (m^{-2})	Number of months censused	Protected TKRC		Heavily grazed neighbouring farm	
			Plot number	Seedlings m-^{2} (monthly means)	Plot number	Seedlings m-^{2} (monthly means)
1989	0.02	6	80	0–1	20	0
1990	0.02	7	80	0–0.6	20	0
1991	0.02	5	80	0.0	20	0
1992	0.02	8	80	0–0.6	20	0
1993	0.02	4	80	0–5.6	20	0
1994	0.02	3	80	0–0.6	20	0
Nov 95	0.02	1	80	0.0	20	0
Jul 96	0.02	1	80	0.0	20	0
Nov 96	0.02	1	80	85.0 SD 169 [a]	20	0
Feb 97	1.00	1	25	8.1 SD 7.8	6	0.2 SD 0.4 [b]
Nov 98	1.00	1	100	9.1 SD 12.5	300	0.15 SD 0.5 [b]

[a] The standard deviation calculated for 0.02-m^2 plots is likely to be greater than that at a scale of 1-m^2 plots, because the distribution of *O. sinuatum* seedlings is patchy on small rather than large scales as a result of seed trapping by litter and small succulents; [b] 10–60 m from boundary fence

Effects of Grazing, Protection and Distance from Seed Source on Recruitment

Densities of seedlings and adult plants differed significantly among transects [seedlings $F_{(6,28)}$ = 5.6, P < 0.001, adults $F_{(6,28)}$ = 4.18, P < 0.001], being significantly greater on transects in TKRC and in grazing exclosures on Argentina than in unprotected plots on Argentina (Table 2). Seedling: adult ratios were greater on TKRC than on grazed transects (χ^2 = 62.9, P < 0.001) or in exclosures (χ^2 = 27, P < 0.001) on Argentina, but exclosures on Argentina had significantly higher seedling: adult ratios than grazed plots (χ^2 = 5.1, P < 0.05). The ratio of seedlings to established plants at distances of 30 m and further from the boundary fence were below parity. Seedling abundance on the 35 transects was correlated with canopy cover of established plants on these transects (r = 0.91, D.F. = 33, P < 0.001) which explained 82% of the variation in seedling density. Seedling density in grazed quadrats on Argentina decreased logarithmically with distance from the fence (r = -0.699, D.F. = 23, P < 0.001, Fig. 1). Of the 255 seedlings recorded in grazed transects on Argentina, 214 (84%) occurred within 5 m, and 247 (97%) within 10 m of the TKRC boundary fence.

Table 2. Total and mean (SD) numbers of seedlings and adult plants of Osteospermum sinuatum in 5 transects 20 m x 1 m, and seedling to adult ratios 5 m inside the TKRC livestock exclosure (- 5 m) and at increasing distances across the fence into a grazed sheep ranch. Plots at 20 m distance were in 12 m diameter exclosures established on the grazed ranch in 1989. Within columns, means with shared superscripts do not differ significantly at P <0.05 (one way ANOVA)

Distance (m) from fence	Seedlings		Adults		Seedling: adult ratio
	Total 100 m^{-2}	Mean (SD) Per transect	Total 100 m^{-2}	Mean (SD) per transect	
-5 TKRC	941	188.2 (153.9)[a]	59	11.8 (9.8)[a]	15.9
2	131	26.2 (18.9)[b]	22	4.4 (2.6)[ab]	5.9
5	83	16.6 (12.1)[b]	16	3.2 (3.1)[ab]	5.2
10	33	6.6 (5.6)[b]	16	3.2 (3.9)[ab]	2.1
[20]	361	72.2 (68.3)[a]	61	12.2 (1.9)[a]	5.9
30	6	1.2 (0.8)[b]	11	2.2 (1.9)[ab]	0.5
60	2	0.4 (0.5)[b]	3	0.6 (0.9)[b]	0.7

Discussion

There is growing support for the notion that vegetation in the arid and semiarid regions of Africa is particularly resilient in the face of heavy grazing because the effects of weather events override the effects of management (Ellis and Swift 1988; Behnke and Scooner 1993; Ward et al. 1999). In contrast, studies that focus on plant populations clearly show that, although recruitment and mortality are event-driven, large-seeded plant species that lack soil-stored seed banks and have short dispersal distances are driven to local extinction where grazing prevents seeding for many consecutive years (O'Connor 1991; Milton 1994b). A reduction in palatable perennials, and their replacement by ephemerals or by perennials that are toxic or less acceptable to domestic livestock, may logically be expected to reduce grazing capacity eventually, even if this process takes many decades (Wiegand and Milton 1996; Palmer et al. 1999). One way in which the coexistence of these optimistic and pessimistic viewpoints may be resolved is that the former may describe vegetation already altered by grazing and dominated by animal-dispersed, seed-banking ephemerals, whereas the latter describes vegetation as yet little influenced by pastoralism and that contains a wide range of plant guilds.

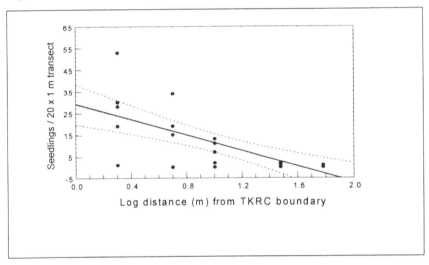

Fig. 1. Decrease in *Osteospermum sinuatum* seedling densities on the Argentina ranch with distance from the seed source on the Tierberg Karoo Research Centre livestock exclosure. Five 20-m transects were placed at distances of 2, 5, 10, 30 and 60 m from the boundary fence

Since 1987 several approaches have been used at TKRC to understand grazing-induced vegetation change and to predict plant population responses to weather and grazing. Initially, we compared population structure of common forage plant species across the fence line (Milton 1994b). Failure of some palatable species to recruit on Argentina was explained as the deterministic result of reduced flowering and seed production of grazed plants on this ranch (Milton and Dean 1990).

Later, we developed an individual-based simulated model incorporating life-history information for the five most common shrub species at TKRC, together with their establishment sites, competitive interactions, seed dispersal distances and flowering, seeding and germination responses to rainfall at various seasons (Wiegand et al. 1995).

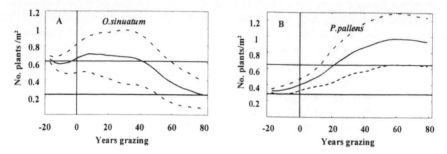

Fig. 2. A,B Results of the simulation model (Wiegand and Milton 1996) showing the mean density of palatable *O. sinuatum* (**A**) and unpalatable *P. pallens* (**B**) averaged for 100 simulation runs with different realistic rainfall scenarios. Simulations started with a range in good condition (Data from Milton and Dean 1990) and the first 20 years were simulated without grazing. Grazing management was modelled as reduced seeding. The *dashed lines* give the mean variation of the simulated plant densities, the *lower horizontal line* indicates the mean density at Tierberg, *and upper horizontal line* shows the mean density at Argentina (Data from Milton and Dean 1990)

Under realistic rainfall scenarios, the expected density of palatable *O. sinuatum* decreased to levels comparable to that in poor condition rangeland within 50 to 70 years (Fig. 2A) while unpalatable *P. pallens* increased rapidly to high levels (Fig. 2B). After 40 years of grazing, when plants that were present when grazing began had died, the expected density of the *O. sinuatum* population declined rapidly due to failure of new recruitment. Because of the stochastic rainfall events, the mean variation in the expected plant density is very high. However, the results of monitoring *O. sinuatum* recruitment suggest an even more pessimistic view than the predictions made by the simulation model that still allowed *O. sinuatum* to recruit at a moderate level under grazing.

O'Connor (1996) found the overriding constraint on seedling recruitment of a perennial forage grass *Themeda triandra* in semiarid savanna to be "seed during appropriate rainfall years". Our monitoring of a major recruitment event for the forage shrub *O. sinuatum* in the arid southern Karoo led to a similar conclusion. Where established plants were rare or grazed too frequently to flower, the density of *O. sinuatum* seedlings remained low, and established plants did not appear to be replacing themselves. Dispersal distances are short (<10 m per generation), suggesting that intensively grazed areas are population sinks. The net result was that the recruitment "event" was a "non-event" on the intensively grazed ranch. However, complete protection of established plants from grazing by domestic livestock and wildlife in exclosures led to a fourfold increase in the density of established plants in 10 years and recruitment of between 10 and 60 times as many seedlings per unit area during the recruitment event.

In such situations, managers of good condition rangeland have two options: (1) to manage livestock for maximum short-term gain regardless of the predictable change in vegetation composition and decrease in long-term carrying capacity, and (2) to manage livestock for sustainable long-term production by endeavouring to maintain viable populations of the major forage plant populations. However, what are the implications of our findings for option (2) that envisages sustainable management of arid rangeland? The provision of unused reserves (or buffers or refuges) has been recognized by modellers (Owen–Smith 1998) and managers (de Bruyn and Scogings 1998) as a practical way of protecting range management units so as to provide forage reserves for animals in dry periods and allow plants to seed and recruit during productive periods. This can be achieved in wildlife management areas by limiting provision of permanent water points (Walker et al. 1987; Owen–Smith 1998), and in rangeland by fencing or herding. However, these options are not always economically feasible or socially acceptable. The effectiveness of such buffers in reseeding sink populations will clearly depend on the dispersal distances of the plant species and the distribution of the refuges. When dispersal distances are short, many small refuges (such as provided by thorny and toxic plants) may be more effective than seed plantations for ensuring the persistence of key forage species in rangelands. There is a perception that bush clearing will improve rangelands invaded by thorny and toxic plants by increasing accessibility and forage production. The possibly beneficial function of defended plants as refuges for palatable species appears unknown, but possibly outweighs their disadvantages in arid regions.

Acknowledgements. This study was supported by a series of grants from the Foundation for Research Development to SJM and from the Umweltforschungszentrum to TW. The owners of Tierberg (C. Hobson, H. Wright, J. Kitzhoff) and of Argentina ranches (W. Niehaus, J. Kitzhoff) have freely permitted research to be carried out on their land over many years.

References

Barnard SA (1987) The influence of sowing depth on the germination of Karoo shrubs and grasses. J Grassl Soc South Afr 4:123–126

Behnke RH, Scoones I (1993) Rethinking range ecology: implications for rangeland management in Africa. In: Behnke RH, Scoones I, Kerven C (eds) Range ecology at disequilibrium. Overseas Development Institute, London, pp 1–30

Bucher EH (1987) Herbivory in arid and semi-arid regions of Argentina. Rev Chil Hist Nat 60:265–273

de Bruyn TD, Scogings PF (1998) Policy-making for the sustainable use of southern African communal rangelands: conclusions of a symposium and workshop. In: de Bruyn TD, Scogings PF (eds) Communal rangelands in southern Africa: a synthesis of knowledge. University of Fort Hare, Alice, South Africa, pp 280–291

Dean WRJ, Macdonald IAW (1994) Historical changes in stocking rates of domestic livestock as a measure of semi-arid and arid rangeland degradation in the Cape Province, South Africa. J Arid Environ 26:281–298

Eldridge DJ, Westoby M (1991) Recruitment and survival in *Atriplex vesicaria* populations in semi-arid western New South Wales, 1977–87. Austr J Ecol 16:309–314

Ellis JE, Swift DM (1988) Stability of African pastoral ecosystems: alternate paradigms and implications for development. J Range Manage 41:450–459

Esler KJ (1993) Vegetation pattern and plant reproductive processes in the succulent Karoo. PhD Thesis, University of Cape Town, Cape Town

Fusco M, Holechek J, Tembo A, Daniel A, Cardeas M (1995) Grazing influences on watering point vegetation in the Chihuahuan desert. J Range Manage 48:32–38

Grice AC, Barchia I (1995) Changes in grass density in Australian semi-arid woodlands. Rangeland J 17:26–36

Hanekom D (1996) Land reform and rangeland management. In: Kerley GIH, Haschick S, Fabricius C, la Cock G (eds) Proc of the 2nd Valley Bushveld Symposium. Grassland Society of Southern Africa Special Publication, Scottsville, South Africa, pp 3–4

Hoffman MT, Todd S, Ntshona Z, Turner S (1999) Land degradation in South Africa. Department of Environmental Affairs and Tourism, Pretoria, South Africa, 245 pp

Milton SJ (1992) Effects of rainfall, competition and grazing on flowering of *Osteospermum sinuatum* (Asteraceae) in arid Karoo rangeland. J Grassl Soc South Afr 9:158–164

Milton SJ (1994a) Small scale re-seeding trials in Karoo rangeland: effects of rainfall, clearing and grazing on seedling survival. Afr J Range Forage Sci 11:54–58

Milton SJ (1994b) Growth, flowering and recruitment shrubs in grazed and in protected rangeland in the arid Karoo. Vegetatio 111:17–27

Milton SJ (1995) Spatial and temporal patterns in the emergence and survival of seedlings in arid Karoo shrubland. J Appl Ecol 32:145–156

Milton SJ, Dean WRJ (1990) Seed production in rangelands of the southern Karoo. S Afr J Sci 86:231–233

Milton SJ, Dean WRJ (1993) Selection of seeds by harvester ants (*Messor capensis*) in relation to condition of arid rangeland. J Arid Environ 24:63–74

Milton SJ, Hoffman MT (1994) The application of state-and-transition models to rangeland research in arid succulent and semi-arid grassy Karoo, South Africa. Afr J Range Forage Sci 11:18–26

Milton SJ, Dean WRJ, Kerley GIH (1992) Tierberg Karoo Research Centre: history, physical environment, flora and fauna. Trans Roy Soc S Afr 48:15–46

Milton SJ, Dean WRJ, du Plessis MA, Siegfried WR (1994) A conceptual model of rangeland degradation. BioScience 44:70–76

O'Connor TG (1991) Local extinction in perennial grasslands: a life-history approach. Am Nat 137:753–773

O'Connor TG (1996) Hierarchical control over seeding recruitment of the bunch grass *Themeda triandra* in a semi-arid savanna. J Appl Ecol 33:1094–1106

O'Connor TG, Roux PW (1995) Vegetation changes (1949–71) in a semi-arid, grassy dwarf shrubland in the Karoo, South Africa: influence of rainfall variability and grazing by sheep. J Appl Ecol 32:612–262

Owen–Smith N (1998) Dynamics of herbivore-vegetation systems and the overgrazing issue. In: de Bruyn TD, Scogings PF (eds) Communal rangelands in southern Africa: a synthesis of knowledge. University of Fort Hare, Alice, South Africa, pp 124–134

Palmer AR, Novellie PA, Lloyd JW 1999 Community patterns and dynamics. In: Dean WRJ, Milton SJ (eds) The Karoo: ecological patterns and processes. Cambridge University Press, Cambridge, pp 208–223

Pierson EA, Turner RM (1998) An 85-year study of Saguaro (*Carnegia gigantea*) demography. Ecology 79:2676–2693

Skarpe C (1990) Shrub layer dynamics under different herbivore densities in an arid savanna, Botswana. J Appl Ecol 27:873–885

Steinschen AK, Görne A, Milton SJ (1996) Threats to the Namaqualand flowers: outcompeted by grass or exterminated by grazing? S Afr J Sci 92:237–242

Todd S (1997) The effects of heavy grazing on plant species diversity and community composition in a communally managed, semi-arid shrubland, Namaqualand, South Africa. MSc Thesis, University of Cape Town, Cape Town

Van der Heyden F, Stock WD (1996) Regrowth of a semi-arid shrub following simulated browsing: the role of reserve carbon. Funct Ecol 10:647–453

Walker BH, Emslie RH, Owen–Smith RN, Scholes RJ (1987) To cull or not to cull: lessons from a southern African drought. J Appl Ecol 24:381–401

Ward D, Ngairorue BT, Kathena J, Samuels R, Ofran Y (1999) Land degradation is not a necessary outcome of communal pastoralism in arid Namibia. J Arid Environments 40: 357–371

Westoby M, Walker B, Noy–Meir I (1989) Opportunistic management for rangelands not at equilibrium. J Range Manage 42:266–274

Wiegand T, Milton SJ (1996) Vegetation change in semiarid Karoo rangelands: simulating probabilities and time scales. Vegetatio 125:169–183

Wiegand T, Milton SJ, Wissel Ch (1995) A simulation model for a shrub-ecosystem in the semi-arid Karoo, South Africa. Ecology 76:2205–2221

Vegetation Degradation in Northeastern Jordan

Othman Sharkas

Keywords. Desertification, rangeland degradation, invader plants, segetal plants, halophyte plants, overgrazing

Abstract. This chapter provides examples of the extent to which Jordanian farmers (Fellaheen) and Bedouin have contributed to vegetation degradation in northeastern Jordan. Irrigated cultivation of marginal lands, deep ploughing of the fragile rangeland, overgrazing, cutting and uprooting of perennial xerophyte species for fire, and even deliberate burning, have in some areas led to a process of desertification.

It is known that the rangeland in northeastern Jordan used to support large numbers of highly palatable species for grazing including *Artemisia sieberi (*syn. *A. herba-alba), Salsola damascena (*syn. *S. vermiculata), Atriplex halimus, Achillea fragrantissima, Hamada eigii* and *Noaea mucronata*. These species were dominant, provided a high degree of surface cover, and were widely distributed.

At present, these palatable species are no longer found in northeastern Jordan, except in the Surra, Khanasri and Shaumari reserves. The palatable species, however, have been replaced by invader, segetal and thorny plants, such as *Peganum harmala, Anabasis syriaca, Salsola jordanicola, Halthamnus hierochunticus, Xanthium spinosum, Onopordum macrocephalum, Chenopodium album* and *Chenopodium murale*, and others.

This phenomenon can be considered and used as an indicator of vegetation degradation in northeastern Jordan.

Introduction

The main causes of desertification and land degradation in arid and semiarid regions lie in the vagaries of a harsh environment coupled with anthropogenic misuse of the natural land resources. This is especially the case in northeastern Jordan (Tadros 1979; Sharkas 1994). This chapter presents some results of vegetation degradation studies in northeastern Jordan to permit us to better understand some of the causes and processes associated with land degradation.

In Jordan there is virtually no control over grazing activities. This is an outcome of a combination of several socioeconomic, historical and geopolitical factors. These factors can be summarized into five general categories:

1. The nomadic tradition of its non-urban residents.
2. Unregulated passage of nomadic Bedouin from surrounding Arab countries, notably Syria and Iraq.
3. The high value, both economic and status, placed by nomads on having large flocks of sheep and goats.
4. The difficulty of governmental enforcement of policy regarding rural land use.
5. Loss of natural landscapes to both pastoral and agricultural use.

The combined effect of these factors has accelerated land degradation in Jordan over the past 15 years.

Description of the Study Area

General Location

The total land area of the Hashemite Kingdom of Jordan is approximately 9.26 million ha. About 91% of this area has been classified as desert and semidesert, with an annual rainfall of less than 350 mm. Due to the low rainfall and high inter -annual rainfall variability, these areas are not suitable for economic agriculture and their optimal use has been left as grazing (Sharkas 1994). The study area falls within an arid and semiarid climate in northeast Jordan.

The total population of Jordan is about 4.2 million inhabitants (Jordanian Statistical Center 1994). The population in the study area is about 2.8 million inhabitants distributed by region as follows: Amman 1.6 million, Zarqa 275 000, north of Azraq 4261 and south of Azraq 1684, Irbid 751 634 and Mafraq 180 000.

Topographic setting

The study area lies within the steppe and steppe desert in northeast Jordan. It is mainly part of the large desert and steppe desert area known as Badiet esh-Sham. The topography is relatively level with elevations falling within 500–800 m above mean sea level (Table 1).

Climate

The study area can be divided into two main climatic regions as defined by Koeppen.

1. A semiarid climate (Bsh) borders an area of Mediterranean climate. It is characterized by a lower annual average precipitation (100 to 350 mm) which falls almost entirely in winter and early spring.

210 O. Sharkas

2. An arid climate (Bwh) dominates the desert areas. This climate area is characterized by hot summers and cold winters (Shehadeh 1985, 1990; El –Kawasma 1983; Al–Eisawi 1985). Average annual rainfall is less than 150 mm. Flash flooding is common in winter and causes soil erosion problems.

Table 1. Topography of the study area

Location	Altitude (m)	Scale of map 1:50000	Degrees north	Degrees east
Mafraq	699	=	32°21'	36°13'
Al–halabat	627	=	32°05'	36°22'
Dhuleil	577	=	32°07'	36°17'
Surrah reserve	750	=	32°24'	36°10'
East Zarqa	595	=	32°05'	36°06'
El–Baij	687	=	32°22'	36°20'
Al–khanasreh	859	=	32°24'	36°03'
Wadi Al–buttom	580	=	32°48'	36°35'

Soil

The soil in Jordan has been studied and classified by several workers such as Moormann (1959), Grueneberg and Dajani (1964) and El–Rihani (1984, 1987). According to the USDA Soil Taxonomy, the main types of soil in the study area are Xerollic Haplargids, Typic Camborthids (FAO Yermic Cabisol), Typic Calciorthids (FAO Yermic-Haplic Calcisol), and Lithic Torriothents (FAO Leptosol). It can be said that the main characteristics of the soil in northeastern Jordan are its shallowness, high salinity, and very low amounts of organic material (Sharkas 1994).

Vegetation

The natural vegetation cover in northeastern Jordan is a reflection of the different temperature regimes, soil characteristics and types, and elevation. Vegetation cover in the study area was studied and classified by several workers using different approaches and methods (Kasapligil 1955, 1956; Long 1955, 1957; Zohary 1962; Al–Eisawi 1985; Al–Eisawi and Hatough 1987; Sharkas 1994). The main type of vegetation in northeastern Jordan are:
• Steppe vegetation. This vegetation is confined to the Irano–Turanian region and may intrude into the Mediterranean or the Saharo–Arabian regions (Al–Eisawi 1985). The composition of the vegetation varies according to the soil and climatic differences depending on its location with respect to the Mediterranean region.
• Desert (hamada) vegetation.
• Halophyte (saline) vegetation.
• Segetal and thorn vegetation.

Vegetation Degradation in Northeastern Jordan   211

During the past five decades the vegetation cover in the study area has been degraded due to large-scale and intensive human activities such as overgrazing, cutting and uprooting of the perennial xerophyte species for fire, irrigated cultivation, deep ploughing and uncontrolled extension of dry land farming (Sharkas 1994).

Land Use

Water is the limiting factor influencing land use in northeastern Jordan. According to the Jordan Ministry of Agriculture (1985), about 1 million ha in the steppe areas (24.5%) receive annual rainfall between 100–350 mm, and 7 million ha of the steppe desert region (7.5%), known as the Al–Badia, receive less than 100 mm (Abu–Zant et al. 1993). The land use chart (Table 2) shows the land use system in the study area.

Table 2. Land use in northeastern Jordan

Cultivation	Grazing
Rain-fed farming	Grazing land
Traditional farming,	
mostly cereals and legumes	Range Reserves
Mechanized farming,	
mostly barley and wheat	
Irrigated cultivation	Pastures
Vegetables and watermelon, forage plants	Sheep and goats
(*Medicago sativa*), fruit trees	Camels with sheep and goats

While cultivated areas can easily be recognized on aerial photos, seasonal land use patterns are more difficult to identify. Such data can be collected, however, by carrying out interviews and surveys of pastoralists and range officers in each particular area. The land tenure in the study area is divided into five categories:

1. Private,
2. Mulk (land held by the municipality),
3. Miri (land within the municipal area),
4. Amiriyya (unregistered land or state land),
5. Musha'(communal open lands). Most of the pastoral areas are officially state lands.

With time, agriculture has encroached on pastoral lands. Hence, the area available for grazing has declined and accordingly the intensity of land use has increased on the remaining pastoral lands.

Methods and Materials

Sampling took place in September and October 1991, and in March, April and May 1992. Generally speaking, the monitoring of degradation of vegetation cover in the study area is done more easily in grazing areas than on cultivated land. The following variables of vegetation change could be determined; plant cover densities (in %), change in abundance (cover), increase or decrease of grass, herbaceous, palatable, unpalatable, segetal and thorn species, clearing for cultivation and urbanization.

In order to maximize the accuracy of the vegetation and land degradation survey, given the relatively short sampling season, the Braun–Blanquet (1964) method was applied as modified by Wilmanns (1984). Sixtyseven releves were used with different land uses. As example, an area of 20 x 20 m² in rain-fed agriculture and irrigated farming areas was used and in the rangeland 3 x 3 m² releves.

Two line transects were also used to study vegetation cover within and outside of the rangeland reserves, and from mountain tops to valley bottoms such as in the Al–Nawasief region.

Results and Discussion

Most of the rangeland in northeastern Jordan is state land. The Bedouin and peasants (Fellaheen) believe they have full right to use this land as they wish, such as for grazing, cultivation, collecting vegetation, ploughing, and even the right to enter the rangeland reserves. This phenomenon explains the origin of the vegetation degradation problem in the study area. The desertification of the natural vegetation cover is a result of the socioeconomic situation.

In 1994 the study area population numbered 2.8 million inhabitants. As the population increased, demand on natural resources was magnified. Without controls, this led to an imbalance between human needs and the animal grazing population on the one hand, and the vegetation, water and land resources on the other. High population pressure in turn led to an increased demand for meat. This manifested itself in a rapid increase in the population of domestic animals to meet the demand. The construction of wells as watering places for sheep and goats (herds) and the trucking of water by trucks have had negative effects, particularly on the area around the wells. According to Tadros (1979) and Abbas (1989) overgrazing around new unplanned wells and bore holes aggravated the problem, for example in Al–halabat, Dhuleil, Mafraq, and Azraq. As governmental grazing control over these pasturelands is almost non-existent, these low-potential range locations were overused.

Another factor that greatly endangers the grass zones is deep ploughing. The residents there are mostly Bedouins and take over wide areas of land that is owned by the government. They overplough these areas to maintain access and to generate a land-tenure legacy. According to traditional law in Jordan, should they use or be able to keep the land for 15 years, then they acquire rights to this land. This phenomenon encourages the inhabitants to try to capture as much as possible. This is really a major cause of the destruction of these areas. On the other hand, the Bedouin plant new trees and bushes that do not belong or are not native to these areas. The importing of water in tanks enables the shepherd owners to remain in a given area for a longer period as they are no longer dependent on rainfall. As a result, overgrazing occurs.

Cross-border movement of thousands of shepherds from Iraq and Syria is another factor to contend with. Their herds (sheep and goats) cross the northeastern Jordanian border and are fattened over 2 months in preparation for the feast which takes place at the end of the Muslim pilgrimage to Mekkah. This causes overgrazing of natural plants in the area, and is called early heavy grazing. This is also practiced by the Saudis for grazing and fattening their herds. During my field studies, for example, I encountered a Saudi who raised 15 000 sheep and goats in the study area. Most Bedouins have trucks that enable them to bring water supplies and other equipment wherever they move in northeastern Jordan.

Jordanian shepherds also have thousands of herds and remain in this area continually. Thus they have to compete with the foreign shepherds. They also have modern equipment that enables them to farm the land there and helps them to move from one place to another with an adequate water supply at any time. They also invade and threaten the natural reserves such as in Surrah, Khanasreh, Al –Shumari and Al–Azraq.

Processes and Indicators of the Destruction of Pastoral Areas in Northeastern Jordan

Overgrazing

Overgrazing of pasture areas is understandable if one considers that the Jordanian farmer (fellah) is proud of having as many sheep as possible as it is a form of wealth. The indicators of this phenomenon are in the form of dominant invader plants (unpalatable species). For example, the pastoral plants (palatable species) in northeastern Jordan are now decreasing (*Artemisia sieberi, Salsola damascena, Poa bulbosa, Carex pachystylis, Stipa capensis, Erucaria pinnata, Hammada eigii, Salsola vermiculata)*. Unpalatable plants replace them, such as *Anabasis syriaca, Peganum harmala, Herniaria hirsuta, Anabasis setifera, Salsola jordanicola, Onopordum carduiforme, Xanthium spinosa* and *Chenopodium opulifolium.*

Moreover, the annual plants are highly endangered. This has resulted in a decline in animal and bird populations, as they either migrate or disappear. As a result of overgrazing and cutting of natural vegetation, soil erosion by wind or water has become more efficient and effective.

As a result of intensive irrigation of cultivated areas, salinity increased in the target area of study. Table 3 shows the ratio of salt in different soil profiles.

A more recent problem involves urbanization encroachment especially in the rangeland areas of northeastern Jordan.

Cutting for Fuelwood

Most of the nomads in northeastern Jordan cut and uproot bushes for fire wood, domestic cooking and heating, and sometimes for fencing in their tents and herds. Two line transects were made within and outside the fenced Suura rangeland reserve and the Al–Nawasif mountain. They show the decrease in palatable plants and increase in unpalatable plants as well as the arrival and spread of invader and segetal plants in degraded land (Table 3).

Overcultivation

A chronic problem in rain-fed agriculture in northeastern Jordan is the expansion into cultivated areas, particularly irrigated farming, in areas which receive less than 200 mm a^{-1}. This new phenomenon in the study area is due to the farmers importing water by truck.

During the last three decades this phenomenon has spread rapidly and is rated as one of the major causes of accelerating desertification, i.e. ploughing of rangelands with annual rainfall of less than 200 mm. Moreover, there is another big problem in the study area ; the rapid increase and expansion of irrigated farming. Most Jordanian farmers in the study area are not the owners of the land but are tenant farmers, who do not care about land degradation (Table 3). The increased salinity is due, in part, to overpumping from wells accompanied by inadequate formation of groundwater, the use of saline water because there are no alternatives, and mismanagement. Consequently, this has led to the abandonment of the rangelands. Indicators include the increase in soil salinity (Table 3), increase in halophyte as well as segetal and thorn species and finally a decrease in the crop yield. According to Abbas (1989) this causes a loss of agricultural potential in semiarid and arid areas. The marked reduction in productivity also makes the loss of the soil apparent. Whereas in the past, wheat harvests amounted to 100 kg ha^{-1}, they are now only about 10 kg ha^{-1}. In some areas, productivity in northeast Jordan has dropped so severely that many farms have had to be abandoned. Table 3 shows the vegetation and soil degradation gradient in different degrees and land use systems in the study area.

Vegetation Degradation in Northeastern Jordan

Table 3. Summary of the extent of vegetation degradation in northeastern Jordan.

Degradation step	No Damage	Small	Moderate	Strong	Very strong
Dwarf shrubs layer (Xerophytes perennial cover)					
Cover degree (%)	>70	40–70	20–40	<20	<5
Height (cm)	40–80	40–70	20–40	5–10	<5
Regeneration	Good	Moderate	Weak	Very weak	No
Grazing	No	Moderate	Strong	Extreme	Bare soil
Area	Surrah, Al–khnasreh, Al–shaumari reserves	Mafraq and Azraq region	Most of northeaster n Jordan	Everywhere in the study area	
Herbs and grass layer and cover					
Cover degree (%)	>80	60–80	<40	<10	<5
Height (cm)	>30	<20	<10	<5	<3
Palatable species (%)	>70	<40	<20	<10	<5
Unpalatable species	<10	>10	40	>40	>10
Regeneration	Good	Moderate	Weak	Very weak	Very weak
Area	Reserves	Reserves	Mafraq, Duleil and Al–halabat	Mafraq, Duleil and Al–halabat	Al–Azrq
Organic matter					
C/N ratio	High	Moderate	Small	Very small	Very small
Area	No	No	Reserves	Mafraq, Halabat	Degraded soil
Salinity					
Salt content	No	Few	Moderate	High	Very high
Conductivity (mhos cm^{-1})		<1	1–4	4–16	>16
Area	No	Virgin soil	Rain-fed agriculture	Irrigated soil, abandoned land	After irrigated cultivation, abandoned land

Conclusions

Key findings of the socioeconomic, land use, soil and the vegetation studies were:

- Illegal encroachment by herders on protected natural rangeland (Surrah, Al –Khanasreh, Al–Azraq protected areas) has increased.
- The method of deep ploughing, which is used today, has reduced the natural vegetation cover in the northeastern Jordan.

- There has been an increase in salinity in the area of Al–halabat because of the misuse of irrigation systems.
- There has been an increase in soil erosion by water (short-term strong rainfall) and wind.
- In the Al–halabat and Al–Mafraq areas the livestock hinders recovery of palatable species, e.g. *Artemisia sieberi* (syn. *A. herba-alba*), *Salsola vermiculata*,
- *Atriplex halimus* and others. Because of the overgrazing caused by the high density of livestock, the indicator plants (invaders) such as *Anabasis syriaca, A. articulata, A. setifera, Peganum harmala, Chenopodium murale* and others are dominant.
- Because of the high amount of salt in the upper soil the halophyte plants, e.g. *Chenopodium* spp. div., *Anabasis* spp. div., *Tamarix jordanis, Sonchus maritimus, Peganum harmala* and others, spread far in the above-mentioned areas. Thus, at present, the change in vegetation is obvious.
- Cutting and uprooting of plants as well as breaking and deep ploughing have reduced previous by existing resources of perennial xerophytes for fuel wood.

The vegetation investigation highlights several current problems as a consequence of vegetation degradation; floristic depletion, soil degradation by erosion, nutrient depletion, increased salinity etc. A rapid impoverishment of the rural population is evident. The immediate intervention of the Jordanian government, particularly the Ministry of Agriculture (Forestry and soil Department, Department of range Management) is very urgent.

This chapter was intended to provide a much-needed background on land degradation in the area in question. Further research and study is, of course, required in Jordan and neighbouring countries and can contribute to our knowledge of desertification processes.

References

Abbas A (1989) Jordan reports and biotic indicators of desertification. In: Ibrahim F, Mueller–Hohenstein K (eds) Problems of resource management and desertification control. In: DSE, Resource conservation and desertification control in the Near East, Feldafing, Schroetter, Peissenberg

Abu–Zant MM, Abusita, M, Tadros, K (1993) Workshop in rangeland management and conservation in the Arab World and combating desertification. Hashemite Kingdom of Jordan, Amman

Al–Eisawi DM (1985) Vegetation in Jordan. In: Hedid A (ed) Studies in the History and Archaeology of Jordan II. Department of Antiquities. Amman, Jordan and Ba'th Press, Avon, England, pp 45–57

Al–Eisawi DM, Hatough AM (1987) Ecological analysis of the vegetation of Ashaumari Reserve in Jordan. In: Dirasat A learned research Journal. University of Jordan, Dean of Academic Research, Amman, pp 81–94

Braun–Blanquet J (1964) Pflanzensoziologie, Fischer, Stuttgart

Vegetation Degradation in Northeastern Jordan 217

El–Kawasma Y (1983) Climatic water balance in Jordan: characteristics and applications. PhD Thesis. University of Ghent, Ghent

El–Rihani A (1984) Soils in Jordan. In: A.C.S.A.D. Arab Center for the study of arid zone and dray land, 1984. Proc VIII int Forum on soil Taxonomy and Agrotechnology Transfer. Symposium Jordan May 14–25, 1984, ACSADIP-28 Soil Science Division, Amman

El–Rihani A (1987) Study and classification of soil profile in the Mafraq district. Ministry of Agriculture, Amman (in Arabic)

Grueneberg F, Dajani F (1964) The soil survey at El–Jaffr. BfB Hannover (unpublished report)

Jordanian Statistical Center (1994) Hashemite Kingdom of Jordan, Amman

Jordan Ministry of Agriculture (1985) Agriclutural data 1980–1985, Amman

Kasapligil B (1955) Preliminary reports on the ecological reports on the ecological survey of the vegetation types in forest and grazing lands of Jordan, ETAP of FAO of the UN to the Government of the Hashemite kingdom of Jordan. Mimeogr, 188 pp

Kasapligil B (1956) An ecological survey of the vegetation in Relation to forestry and grazing. FAO Report to the Government of the Hashemite Kingdom of Jordan, Report No. 549

Long GA (1955) Plant ecological survey and studies on the possiblity of pasture utilization in the enclosures of Surra and Khnasri (Region of Mafraq, Jordan). FAO of the UN, mimeogr report, Rome, pp 1–31

Long GA (1957) The bioclimatology and vegetation of east Jordan. UNESCO/FAO, Rome

Moormann F (1959) Report to the Government of Jordan, on the soil of east Jordan. FAO, Rome

Sharkas OA (1994) Boden- und Vegetationsdegradierung in Nordjordanien. PhD Thesis, University of Bayreuth, Bayreuth, 305 pp (unpublished)

Shehadeh NA (1985) The climate of Jordan, past and present. In: Symp on the environment of Jordan, 2nd int conf on the history and Archaeology of Jordan, Amman, pp 4–11

Shehadeh NA (1990) The climate of Jordan. Jordan University Press, Amman (in Arabic)

Tadros KI (1979) Combating desertification in Jordan. In: Bishay A, Mc Ginnies W (eds) Advances in desert and arid land technology and development, vol 1. Harwood Academic Publishers, London, pp 135–138

Wilmanns O (1984) Oekologische Pflanzensoziologie 3. Aufl UTB, Quelle Meyer, Heidelberg, 372 pp

Zohary M (1962) Plant life of Palestine, Ronald Press, New York

Zohary M (1973) Geobotanical foundation of the Middle East. Swets and Zeitlinger, Amsterdam

Impact of Grazing on the Vegetation of South Sinai, Egypt

AbdEl–Raouf A. Moustafa

Keywords. Grazing intensity, south Sinai, endemic species, endangered species, species richness

Abstract. The vegetation of southern Sinai is subjected to great disturbance through unmanaged human activities, including overgrazing, overcutting, uprooting, tourism and quarrying. Many plant species are threatened due to the severe impact of grazing and human activities. The present chapter addresses the main question; how does grazing change the vegetation and its structure in the main wadis in south Sinai? The survey was designed for the vegetation of protectorate and adjacent wadis (overview and identification) and to quantify the grazing intensity in the main wadis. Eighteen main localities within and adjacent to the Saint Catherine Protectorate were studied through choosing 54 stands from whole area. In each stand three transects (500 m in length) were distributed randomly to estimate grazing intensity (based on grazing index), total cover percent, and current species status. Three hundred and sixteen species were identified, including 19 endemic species, 10 extremely endangered, 53 endangered and 37 vulnerable species. The grazing intensity was significantly and negatively correlated with species richness, number of endangered species (extremely endangered, endangered, vulnerable), number of endemic species and total plant cover. The results showed that 50% of selected localities was overgrazed, 27.8% had high grazing, 16.7% medium grazing and 5.6% low grazing intensity. Mt. Catherine showed the highest percentage of both species richness and number of endemic species while W.El–Kid and W. Nabq have the lowest values of species richness and no endemic species. At the same time, W. El–kid and W. Nabq showed the highest number of endangered species and a huge reduction in the total cover percentage due to overgrazing activities.

Introduction

Grazing in South Sinai, especially in the Saint Catherine Protectorate, is a tradition and still the first job for young girls who are not yet married. They start their daily trip of grazing early in the morning as a Bedouin tradition until sunset. In the Bedouin rule, every locality (of the selected localities) has its own people to graze, reached by a very easy route between the mountains using the foot track. The average animal number is between 50–60 goats and sheep, that are owned by the whole community (e.g. families share in the herd of animals). In general, grazing never stops with domestic animals, even in dry seasons; sometimes the route is changed inside the area, in the search for more vegetation, but grazing never stops.

Saint Catherine mountains are characterized by a unique type of vegetation due to climate and geomorphologic formations. The vegetation includes a huge number of medicinal plant species, endemic species and rare species as well. The disappearance of large numbers of plant and animal species is known nowadays, both worldwide and for Egypt, due to drastic environmental changes and habitat destruction. In recent years, it was noticed in South Sinai that many plant species are threatened due to the severe impact of grazing and human activities. The continuous overgrazing, overcutting and uprooting (for fuel and medicinal uses and feeding animals) has resulted in the disappearance of pastoral plants, a paucity of trees and shrubs, as well as the disappearance of many rare and endemic species.

Within the framework of both international and national interests for environmental conservation, Egypt has expressed concern about the preservation of the genetic resources of living organisms, particularly plants, animals and microorganisms of economic value, and those threatened with extinction. In 1992, Egypt signed the convention on Biological Diversity Conservation and the Egyptian government declared most of the granite massif (the mountainous area of southern Sinai) as the Saint Catherine Protectorate area. Therefore, a plan was designed to survey the Protectorate and adjacent areas through two missions. These two missions (2 years) aimed to carry out (1) studying the grazing effect on the status of species and describing the change due to various grazing intensities and (2) a vegetation survey (overview and identification) including an estimation of the number of plant species and their categories in the protectorate.

Materials and Methods

The botanical survey in South Sinai was carried out in 18 main localities, and included recording the different vegetation types and associations, and listing and evaluating the presence of threatened, rare, endemic and common species. These localities were divided into two main categories: low and high elevation. The first mission, in 1996, focused on low-elevation wadis and plains which included the El–Agramia plain, the Umm Alawi area, W. Sa'al, W. Solaf, El–Qaa plain, W. El –Nasb, the Nabq area, W. El–Kid, W. Yahmed, W. Lithi and W. Mandar. The second mission, in 1997, focused on the high mountainous area which included Mt. Catherine, Mt. Musa, Mt. Serbal, Mt. Umm Shomer, W. Jibal, W. Isla, Mt. Tarbush and the most important related wadis (e.g. W. Arbae'en, W. El–Faraa Shaqq Musa, W. El–Rutig) as well as some of their tributaries (Fig. 1).

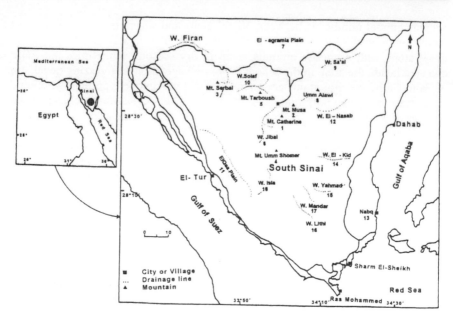

Fig. 1. Map of South Sinai showing the selected 18 main localities

General grazing information was collected by visual estimates over a period of 2 years throughout the 18 main localities. Fiftyfour stands were selected in these 18 main localities while 162 transects (500 m in length), three transects per stand, were distributed randomly to estimate grazing intensity and total cover percentage. In each transect, basic descriptive information, species list and human activities (settlements, grazing, cutting) were recorded for each locality. Quantitative grazing intensity was determined for each locality based on the grazing index, scored from 1 to 6 as follows: 1 = no interference or grazing; 2 = low grazing; 3 = medium grazing, 4 = high grazing, 5 = overgrazing and 6 = no vegetation on the ground due to overgrazing and human activities. In fact, the basis of this index depends on the data collected from each locality and the main points can be summarized as follows:

(1) The number of grazed individuals per plant species (palatable or unpalatable), number of browsed branches in each individual, and height of each grazed species and its vigor and vitality; (2) the number of visits by animals during different seasons; (3) the number of animals, their types (goats, sheep, donkeys and camels and their relative age, young animals or old; (4) the number of traces of browsing by domestic and wild herbivores using 100 randomly distributed (1x1 m) quadrats in each stand; (5) the rate of cutting and uprooting of plants (number of individuals uprooted daily due to human activities in each locality; and (6) the ratio of palatable to unpalatable plant species.

Topographic maps of scale 1:50 000 were used for locating the stands in the mountainous area, and maps of scale 1:250 000 were used in low-elevation areas,

Impact of Grazing on the Vegetation of South Sinai, Egypt 221

as they are distributed in large-scale transects. Täckholm (1974), Zohary (1966, 1972), and Feinbrun–Dothan (1978, 1986) were followed for the identification of plant species, while updating of species names followed Boulos (1995).

Categories of endangered and vulnerable plant species are used to indicate the degree of threat to individual species in their wild habitat (IUCN 1980). Definitions of these categories are:

Vulnerable: Species believed likely to move into the endangered category in the near future if the causal factors continue operating. Included are species of which most or all of the populations are decreasing because of overexploitation, extensive destruction of habitat or other environmental disturbance; species with populations that have been seriously depleted and whose ultimate security is not yet assured; and species with populations that are still abundant but are under threat from serious adverse factors throughout their range.

Endangered: Species in danger of extinction and whose survival is unlikely if the causal factors continue operating. Included are species whose numbers have been reduced to a critical level or whose habitats have been so drastically reduced that they are seemed to be in immediate danger of extinction. This is interpreted to mean including species with populations so critically low that a breeding collapse due to lack of genetic diversity becomes a possibility, whether or not man threatens them.

Extremely endangered: A new category proposed by the author for some endangered species with very few individuals in very few sites.

Results

Species Composition

The result of the botanical survey throughout the 18 localities identified 316 plant species growing in different habitats, most of them perennial, 26.9% (85 from 316) of which are annuals and 15.5% trees and shrubs (10 trees and 39 shrubs). The identified species belong to 56 families: Compositae (40 species), Graminae (30 species), Labiatae (25 species), Caryophyllaceae (21 species), Scrophulariaceae (18 species) and Cruciferae (16 species) being represented by the largest number of species, respectively. Twentyone families are represented only by one species; Equisetacea, Adiantaceae, Sinopteridacea, Salicaceae, Amaranthaceae, Menispermaceae, Guttiferae, Fumariaceae, Moringaceae, Nitrariacea, Polygalaceae, Pistaceae, Salvadoraceae, Rhamnaceae, Avicenniaceae, Globulariaceae, Acanthaceae, Plantaginaceae, Campanulaceae, Palmae and Typhaceae. From the identified species, 19 are endemic, 10 extremely endangered, 53 endangered and 37 vulnerable. Eleven species are newly recorded in certain areas; *Silene arabica, Diplotaxis erucoides, Reseda arabica, Andrachne telephioides, Chrozophora tinctoria, Scrophularia hypericifolia, Conyza bonariensis, Lolium rigidum, Cyprus conglomeratus, Ephedra alata, Ephedra* pachyclada, and three are new for Sinai; *Galium parisiense, Centaurea fururacea,* and *Echinops hussonii.* Three species are Pteridophyta (*Equisetum ramosissimum, Adiantum capillus-veneris,* and *Cheilanthes pteridioides*), and another three are Gymnospermae (*Ephedra alata, E. aphylla,* and *E. pachyclada*).

Grazing Intensity and Species Richness

The grazing intensity is expressed by grazing index (score 1–6 points), is based mainly on the grazing and human activities that are related to the grazing pattern in the area (Fig. 2). In general, most of the area showed massive signs of overgrazing intensity (nine localities) while only one locality (Mt. Serbal) has a low degree of grazing (Fig. 2, due to its rugged and high elevation, the low number of herds grazing in this area, and centralization of Bedouins far from the top of Mt. Serbal. Three main localities have a medium degree of grazing (Umm Shomer, Mt. Tarboush, Umm Alawi) and five localities have high grazing intensity (Mt. Catherine, El–Agramina, W. Sa'al, Nabq and W. Isla).

The relation between grazing intensity and species richness was a negative correlation ($r = -0.318$), where species richness increased in localities with low and medium grazing intensity and reached maximum values at a high grazing locality (Mt.Catherine), decreasing at overgrazed areas such as W.Kid, W. Lithi and W. Mander (Fig. 3).

Impact of Severe Grazing

Many results can be seen as direct effects of overgrazing in the area, such as an increasing number of endangered species that are included (extremely endangered, threatened, and vulnerable), reduction in the total number of endemic species, and reduction in the total cover percentage in all plant species. The percentage of all endangered species was 31.6% (100 species), with the highest percentage found in Kid and Nabq and the lowest values at El–Agramia, Umm Alawi (Fig. 4).

In spite of W. Lithi, W. Mandar and Yahmad have intensive pressure of overgrazing and cutting, but the percentage of endangered species is not very high, due to the pressure of grazing being concentrated on two species, *Acacia* and *Caligonium.* Figure 5 shows the total percentage of endangered species, including the percentage of extremely endangered, endangered, and vulnerable in relation to grazing intensity, scored by the grazing index.

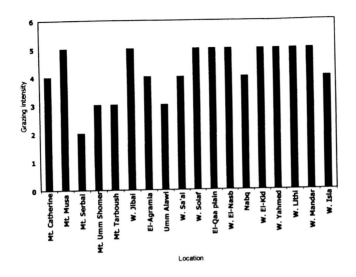

Fig. 2. Variation in grazing intensity throughout the 18 selected localities in Souith Sinai. *1,* no grazing; *2,* low grazing; *3,* medium grazing; *4,* high grazing; *5,* overgrazing; *6,* no vegetation due to overgrazing

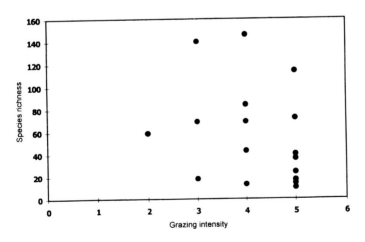

Fig. 3. Relationship between grazing intensity scored by grazing index and species richness in the main 18 localities

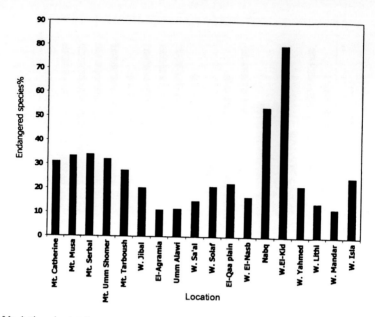

Fig. 4. Variation in total percentage of endagered plant species (extremely endagered, endagered, and vulnerable species) in the selected 18 localities

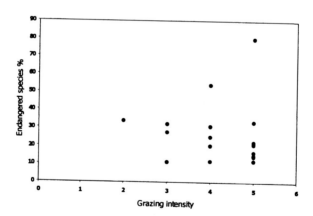

Fig. 5. Relationship between grazing intensity scored by grazing index and total percentage of endagered species (extremely endagered, endagered, and vulnerable)

The total number of endemic species to all Sinai is 37 (Boulos 1995) while the endemics recorded in the observed localities were 19 species. These recorded species are concentrated in Mt. Catherine (14 species), Umm Shomer (11 species), Mt. Musa (9 species), W. Isla (7 species), Mt. Serbal (6 species). Some localities have a small number of endemic species such as Agramia, El–Nasab and W. Isla. Other wadis showed no record of endemic species, due to the huge pressure of

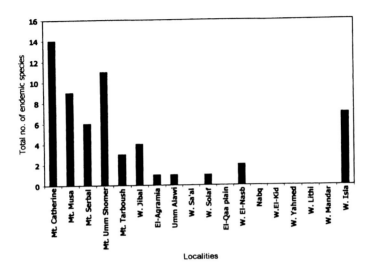

Fig. 6. Number of endemic species occurring in each locality

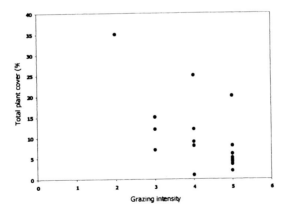

Fig. 7. Total plant cover percentage in relation to grazing intensity thruoughout the 18 main localities in South Sinai

grazing and cutting effects (e.g. Nabq, El–Kid, W. Yahmad, Lithi, and Mandar); (Fig. 6). The relationship between number of endemic species and grazing intensity was a negative correlation ($r = -0.4032$). The total plant cover in most localities did not exceed that 15%, due to grazing and cutting for feeding animalsat home (Fig. 7). The highest plant cover can be seen at gorge habitats of Mt. Serbal and Mt. Catherine, followed by Mt. Musa. The plant cover correlated negatively with grazing intensity ($r = -0.615$), where a huge reduction in plant cover occurred at most localities with high grazing and overgrazing effects.

Discussion and Conclusion

The flora of Sinai comprises nearly 900 species distributed through 250–300 associations and communities (Danin 1986), and is characterized by the dominance of four families: Compositae, Zygophyllaceae, Leguminosae and Labiatae, which are represented by a large number of endemic species (Moustafa 1990). The most characteristic trees and shrubs include *Crateagus* x *sinaica, Ficus palmata, Rhamnus disperma, Cotoneaster orbicularis, Pistacia khinjuk, Colutea istria,* and *Acacia tortilis* subsp.*raddiana* (Danin 1983).

The change in flora and vegetation due to grazing can be seen by comparing the number of species occurring in the same area under more or less the same conditions. In fact, the scarcity and irregularity of rainfall in Sinai and the type of plant species (annuals or perennials) in Sinai and Egypt, make this kind of comparison impossible to be measured over a few years. However, Moustafa and Kamel (1996) found that the number of plant species (annual and perennials) increased, associated positively with the amount of rainfall. However, their study concerned only Mt. Catherine, Mt. Musa, and surrounding wadis, not the whole South Sinai. Their highest record of species in 3 successive years (1992–1994) was 221 species. In addition, Danin (1983) stated that the Saint Catherine area has 420 species in the district of the Saint Catherine mountains. In contrast, the present study identified 316 species recorded in 18 localities that represent most of the south Sinai area, not only the Saint Catherine area.

However, severe grazing is one of the main threats affecting the vegetation and changing its structure (Fig 8). Its main effects can be seen in the results of overgrazing through the following points:

(1) Threat to endemic and rare species (e.g. *Primula boveana,* and *Rosa arabica*) and increase in the number of endangered species associated with grazing intensity. (2) Paucity of trees in the studied localities, such as *Acacia, Crataegus* and *Lycium*. (3) Disappearance of pastoral plants and difficulty in finding some species which are commonly used in folk medicine (e.g. *Salvia* species). (4) Reduction in the total plant cover percent and increase in the number of unpalatable plant species such as *Artemisia judaica, Anabasis articulata* and *Fagonia mollis*. (5) Increase in the number of grasses due to importing animals (goats and sheep) and their fodder. (6) Change in soil surface and moisture due to

the movement of animals and the disturbance caused by daily tracking in the same area.

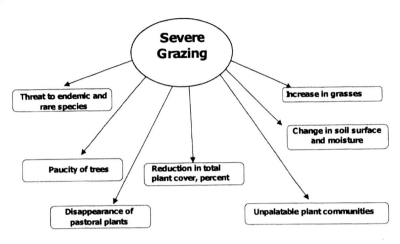

Fig. 8. Output results of severe grazing affecting the vegetation in South Sinai

The change in vegetation in the wadis and mountains due to grazing can be summarized in the following points:

- Mt. Catherine and Mt. Musa showed a high degree of grazing, as a result of which more than 30% of their flora is threatened. In addition, the vegetation has been changed to be dominated mainly by unpalatable assemblages such as *Phlomis aurea, Artemisia herba-alba, Echinopis spinosissimus, Fagonia mollis* and *Alkanna orientalis*.
- The vegetation in W. Kid, W. Yahmed, W. Lithi and W. Mandar has been dramatically reduced to less than 5% and is represented by a small number of perennial species (e.g. *Acacia tortilis* and *Haloxylon salicornicum)* associated with high disturbance of the soil surface and loss of soil moisture as well.
- Grazing in Mt. Serbal is low (mainly by feral donkeys), therefore the general status of its vegetation is still good except for the number of trees of *Pistacia* which recorded by Danin et al. (1985) as thousands in the 1980s and now decreased to a few hundreds.
- Palatable species in the wadi El–Qaa plain, W. ElNasb, W.Sa'al, and W. Solaf (e.g. *Pennisetum, Panicum* and *Crotalaria*) and moderately palatable species such as *Achillea, Retama* and *Lycium* suffer from high pressure of grazing and cutting for feeding animals.

References

Abd El–Wahab RH (1995) Reproduction ecology of wild trees and shrubs in Southern Sinai, Egypt. MSci, Suez Canal Univ, Ismailia, Egypt, 110 pp

Boulos L (1995) Flora of Egypt – checklist. Al–Hadara Publishing, Cairo, Egypt, 283 pp

Danin A (1983) Desert vegetation of Israel and Sinai. Cana Publishing, Jerusalem, 148 pp

Danin A (1986) Flora and vegetation of Sinai. Proc R Soc Edinb 89B: 159–168

Danin A, Shmida A, Liston A (1985) Contribution to the flora of Sinai III – checklist of the species collected and recorded by the Jerusalem team. Willdenowia 15: 255–322

Feinbrun–Dothan N (1978) Flora Palaestina, Part III The Israel Academy of Sciences and Humanities, Jerusalem

Feinbrun–Dothan N (1986) Flora Palastina, Part IV The Israel Academy of Sciences and Uumanities, Jerusalem

IUCN (1980) World conservation strategy, Threatened plants committee secretariat of IUCN, IUCN, UNEP, WWF

Moustafa AA (1990) Ecological gadients and species distribution on Sinai Mountains. PhD Thesis, Suez Canal Univ., Ismailia, Egypt, 115pp

Moustafa AA, Kamel WM (1996) Ecological notes on the floristic composition and endemic species of Saint Catherine area, South Sinai, Egypt. Egypt J Bot 35: 177–200

Täckholm V (1974) Students' flora of Egypt. 2nd edn. Cairo University, Beirut, 888pp

Zohary M (1966) Flora Palaestina, Part I The Israel Academy of Sciences and Humanities, Jerusalem

Zohary M (1972) Flora Palaestina, Part 2: The Israel Academy of Sciences and Humanities, Jerusalem

Arid Rangeland Management Supported by Dynamic Spatially Explicit Simulation Models

Florian Jeltsch, Thomas Stephan, Thorsten Wiegand and Gerhard E. Weber

Keywords. Farm management, grazing threshold, land use, long-term dynamics, stochastic modelling, vegetation dynamics

Abstract. In arid regions, the effects of grazing management on natural communities of long-lived plants generally take years or even decades to become evident. Event-driven dynamic behaviour, disturbances, unpredictable and low rainfall and complex interactions between species make it difficult to gather sufficient understanding of vegetation dynamics for developing guidelines for sustainable management of arid rangelands. This is further complicated by the importance of spatial scales and patterns, e.g. patchiness of rainfall, heterogeneous grazing behaviour of domestic livestock or distances between artificial watering points.

Simulation models that consider the essential processes determining vegetation dynamics offer scope for quantitatively exploring long-term vegetation dynamics of arid rangelands. If these models are spatially explicit, they additionally allow for the investigation of spatial processes, such as competition or dispersal, and patterns such as landscape features or structures imposed by management (boreholes, paddocks etc.).

This chapter, discusses the promises and limitations of spatially explicit simulation models as (often neglected) tools for rangeland management. We focus on model examples from rangelands in southern Africa, namely a set of models simulating cattle grazing in the southern Kalahari and a model simulating a karakul sheep farm at the border of the Namib desert (Namibia).

Results of the Kalahari models show the existence of a grazing threshold that determines the long-term sustainability of livestock grazing. This threshold depends on rainfall, grazing intensity and grazing heterogeneity. Its effect is illustrated with the spatial vegetation dynamics around artificial watering points. The model of the karakul sheep farm is used to investigate a successful example of sustainable rangeland management under harsh conditions.

Introduction: Problems in Arid Rangeland Management

According to the United Nations (UNEP, Agenda 21), approximately 1/6 of the world's population, 70% of all drylands with a total area of 3.6 billion hectares (ha) and a quarter of the total land surface of the Earth is endangered by desertification (Kruger and Woehl 1996). The most obvious consequences are the increasing poverty and the increasing damage to 3.3 billion ha of rangeland. In 1987 it was estimated that every year 27 million ha of land are lost to desert or zero economic productivity (Hellden 1991). Even though some of these numbers are probably too high, their magnitude illustrates the importance of improving strategies and measurements of rangeland management.

Especially in the arid and semiarid–henceforth "arid" for short–regions of Africa, overexploitation caused by a drastic increase in the population has continually degraded the environment (Kruger and Woehl 1996). Despite the numerous efforts to improve this situation, arid Africa suffers a decrease in per capita production of 2% every year (Pieri and Steiner 1996). This threatening situation emphasizes the importance of a sustainable management of the scarce resources of arid Africa.

Given this background of degradation and desertification the question arises whether arid ecosystems are specifically fragile and unstable. This question is difficult to answer and controversially discussed in the literature. It is widely agreed that arid systems are typically event-driven (Westoby et al. 1989; Walker 1993) and that a characteristic feature of dryland ecosystems is their changeability in response to rainfall. Additional variation is caused by the high variability of rainfall in time and space. Thus, in arid ecosystems it is extremely difficult to establish whether an area is suffering a progressive, long-term decline in biodiversity and productivity, and hence desertification, or whether it is merely suffering a short–term drought from which the land may recover if the human impact is reduced or removed. Consequently, under these conditions it is almost impossible to evaluate management decisions and to make predictions of future developments; and if predictions are limited, how can we manage an almost unpredictable system?

In order to answer this question, it is necessary to learn more about the natural long-term dynamics of arid ecosystems. The essence of the approach to management in arid ecosystems is that it must be longterm, i.e. it should match the timescale on which the biological world operates; but how can we learn more about these systems on biological timescales that may exceed decades, given the current lack of long-term information? How can we identify the essential ecological processes and how can we maintain them? How can we distinguish environmental variability noise from signals of ecological change and degradation?

It is evident that answers to these questions are not straightforward and easy to find. However, modern natural sciences offer several tools that, separate or in combination, can help to deal with problems that are typical in arid systems. Suitable tools to deal with these questions are (1) long-term investigations and monitoring, (2) comparative approaches including different systems or ecological gradients, (3) process-oriented field research, (4) interdisciplinary approaches combining experts in the field of nature conservation, ecology, geology and geography as well as socioeconomics and (5) the development and application of computer models at different levels and scales. In the following we will highlight the latter point with two examples from southern Africa. The first example deals with land use and degradation in the southern Kalahari, the second analyses a sustainable management strategy of a karakul sheep farm at the border of the Namib desert. Both examples are based on computer simulation models that include the available knowledge of local field experts.

Spatially Explicit Simulation Models

Computer simulation models can link current knowledge of processes driving vegetation dynamics and allow systematic investigations of the interactions between all relevant factors and their logical consequences. Also, simulation modelling is a powerful tool for controlled experimental manipulation for a large number of environmental conditions and over long time spans (Thiéry et al. 1995; Wiegand et al. 1995; Jeltsch et al. 1996, 1997a,b, 1998, 1999; Weber et al. 1998). If these models are spatially explicit, they additionally allow for the investigation of spatial processes, such as competition or dispersal, and patterns, such as landscape features or structures imposed by management (boreholes, paddocks etc.). Most spatially explict models subdivide an area into spatial subunits (e.g. patches or grid cells), the size or shape of which is determined by the question to be addressed. The subunits are characteristically based on typical biological or management scales of the modelled system, e.g. the size of individual plants, types of plant interactions or, as will be shown here, grazing paddocks as management units of a farm (Wiegand et al. 1995; Jeltsch et al. 1996, Stephan et al. 1998, Weber et al. 1998). Spatially explicit simulation models focus on processes and mechanisms considered crucial for driving system dynamics. Although availability of long-term field data on plant community dynamics in arid rangelands is usually poor, many driving processes can be observed on shorter timescales. The basic idea of a model-based bottom-up approach is to incorporate such short-term knowledge in the form of rules into a computer simulation model. In order to investigate vegetation dynamics, the model simulates the fate and the interactions of individual plants or plant assemblages within the rangeland. In the case of farm paddocks as spatial subunits, the model simulates the dynamics of palatable and vital biomass under variable rainfall conditions and grazing.

By using long-term climatic data and plausible management scenarios, the models extrapolate from the known short-term behaviour of individual plants, plant assemblages or farm paddocks to long-term rangeland dynamics. An important advantage of spatially explicit simulation models is the inclusion of biological information on the modelled processes in the form of rules rather than mathematical equations. Especially in more complex problems, this allows the direct inclusion of expert knowledge that is not necessarily restricted to hard data.

Example 1–Degradation in the Semiarid Savanna of the Southern Kalahari

On all continents, rangeland utilization and overgrazing by domestic livestock have resulted in changes in plant species composition that reduce carrying capacity for these animals (Schlesinger et al. 1990; Friedel 1991; Dean and Macdonald 1994). For example, shrub encroachment as a form of rangeland degradation is reducing the carrying capacity of the Kalahari and other semiarid savannas or grasslands in southern Africa. Shrub encroachment is the progressive increase of unpalatable shrubs and/or stunted trees at the expense of palatable herbaceous vegetation (Walter 1954; Leistner 1967; Skarpe 1990, 1991; Archer and Smeins 1991). Because vegetation changes such as shrub encroachment are relatively slow processes, and because animals are stocked at low densities in these areas, field experiments to determine stocking rates that avoid degradation under various rainfall scenarios are almost impossible to replicate. Therefore we used a spatiallyexplicit simulation model to investigate the shrub-grass dynamics of the southern Kalahari under various realistic rainfall scenarios and stocking rates of domestic livestock (Jeltsch et al. 1997a,b).

Simulations are based on daily rainfall data of the modelled localities (South African Weather Bureau, unpubl. data; Zucchini et al. 1992) in the South African part of the southern Kalahari bordering Botswana. Annual rainfall ranges from 170 to 450 mm. The model subdivides an area of 50 ha into 20 000 grid cells of a size of 5 x 5m. This size corresponds with observed maximum scales of direct shrub-grass interaction (Jeltsch et al. 1997a,b). On the basis of these grid cells and a 1-year time step, different submodels are distinguished, simulating moisture availability in a top and subsoil layer, vegetation dynamics and biomass production, grazing by cattle and grassfires. In each grid cell, actual soil moisture availability in two soil layers is calculated on the basis of annual precipitation, water losses such as evaporation and runoff, and moisture uptake by the local vegetation. Moisture levels and vegetation cover determine annual grass biomass production in the modelled area which is the basis for both cattle grazing and grass fire fuel. The submodel grazing simulates the spatially explicit reduction of herbaceous biomass by cattle and the corresponding trampling effects. Varying spatial grazing behaviour of cattle is simulated by modifying the rules for choosing individual patches (i.e. grid cells) for grazing (Weber et al. 1998). Grazing and trampling pressure on the herbaceous component in a grid cell eventually influence potential biomass production and local occurrence of the herbaceous vegetation. Finally, in the submodel grass fire, the amount of remaining grass fuel determines whether and with what probability grass fires occur in the modelled area. Grass fires cause a small mortality risk for shrubs and modify potential biomass production.

As a first test, the model was applied to artificial watering points in the southern Kalahari (Jeltsch et al. 1997a). The effect of grazing is particularly marked around boreholes in southern Africa (Perkins and Thomas 1993), Australia (Andrew 1988) and southwestern North America (Archer and Smeins

1991; Fusco et al. 1995), where grazing animals are concentrated in high densities for long periods. In general, distinct zones of bare soil, ephemeral plants, woody shrubs and grasses (the piosphere) may be distinguished from the waterpoint to the outer edge of the intensively used area (Andrew 1988; Pickup 1994). In the model, an area of 3.2 km length and 0.2 km width represents a transect from a waterpoint to the outer piosphere. A distance of 3 km corresponds with the area over which cattle preferentially forage (Tolsma et al. 1987), with grazing intensity decreasing with increasing distance from the borehole.

Model results agree well with vegetation patterns occurring in the field. A locality with 385 mm mean annual precipitation was simulated in the example of Figure 1, which shows a distinct formation of piosphere zones within less than 10 years. An area of bare soil and annuals (approximately 150 m-sacrifice zone) is surrounded by a zone dominated by shrubs (approximately 1 km with increasing tendency). This zone of shrub encroachment is surrounded by the typical pattern of grass-shrub as given in the ungrazed savanna. This clear zonation is known from field studies in the Kalahari (Martens 1971; Tolsma et al. 1987; Thomas and Shaw 1991; Perkins and Thomas 1993). The good agreement with these field studies increases confidence in the model and allows more general investigations of the effect of different grazing intensities on shrub encroachment and savanna degradation. Simulation experiments exploring the average shrub cover after 5, 20 and 50 years of grazing under different utilization intensities suggest the existence of a grazing threshold (Fig. 2: same locality as in Fig. 1). Increasing grazing pressure leads to a sudden increase in shrub cover once a specific threshold is exceeded. This threshold of grazing intensity is determined by long-term mean annual rainfall. Below the threshold, i.e. at stocking rates which cause a lower grazing pressure, there is almost no visible effect on the shrub cover (Jeltsch et al. 1997b). The results of the model agree well with the results of a 5-year grazing experiment (Skarpe 1990) at a site in the Botswana Kalahari with a mean annual precipitation of 300 mm (Fig. 2; for details see Jeltsch et al. 1997b).

It is remarkable that stocking rates recommended by field experts (Fourie et al. 1985) for the given locality are in the range of the detected grazing threshold (Fig. 2). The simulation experiments suggest that, although the stocking rates currently recommended by pasture scientists are unlikely to lead to shrub encroachment within 20 years, they have a high probability of bringing about shrub encroachment within a century (Fig. 2).

Even though this result is impressive one has to be careful with the exact quantification of the grazing threshold. Figure 3 shows that small-scale grazing heterogeneity can have a strong effect on the grazing threshold (Weber et al. 1998). In these simulations we distinguish three grazing scenarios which are referred to as high, moderate and low grazing heterogeneity. Under low heterogeneity, grazing pressure is normally distributed for all grass patches (i.e. grid cells dominated by grasses). Contrarily, under higher heterogeneity, grazing pressure is bimodal with high frequencies of extreme defoliation severities, and only a very small fraction of grass patches shows defoliation severities close to mean defoliation severity. Frequencies of grass patches with extremely low or extremely high defoliation intensities increase with grazing heterogeneity. Figure

3 shows that under higher heterogeneity, shrub encroachment starts at lower utilization intensity, and at given critical stocking rates, shrub encroachment is faster than under lower heterogeneity. After 50 years, stocking at 10 ha lsu-1 results in 60% shrub cover under high heterogeneity, whereas under low heterogeneity, shrubs remain close to the initial level.

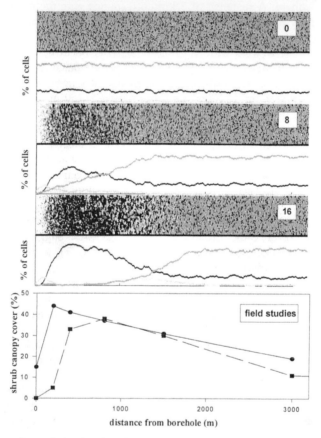

Fig. 1. Comparison of simulated and real pattern formation around watering points in the Kalahari. *Top* Simulated spatial vegetation dynamics around an artificial waterhole (385 mm mean annual precipitation; under locally recommended stocking rate of 11 ha/lsu). The depicted area of 3.2×0.2 km represents a transect from a waterpoint (*left side*) to the outer piosphere. The size of the individual "cells" is 5×5 m. For different years (0, 8, and 16 years of grazing) the two-dimensional vegetation pattern is shown together with a one-dimensional projection. The latter shows vegetation cover by different life forms (i.e. % of dominated cells) in a 50-m-wide strip versus distance from the waterhole. For initial conditions the simulation was run for 20 years without grazing under the given rainfall scenario. Colour legend: *black* shrubs; *dark grey* perennial grasses and herbs; *light grey* annuals; *white* empty space. *Bottom* Shrub cover data for two artificial waterholes in the southern Kalahari under conditions that are comparable to the simulated scenario. (Perkins and Thomas 1993)

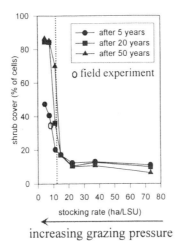

Fig. 2. Comparison of simulation results of shrub cover versus stocking rate after 5, 20 and 50 years of grazing at a rainfall site of 385 mm mean annual precipitation. The *dotted line* gives the recommended stocking rate for the site. (After Fourie et al. 1985). The field experiment was conducted by Skarpe (1990) (see text)

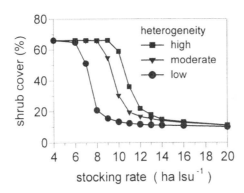

Fig. 3. Means of shrub cover by stocking rate after 50 years of grazing under three different levels of small-scale grazing heterogeneity

Qualitative empirical evidence suggests that the grazing impact of domestic herbivores in arid rangelands does show considerable small-scale heterogeneity (for details see Weber et al. 1998). Given the fact that respective quantitative knowledge is very poor, the sensitivity of the grazing threshold to grazing heberogeneity indicates a clear and important need for additional research.

Example 2–Sophisticated Strategy of Farm Management in an Arid Environment

In contrast to the previous example, we used a top-down simulation approach to analyze the sustainable management strategy of a farm (Stephan et al. 1998). The farm Gamis is located 250 km southwest of Windhoek at the edge of the Namib desert. Mean annual precipitation is 180 mm. Extensive pastoral agriculture is supposed to be the suitable type of land use in this ecologically sensitive region where much of the land is now degraded due to overgrazing in the past. At Gamis farm, karakul sheep are kept for fur production; 30 000 ha of farmland are divided into 98 paddocks, approximately 60 of them are used for the grazing of up to three production flocks. A sophisticated rotational grazing system has prevented degradation, and ensured the farm's economic survival. This is remarkable because most neighbouring farms are economically and ecologically degraded. The simulation model is constructed in order to understand the basis of this strategy's success and to apply this knowledge to other situations. Instead of modelling detailed vegetation dynamics, the model simulates rainfall and vegetation response at the paddock level according to long-term observations of the farmer. This expert knowledge, as well as the grazing strategy, are included in the model in the form of rules, similar to a dynamic expert system.

We distinguish four different life forms (trees, shrubs, annuals and perennial grasses) and five habitats with different soil types (river, slope, brackish, stones, slate). One paddock can contain all five soil types. Typical distributions of the life forms on the different habitats under different rainfall conditions are known from the farmer's long-term observations. Trees, shrubs and perennial grasses are characterized by two components: palatable and vital biomass; annuals are characterized by palatable biomass only. Vital biomass describes the potential of the vegetation to produce palatable biomass and depends on the rainfall and the intensity of grazing during the past years. It represents the "memory" of the vegetation and considers, for example, that production of palatable biomass will be relatively low after a long period of drought or heavy grazing stress. Grazing is modelled by reducing palatable as well as vital biomass, thus affecting palatable biomass production in subsequent years. The model keeps track of palatable and vital biomass of all life forms in each paddock and simulates their temporal development in accordance with rainfall and grazing intensity. It considers a spatially explicit distribution of rainfall and distinguishes between bad, mean and good rainfall for each year and paddock.

Rules for the yearly growth of vital biomass and the production of palatable biomass under different rainfall conditions and different grazing intensities are based on the farmer's long-term experience. The grazing system is based on rotative 2-week grazing followed by at least 2 months of resting. Additionally, one third of all paddocks remain ungrazed during the growing season (Sept.–May). In the case of insufficient rainfall conditions, these basic rules are modified. If rainfall is not sufficient for 1 year, "spare" paddocks are used, resting time is decreased and lamb raising is ceased or translocated to rented rangeland. After 2 years with bad rainfall, adult sheep numbers are reduced and all paddocks are grazed without resting to distribute the grazing pressure equally over all paddocks. In the case of severe droughts (more than 2 years with bad rainfall), all sheep are translocated to rented rangeland.

Variations of the grazing rules allow the comparison of different strategies. Here, we investigated the ecological and economical sustainability of stocking strategies differing in the management rule for the additional resting of paddocks. We used three different strategies, namely no additional resting, Gamis strategy (additional resting in years with sufficient rain) and full resting, which also takes into account years with insufficient rain. The economic success of the strategies is measured by the number of sheep that can be kept on the farm when the stocking rate is adapted to the carrying capacity. By doing so, economic success directly depends on the ecological state of the farmland.

The simulation experiments indicate (Fig. 4) that no additional resting is superior at first, but then falls back behind the Gamis and the full resting strategy. However, in the long run, the Gamis strategy is the most beneficial. From the ecological perspective it preserves farmland, with carrying capacity remaining almost constant over two centuries. This also ensures the economic superiority of the Gamis strategy, despite its inferiority to less protective strategies over the first decades; indeed, it might even take more than a human life time to make up for these initial losses. The no additional resting strategy is unsustainable in the long term, and despite its initial success, could finally destroy vital biomass of perennial plants and thus result in degraded land with very low carrying capacities.

Discussion

These examples of spatially explicit simulation models show that this type of computer model is well adapted to ecological problems inherent to arid rangelands. Especially the problem that large temporal scales have to be considered in arid rangeland management is difficult to solve without the help of models. This problem is even aggravated by the fact that certain events, such as extraordinary rainfall conditions or coincidences of high grazing pressure followed by a prolonged period of drought, can lead to ecological changes that persist for decades. In addition to the temporal aspects, the Kalahari model (example 1) shows that spatial aspects such as small-scale grazing heterogeneity significantly influence overall rangeland dynamics on time scales of several years or decades. The model indicates the existence of a grazing threshold with respect to shrub encroachment and savanna degradation. This threshold is largely determined by the long-term mean annual precipitation and the stocking rate but also depends on the heterogeneity of rangeland utilization on a spatial scale of a few meters. These findings, i.e. the existence of such a threshold as well as its dependence on the spatial rangeland utilization, open necessities and possibilities for long-term management strategies that lead to an ecologically and economically sustainable land use.

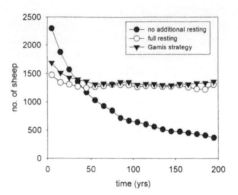

Fig. 4. Analysis of different grazing strategies with regard to resting of grazing paddocks. The *y-axis* shows the number of sheep kept on the farm related to carrying capacities. The *data points* are floating averages over 10 years taken from 100 simulations, each with a stochastic rainfall scenario

The main benefit of the second modelling example is the possibility to generalize an existing sustainable strategy. However, without the model, it is difficult to say whether the applied strategy is really the best choice or whether other strategies could be of more benefit. Without the model it would also be difficult to apply this strategy to other farms or systems with slightly different conditions. The model helps to identify the key elements of the successful strategy and to implement them in other examples. To this end, it is planned to develop a user-friendly simulation tool based on the Gamis model which will allow rangeland managers and students to test alternative strategies and conduct simulation experiments under a large variety of conditions. The main aim of this approach is to improve the understanding of rangeland systems on larger temporal scales and to help to question and to test the intuitive models of rangeland dynamics that are based on personal experience only.

In practice, rangeland management is necessarily based on an intuitive model (Westoby et al. 1989). Each farmer or land manager has a clear idea (i.e. a model) about the functioning of his own system. This model is a philosophic system of concepts, generalizations or assumptions rather than a qualitative model and is used to predict how the system reacts in response to climate, disturbance and management. The understanding that a farmer or manager has of the ecosystem is the basis for defining the management strategy. Clearly, this understanding depends on the current knowledge and on the experience of the particular farmer or manager, and on the period of time they could observe the system–a time span which usually does not exceed 20 or 30 years. Consequently, a lack of long-term experience under rare and unpredictable driving events that may occur only a few time in a century, may lead to an inappropriate conceptual model which would cause non-sustainable management decisions. Walker (1993) stated that "undesirable changes in rangelands have mostly been brought about by applying the wrong ecological model(s) of rangeland dynamics". Therefore, an understanding of long-term dynamics of arid or semiarid plant communities is indispensable for developing strategies for sustainable management.

Spatially explicit simulation models can help to improve this understanding by (1) structuring available information and knowledge and thus identifying knowledge gaps, (2) linking these information and detecting logical implications and consequences as well as developing prognoses, (3) extrapolating local information that were collected over short periods of time to larger temporal and spatial scales and (4) allowing simulation experiments that are not possible in the field either because of logistic, temporal or economic problems. However, these aims can only be achieved if models are built in collaboration with field ecologists and rangeland managers as well as experts from other fields such as geography, socioeconomy or remote sensing. In this context, models can work as catalysts of interdisciplinary research leading to an improved understanding and predictions in almost "unpredictable" systems.

References

Andrew MH (1988) Grazing impact in relation to livestock watering points. Trends Ecol Evol 3: 336–339

Archer S, Smeins FE (1991) Ecosystem-level processes. In: Heitschmidt RK, Stuth JW (eds) Grazing management. Timber, Oregon, pp 109–139

Dean WRJ, Macdonald IAW (1994) Historical changes in stocking rates of domestic livestock as a measure of semi-arid and arid rangeland degradation in the Cape Province, South Africa. J Arid Environ 26: 281–298

Fourie JH, van Niekerk JW, Fouché HJ (1985) Weidingskapasiteitsnorme in die Vrystaatstreek. Glen Agric 14: 4–7

Friedel MH (1991) Range condition assessment and the concepts of thresholds: view point. J Range Manage 44: 422–426

Fusco M, Holechek J, Tembo A, Daniel A, Cardenas M (1995) Grazing influences on watering point vegetation in the Chihuahuan desert. J Range Manage 48: 32–38

Hellden U (1991) Desertification: time for an assessment? Ambio 20:372–383

Jeltsch F, Milton S, Dean WRJ, van Rooyen N (1996) Tree spacing and coexistence in semi-arid savannas. J Ecol 84: 583–595

Jeltsch F, Milton S, Dean WRJ, van Rooyen N (1997a) Simulated pattern formation around artificial waterholes in the semi-arid Kalahari. J Veg Sci 8: 177–189

Jeltsch F, Milton S, Dean WRJ, van Rooyen N (1997b) Analysing shrub encroachment in the southern Kalahari: a grid-based modelling approach. J Appl Ecol 34: 1497–1509

Jeltsch F, Milton SJ, Dean WRJ, Moloney K (1998) Modelling the impact of small-scale heterogeneities on tree-grass coexistence in semi-arid savannas. J Ecol 86: 780–794

Jeltsch F, Milton SJ, Moloney K (1999) Detecting process from snap-shot pattern-lessons from tree spacing in the southern Kalahari. OIKOS 85: 451–467

Kruger AS, Woehl H (1996) The challenge for Namibia's future: sustainable land use under arid conditions. Entwicklung Ländlicher Raum 4: 16–20

Leistner OA (1967) The plant ecology of the southern Kalahari. Mem Bot Surv South Africa 38: 1–172

Martens HE (1971) The effect of tribal grazing patterns on the habitat in the Kalahari. Botswana Notes Rec 1: 234–241

Perkins JS, Thomas DSG (1993) Environmental responses and sensitivity to permanent cattle ranching, semi-arid western central Botswana. In: Thomas DSG, Allison RJ (eds) Landscape sensitivity. Wiley, London, pp 273–286

Pickup G (1994) Modelling patterns of defoliation by grazing animals in rangelands. J Appl Ecol 31: 231–246

Pieri C, Steiner KG (1996) The role of soil fertility in sustainable agriculture with special reference to Sub–Saharan Africa. Entwicklung Ländlicher Raum 4: 3–6

Schlesinger WH, Reynolds JF, Cunningham GL, Huenneke LF, Jarrell WM, Virginia RA, Whitford WG (1990) Biological feedbacks in global desertification. Science 247: 1043 –1048

Skarpe C (1990) Structure of the woody vegetation in disturbed and undisturbed arid savanna, Botswana. Vegetatio 87: 11–18

Skarpe C (1991) Impact of grazing in savanna ecosystems. Ambio 20: 351–356

Stephan T, Jeltsch F, Wiegand T, Wissel C, Breiting HA (1998) Sustainable farming at the edge of the Namib: analysis of land use strategies by a computer simulation model. In: Usó JL, Brebbia CA, Power H (eds) Ecosystems and sustainable development. Advances in ecological sciences 1. Computational Mechanics Publication, Southampton, pp 41–51

Thiéry JM, D'Herbés JM, Valentin C (1995) A model simulating the genesis of banded vegetation patterns in Niger. J Ecol 83: 497–507

Thomas DSG, Shaw PA (1991) The Kalahari environment. Cambridge Univ Press, Cambridge

Tolsma DJ, Ernst WHO, Verwey RA (1987) Nutrients in soil and vegetation around two artificial waterpoints in Eastern Botswana. J Appl Ecol 24: 991–1000

Walker BH (1993) Rangeland ecology: understanding and managing change. Ambio 22: 80–87

Walter H (1954) Die Grundlagen der Weidewirtschaft in Südwestafrika. Ulmer, Stuttgart

Weber GE, Jeltsch F (1998) Spatial aspects of grazing in savanna rangelands – a modelling study of vegetation dynamics. In: Usó JL, Brebbia CA, Power H (eds) Ecosystems and sustainable development. Advances in ecological sciences 1. Computational Mechanics Publication, Southampton, pp 427–437

Weber GE, Jeltsch F, van Rooyen N, Milton SJ (1998) Simulated long-term vegetation response to spatial grazing heterogeneity in semi-arid rangelands. J Appl Ecol 35: 687 –699

Westoby M, Walker BH, Noy–Meir I (1989) Opportunistic management for rangelands not at equilibrium. J Range Manage 42: 266–274

Wiegand T, Milton SJ, Wissel C (1995) A simulation model for a shrub-ecosystem in the semi-arid Karoo, South Africa. Ecology 76: 2205–2221

Zucchini W, Adamson P, McNeill L (1992) A model of Southern African rainfall. South Afr J Sci 88: 103–109

Part IV:

Desertification Processes and Monitoring

Remote Sensing of Surface Properties. The Key to Land Degradation and Desertification Assessments

Joachim Hill

Keywords. Spectral mixture modelling, vegetation cover, soil organic carbon, erosion cell mosaic, long-term studies, Landsat time series, desertification monitoring, resource assessments

Abstract. Since the International Convention on Desertification of the United Nations has come into force, the need to measure degradation processes in these areas has further increased. While standard methods for undertaking such measurements are imperfect or expensive, remote sensing with air- or space-borne sensor systems offers considerable potential. It provides a comprehensive spatial coverage, is intrinsically synoptic, collects objective, repetitive data and is thus ideally suited for monitoring environmentally sensitive areas. However, the relatively low spatial resolution provided by remote sensing satellites implies that one needs to exploit the spectral contrast between soil and vegetation and their intermediate states to develop surrogate measurements of degradation. Relationships between climate and environmental conditions, which may be of primary importance at a global scale, become problematic when local factors such as topography, lithology and soil properties determine the redistribution of water available for plant growth. Suitable remote sensing approaches must therefore concentrate on both, vegetation and soil-related assessments of ecosystems.

Introduction

Remote sensing is commonly introduced as the science to collect information about objects without coming into physical contact with them; in Earth observation, the most important medium to transmit this information is electromagnetic radiation in the optical and microwave region. Numerous textbooks can be found which give a thorough description of the physical background, principles of image interpretation, digital image processing and specific sensors and applications (e.g. Elachi 1987; Lillesand and Kiefer 1987; Asrar 1989; Richards 1993; Schott 1996; Schowengerdt 1997). In this chapter, primary importance will be attributed to more closely analyze which relationships between parameters of interest and corresponding electromagnetic signals from dryland ecosystems exist.

Remote Sensing of Arid Ecosystems

When compared to the early 1970s, when the first Landsat system was placed into a space orbit, remote sensing systems exhibit a large variety of spectral, spatial and temporal parameters (Kramer 1996), and it depends on the user requirements which system is to be used (Fig. 1). If a frequent repetitive coverage with relatively low spatial resolution is desired (e.g. for meteorology) one would certainly be inclined to base the approach on the AVHRR system available from the polar-orbiting satellites of the NOAA series, or even on data from geostationary satellites such as METEOSAT or GOES. Alternatively, one might look for the highest spatial and spectral resolution available, even at the expense of relatively low repetition rates, and would thus choose one of the available Earth observation satellite systems (Landsat, SPOT, IRS).

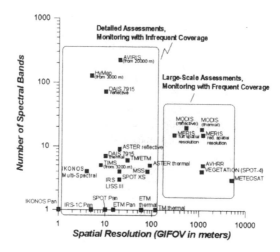

Fig. 1. Important operational, planned, and experimental remote sensing systems in a two-dimensional parameter space determined by spectral and spatial resolution.

As most information on the composition of surface elements is provided by passive remote sensing systems we shall limit our discussion to sensors that measure electromagnetic radiation either emitted directly or reflected by the objects. The focus is therefore on air-borne and space-borne multi-spectral systems, including high spectral resolution sensors (i.e. imaging spectrometers), but also on colour aerial photographs which still provide a cost-attractive and

Remote Sensing of Surface Properties 245

efficient tool for mapping important ecosystem parameters in arid and semiarid environments (e.g., Hill et al. 1998b).

The Importance of Surface Properties

It is widely agreed that environmental change in arid, semiarid and dry subhumid ecosystems is not necessarily driven by climatological variables but triggered by processes which result from adverse human impact (e.g. Mainguet 1994; Perez –Trejo 1994). Traditionally, albedo and vegetation cover have been considered the most important remotely sensed indicator variables to characterize the state of ecosystems under the threat of global environmental change. More recently, however, the awareness has grown that more specific information on surface properties is required to understand processes on local to regional scales (e.g. Yair 1994; Hill and Peter 1996).

Vegetation attributes are usually described by structure, dynamics and taxonomic composition, of which taxonomy is the least important of the three. The classification which is most compatible with remote sensing relates to the projected foliage cover (PFC) and the life form of the tallest stratum (Graetz 1990). However, since arid or semiarid ecosystems are dominated by sparse vegetation, the key issue is to obtain accurate estimates of vegetation abundance which are not biased by the spectral contribution of background components (i.e. litter and substrate). At the same time, the soil surface itself should be as much an object of attention as is the vegetation. At the desert fringe, which is characterized through an enormous interannual variability in rainfall rather than long-term averages, specific surface properties with implications for water concentration, infiltration and leaching are becoming primary indicators for assessing ecosystem conditions. Soil texture, in particular, is one of the most important factors for the water balance as it controls infiltration and the capillary rise of water (Fig. 2). In fine-textured (i.e. loamy) soils, rainfall can hardly infiltrate and is therefore subject to rapid and substantial evaporative losses. More coarsely textured soils (e.g. sandy or stony substrates) permit water to rapidly infiltrate into greater depth where it is well protected against evaporation after the topsoil layer dries out; additionally, almost all infiltrated water is readily available to plants, and salt concentrations in the soil remain low due to the higher leaching efficiency. Relationships between climate and environmental conditions, which may be of primary importance at a global scale, thus become problematic when local factors such as topography, lithology and soil properties determine the redistribution of water available for plant growth (Yair 1994).

Suitable remote sensing approaches must therefore concentrate on both vegetation and soil-related assessments. Also, it must be recognized that the instantaneous state of the surface may fluctuate seasonally or annually in response to rainfall, illustrating the dynamic nature of the system. Repeated observations over longer time periods are then indispensable to assess significant changes. Here, retrospective studies might be as important as continuously monitoring environmental change by remote sensing (Graetz 1996).

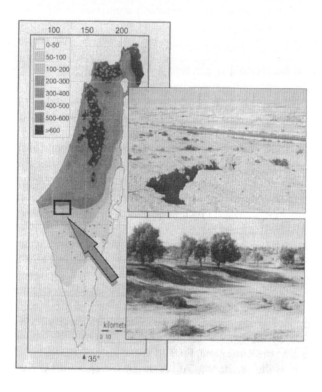

Fig. 2. Loess- and sand-covered areas at the desert fringe in the Northern Negev, Israel. Notwithstanding the possible interference of human actions, the two photographs demonstrate the importance of substrate properties for water redistribution and availability, and plant growth (Rainfall map from Kadmon and Danin 1997)

Remote Sensing Data Interpretation

Optical remote sensing systems measure the spectral properties of surfaces; they cannot measure specific resources or land degradation directly. The major problem associated with their use is now to quantitatively interpret a measured signal that has interacted with remote objects in terms of the properties of these objects. In order to assess land resources or land-degradation processes, it is necessary to define diagnostic indicators. These may be primary (e.g. high salt content in salinized soils) or secondary indicators which are produced by the problem (e.g. reduced vigour of vegetation). It would, of course, be appropriate to discuss some issues in more detail, such as principles of soil and vegetation reflectance, limitations of conventional vegetation indices, detection thresholds for photosynthetic vegetation, the influence of specific surface constituents in arid regions, and spectral characteristics related to non-photosynthetic vegetation and variable surface conditions. However, with regard to the scope of this chapter the reader is referred to specific summaries, such as, for example, Hill and Peter (1996). Here, more emphasis will be given to interpretation concepts and examples from case studies suited to illustrate some of the more general concepts.

Methodological approaches to derive vegetation structural parameters from spectral variables range from empirical approaches to the inversion of physically based models. In arid ecosystems, uncertainties in measuring vegetation abundance can only be minimized by accounting for the reflective properties of background materials (e.g. Siegal and Goetz 1977). Conversely, natural vegetation can significantly mask and alter the spectral response of the ground, and attention must be given to the spectral characteristics of non-green plant components and associated litter (Elvidge 1990). Therefore, as a compromise between empirical indices and more physically based reflectance models, it seems well justified to use relatively simple models which can be successfully inverted.

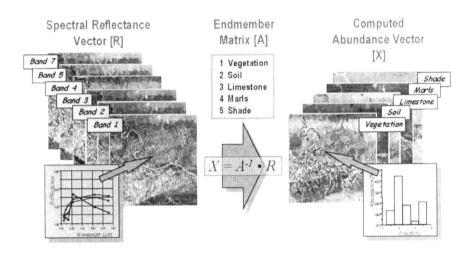

Fig. 3. Scene model for estimating vegetation and substrate properties (i.e. abundance of plant and soil/rock components) based on the inversion of a linear mixture model

One of the these approaches attempts to computationally decompose multispectral measurements with regard to a finite number of pure spectral components, i.e. end-members (Fig. 3). The method has become known as spectral mixture analysis (e.g. Adams et al. 1989; Smith et al. 1990; Craig 1994), and it assumes that most of the spectral variation in multispectral images is due to mixtures of a limited number of surface materials, and that these mixtures can, in first approximation, be described as a result of additive (linear) spectral mixing (i.e. where each photon contacts only one type of surface material). The unknown vector of material abundances (X) is then computed by

$$X = A^{-1} \cdot R$$

i.e. essentially by multiplying a multispectral measurement (R) obtained from air-borne or satellite sensors with the inverted end-member matrix A (e.g. Schowengerdt 1997). Many studies have shown that it is to be considered one of the most promising approaches for computing the proportional abundance of materials that occur within a specific surface area (i.e. pixel). Alternatively, only the use of physically based reflectance models might open the avenue to quantitatively retrieve selected model parameters (i.e. cover, LAI, pigment concentrations) through iterative inversion procedures which also may more easily account for non-linearities (e.g. Baret and Jacquemoud 1994; Pinty et al. 1996).

Case Studies

The spectral mixing paradigm has been successfully used for a wide range of applications in ecology, land use and land-degradation studies (e.g. Adams et al. 1989; Smith et al. 1990; Hill et al. 1995). Due to its direct response to preselected component spectra (i.e. typical desert shrubs and specific substrates) and its increased robustness against soil colour differences in the absence of dense vegetation, the method appears particularly suited to overcome the conceptual limitations of two-band vegetation indices, such that we can expect improved estimates of vegetation abundance in areas with sparse cover of mainly woody plants. Additionally, it is to some extent possible to account for fallen dead material (litter) and the ubiquitous crust of cryptogams, algae, fungi and bryophytes, that play significant roles in the redistribution and nutrients at the soil surface (e.g. Hill et al. 1998b)

Vegetation Assessments and Monitoring

Graetz and Gentle (1982) and Pech et al. (1986) have already used spectral mixture modelling to estimate plant cover in arid Australian landscapes. Later, Smith et al. (1990) successfully applied the technique to Landsat-TM images for mapping vegetation cover in semiarid parts of California, and Hill et al. (1995) have obtained improved vegetation estimates in a Mediterranean study site. Whether and to which extent these methods can be applied to process and interpret long term series of satellite imagery has recently been illustrated in more detail through a case study developed for the island of Crete, Greece (Hill et al. 1998a). In this study, a long time series of Landsat TM images was used which covers the time span between 1984 and 1996. The data set was further extended by adding Landsat-MSS data dating back until 1976, so that the complete data set covered an observation period of 20 years.

It was demonstrated that degradation trends are limited to specific areas, while extended parts of the central highlands remained unaffected. This is found in good agreement with the fact that most of the increasing grazing activities are, and have been, relying on the use of access roads for transporting animals and additional food supplies. Vegetation dynamics for most of the remaining areas exhibit patterns typical for a steady-state equilibrium, as they are known from large areas in the Mediterranean which, for several decades, have been excluded from land use pressure (e.g. Tabarant 1999). However, the fact that it has been possible to identify a degradational reduction of vegetative cover in specific locations with continued grazing pressure implies that remote sensing can also provide substantial contributions to a more conscious management of precious land resources.

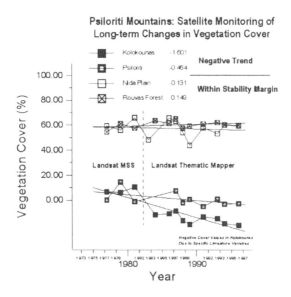

Fig. 4. Regression-based trend analysis (1976–1996) of satellite-derived estimates of proportional vegetative cover for selected reference areas in the Psiloriti highlands of central Crete. The use of this approach (i.e. mapping the regression slope for each individual pixel) permitted identification of the areas where degradation processes have caused a significant loss of perennial plant cover.

Besides the ongoing development of high spectral resolution air-borne scanner systems, it has been almost neglected that high-quality scanning devices allow multispectral images to be generated from aerial photographs. These can then be analyzed by using classical image processing techniques, and, in the case of spectral reference measurements being available, it is even possible to calibrate colour photographs to reflectance. An example from a study on disturbance and regeneration of biotic crusts in a sandy arid ecosystem of the Negev Desert in

Israel (Hill et al. 1998b) shows how digitized aerial photographs are used successfully to differentiate the proportional cover by photosynthetic and woody vegetation components (Fig. 5). The methodological approach is based on the spectral mixture analysis which, after stratifying the aerial photographs into vegetated and non-vegetated spatial domains, provides the possibility to use reflectance spectra of specific components measured in the field.

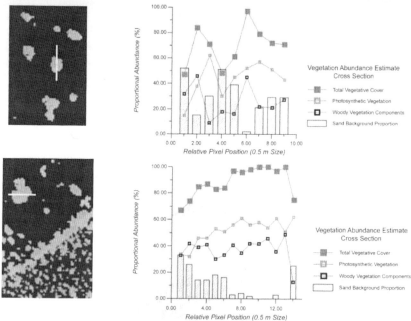

Fig. 5. Cross-section through individual desert shrubs in a sandy arid ecosystem of the Negev Desert (Nizzana, Israel). The diagrams, which are based on the spectral unmixing of true-colour aerial photographs, show the proportional cover of photosynthetic and woody vegetation against the sandy background (Hill et al. 1998b)

Substrate Properties

Not all studies of dryland ecosystems have relied solely on vegetation cover as an indicator of land suitability or degradation. Escadafal et al. (1994) and Hill et al. (1995), for example, were able to use also parameters indicative of soil conditions and hence to apply concepts taken from pedology and geomorphology to obtain qualitative indications for soil degradation or erosion processes. Spectral mixture analysis allowed the latter to estimate relative amounts of rock fragments and soil particles on the surface. Since, within a specific context, soil erosion leads to an increase in rock cover in source areas and the accumulation of soil material as colluvium elsewhere, rock content can be used as an indicator of degradation. This corresponds to defining the erosional state of soils as a function of the mixing ratio between developed soil substrates and parent material components which, of course, need to be spectrally distinct from each other (Hill et al. 1995).

Based on hydrological concepts, Pickup and Chewings (1988) had already earlier developed the erosion cell approach whereby the landscape is divided into the production zone, where there is a net soil loss, a transfer zone with intermittent erosion and deposition and a sink where accumulation occurs. The two approaches can be combined with quantitative estimates of soil properties such as organic carbon, thereby leading to increasingly differentiated possibilities to analyze environmental conditions in dryland ecosystems. The objective of differentiating between favourable sink areas and active erosion and transport zones then condenses on a detection problem, which is to identify the organic matter content of dryland soils based on spectral indicators. Some of the classical relations between organic carbon content and spectral variables were based on the reflectance in the visible range. Traditionally, one has tried to analyze the relationships to soil colour and soil brightness in the visible range of the solar spectrum. Hill and Schütt (2000) proposed a method which can be applied to high-resolution spectra as well as to the spectral resolution of Landsat TM. It can thus be applied to real images, provided the data have been corrected for atmospheric effects. Basically, organic matter content (in wt.%) is estimated through a multiple linear regression between soil organic carbon and the coefficients of a parabolic curve fit to the soil reflectance spectrum between 0.45 and 1.6 μm; they could use this model to successfully map the organic carbon content in their study site based on atmospherically corrected Landsat TM images (Fig. 6).

Similarly, based on digitized and calibrated high spatial resolution true-colour aerial photographs, Hill et al. (1998b) could demonstrate that spectral unmixing of reflectance signatures from the non-vegetated domain allowed the successful identification and spatial differentiation of substrates and biological crusts in a sandy arid ecosystems.

Concluding Remarks

Assessing and monitoring desertification by conventional means, in particular, has traditionally been lacking in standardization because of the range of criteria and indicators. The various data sources available through remote sensing offer the possibility of gaining environmental data over both large areas and relatively long time periods. Resource assessments and continuous monitoring of environmental parameters are complementary issues to be observed for a sustainable management of dryland ecosystems.

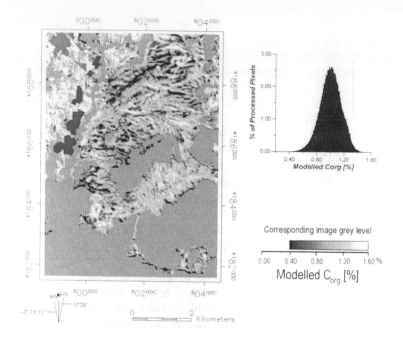

Fig. 6. Average soil organic carbon concentrations derived from applying a laboratory-calibrated regression model to specifically preprocessed Landsat Thematic Mapper images in a study site in SE Spain. Seminatural lands, irrigated areas and townships are masked in *different shades of grey*. (Hill and Schütt 2000)

The advantages of remote sensing result from its synoptic nature, comprehensive spatial information and objective, repetitive coverage. While remote sensing was initially used primarily for resource mapping and inventory, it turns out that monitoring and predictive modelling is becoming more important and successful. Advanced scene models, such as the spectral mixing paradigm or invertible physically based analytical models, can also be used to derive quantitative estimates and improved indicators for land resources and degradation processes. The integrated interpretation of the satellite-derived information layers, available climatic records and results from detailed field studies may then provide a new perspective to understand environmental change in arid ecosystems.

Finally, the success of integrated studies will depend more on the interdisciplinary co-operation between experts than on further instrumental and methodological advances, in particular because remote sensing as a science has achieved the provision of highly diversified and advanced level expertise, aiming at identification and quantification rather than relative discrimination.

References

Adams JB, Smith MO and Gillespie AR (1989) Simple models for complex natural surfaces: a strategy for the hyperspectral era of remote sensing. Proc IGARSS '89 Symp, July 10–14, Vancouver, Canada, pp 16–21

Asrar G (ed) (1989) Theory and applications of optical remote sensing. Wiley Interscience, New York

Baret F and Jacquemoud S (1994) Modeling canopy spectral properties to retrieve biophysical and biochemical characteristics. In: Hill J and Mégier J (eds) Imaging spectrometry – a tool for environmental observations. EUROCOURSES: remote sensing, vol 4. Kluwer, Dordrecht, pp 145–167

Craig MD (1994) Minimum–volume transforms for remotely sensed data. IEEE Trans Geosci Remote Sens 32: 542–552

Elachi C (1987) Introduction to the physics and techniques of remote sensing. Wiley, New York

Elvidge CD (1990) Visible and near infrared reflectance characteristics of dry plant materials. Int J Remote Sens 11:1775–1795

Escadafal R, Belgith A and Ben Moussa A (1994) Indices spectraux pour la télédétection de la dégradation des milieux naturels en Tunisie aride. Proc 6th Int Symp on Physical Measurements and Signatures in Remote Sensing, 17–24 January 1994, Val d'Isere, France, ESA Publ, pp 253–259

Graetz RD (1990) Remote sensing of terrestrial ecosystem structure: an ecologist's pragmatic view. In: Hobbs RJ and Mooney HA (eds) Remote sensing of biosphere functioning. Springer, Berlin Heidelberg New York, pp 5–30

Graetz RD (1996) Empirical and practical approaches to land surface charaterization and change detection. In: Hill J, Peter D (eds) The use of remote sensing for land degradation and desertification monitoring in the Mediterranean basin. State of the art and future research. EUR 16732 EN. Office for Official Publications of the European Communities, Luxembourg, pp 9–21

Graetz RD and Gentle MR (1982) The relationships between reflectance in the Landsat wavebands and the composition of an Australian semi-arid shrub rangeland. Photogramm Engin Remote Sens 48:1721–1730

Hill J and Peter D (eds) (1996) The use of remote sensing for land degradation and desertification monitoring in the Mediterranean Basin. State of the art and future research, EUR 16732 EN. Office for Official Publications of the European Communities: Luxembourg

Hill J and Schütt B (2000) The use of earth observation satellites for mapping complex patterns of erosion and stability in dry Mediterranean ecosystems. Remote Sens Environ (in press)

Hill J, Mégier J and Mehl W (1995) Land degradation, soil erosion and desertification monitoring in Mediterranean ecosystems. Remote Sens Rev 12:107–130

Hill J, Hostert P, Tsiourlis G, Kasapidis P and Udelhoven T (1998a) Monitoring 20 years of intense grazing impact on the Greek island of Crete with earth observation satellites. J Arid Environ 39:165–178

Hill J, Udelhoven T, Schütt B and Yair A (1998b) Differentiating biological soil crusts in a sandy arid ecosystem based on hyperspectral data acquired with DAIS-7915. In: Schaepman M, Schläpfer D and Itten K (eds) Proc 1st EARSeL Workshop on Imaging

Spectrometry, RSL, University of Zurich, Switzerland, 6–8 Oct. 1998, EARSeL, Paris, pp 427–436

Kadmon R and Danin A (1997) Floristic variation in Israel: a GIS analysis. Flora 192:341–345

Kramer HJ (1996) Observation of the earth and its environment: survey of missions and sensors. Springer, Berlin Heidelberg New York

Lillesand TM and Kiefer RW (1987) Remote sensing and image interpretation. John Wiley, New York

Mainguet M (1994) Desertification, natural background and human mismanagement, 2nd edn. Springer, Berlin Heidelberg New York

Mainguet M (1999) Aridity. Droughts and human development. Springer, Berlin Heidelberg New York

Pech RP, Davis AW and Graetz RD (1986) Reflectance modelling and the derivation of vegetation indices for an Australian semi-arid shrubland. Int J Remote Sens 7: 389

Perez–Trejo F (1994) Desertification and land degradation in the European Mediterranean, EUR 14850 EN. Office for Official Publications of the European Communities, Luxembourg

Pickup G and Chewings VH (1988) Forecasting patterns of soil erosion in arid lands from Landsat MSS data. Int J Remote Sens 9, 1:69–84

Pinty B, Verstraete M, Gobron N and Iaquinta J (1996) Advanced modelling and inversion techniques for the quantitative characterisation of desertification. In: Hill J and Peter D (eds) The use of remote sensing for land degradation and desertification monitoring in the Mediterranean basin. State of the art and future research. EUR 16732 EN. Office for Official Publications of the European Communities, Luxembourg, pp 169–180

Richards JA (1993) Remote sensing digital image analysis – an introduction. Springer, Berlin Heidelberg New York

Schott JR (1996) Remote sensing: the image chain approach. Oxford University Press, New York

Schowengerdt RA (1997) Remote sensing. Models and methods for image processing. Academic Press, San Diego

Siegal BS and Goetz AFH (1977) Effect of vegetation on rock and soil type discrimination. Photogramm Engin Remote Sens 43:191–196

Smith MO, Ustin SL, Adams JB, Gillespie AR (1990) Vegetation in deserts I: a regional measure of abundance from multispectral images. Remote Sens Environ 32:1–26

Tabarant F (1999) Apport de la télédétéction et de la modélisation à l'etude de la dynamique de production d'un écosysteme méditerranéen de chêne vert (Quercus ilex) dans le sud de la France. Laboratoire d'Ecologie Vegetale, Universite Orsay, Paris Sud

Thomas DSG and Middleton NJ (1994) Desertification. Exploding the myth. John Wiley, Chichester

Yair A (1994) The ambiguous impact of climate change at the desert fringe: Northern Negev, Israel. In Millington AC and Pye K (eds) Environmental change in drylands: biogeographical and geomorphological perspectives. John Wiley, Chichester, pp 199–227

Evaluation of Potential Land Use Sites in Dry Areas of Burkina Faso with the Help of Remote Sensing

Martin Kappas

Keywords. Soil conditions, detection of soil and vegetation patterns, Sahel, remote sensing

Abstract. Water stress is one of the major growth-limiting factors in the Sudano–Sahelian Zone of West Africa. Due to the complex interactions of multiple growth restricting factors, a scientific quantification of water deficits on plant growth is complicated. Deterministic plant growth models (e.g. SWASUC model, CERES–Millet model) are able to describe the interactions between growth-determining factors and the environment. The development of simulation models would therefore be helpful in the classification of potential production zones for the northern part of Burkina Faso. Before plant growth models can be built up and applied, local measurements, calibration and verification are required. Many simulation results from models (see Fechter 1993) indicated that growth restriction factors other than water stress, i.e. low soil fertility or soil properties which affect soil moisture, had a significant influence on yield formation. The exploration of soil properties is of fundamental importance in the African Sahel to detect possible agricultural sites. By the combination of relief, soil types and soil water content sites for potential land use are derived. The derived field data have to be transfered to patterns in aerial photos and spectral signatures of satellite images (mainly Landsat TM and Spot XS) in order to link the field data to remote sensing information. For integrating field data to remote sensing data, robust parameters have to be found, and advanced RS techniques like RGB to USGS Munsell Color transformation or the calculation of specific indices (i.e. redness index or brightness index) have to be used.

Introduction

In addition to the described problems of agricultural production, an overall decline in the agricultural and pastoral output of the Sahelian region has occurred. This decline is a result of many interacting factors. Population growth of 3.6% and a decrease in crop yields have created a situation where cultivated areas have been expanded to include more marginal regions, and fallow periods have been shortened. FAO statistics showed that millet-planted areas in Burkina Faso increased, while the average yield decreased. The present method of land use is a major factor in the destruction of the sensitive ecosystem of the Sahel. One attempt to manage both land conservation and sufficient food production is to reduce production in regions with environmental limitations (i.e. marginal areas) and to increase productivity in regions that have adequate resources for sustainable yields. The question is how to evaluate potential sites and to transfer the single-test-site data to a larger area. Potential sites for agriculture can be interpreted by evaluation of soil types and soil conditions as grain size of soils and soil moisture. Field work and laboratory analysis produce a quantity of data to evaluate possible sites, but these data are valid only for small areas (comparable with point data of precipitation measurements). To evaluate potential sites from remote sensing data, robust surface parameters have to be specified to extract them from the spectral mixture of the single sensor data.

Location of the Study Area

The study area covers a region in northern Burkina Faso and is spatial by limited to the watershed of the River Gorouol (see Fig. 1). This region belongs to the West African Sahel and is viewed as an eco-climatically autonomous region, defined by its highly variable precipitation and characteristic seasonal therophyte vegetative communities. The whole region is framed within the 200- and 500-mm isohyets and belongs therefore to the Sahel zone proper. The long-term amount of precipitation at the station Dori (14°02 N, 0°02 W) from 1922 to 1996 is shown in Fig. 2. The Sahelian landscape of the Gorouol watershed is dominated by a herbaceous stratum and is mainly used as rangeland by seminomadic societies. The region is characterized by dune formations that stretch from east to west. There were two major stages of dune formations during the late Quatenary; the first occurred during the Wurm Regression around 40 000 b.p. and the second is associated with a period of exeptional aridity around 20 000 b.p. (Ogolian). The dunes and aeolian sands cover a large part of the area. The growing of millet (*Pennisetum* spp.) is mostly limited to these areas and millet is cultivated under extensive, traditional farming systems. A more detailed presentation of the study area, the data set, and the laboratory analysis of soils is given in Kappas and Wandelt (1996) and Kappas (1997).

Basic Requirements for Land Use, Especially Pearl Millet Stands

Millet (*Pennisetum* spp.), being a drought-tolerant crop, is in general unpretending and tolerates soils with low nutritive values, but responds to nitrogenous and potash fertilizers. Fertilizing seldom occurs in the Sahel of Burkina Faso. The Extraction of nutrients (P_2O_5, N, K_2O CaO, MgO) during one growing period is

Evaluation of Potential Land Use Sites in Dry Areas of Burkina Faso 257

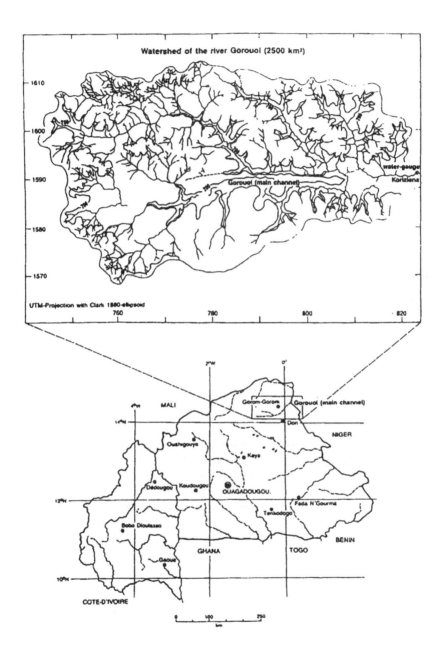

Fig. 1. Location of the study area

modest, but after a period of 4 years' intensive cropping without fallows, potash fertilizing is required. The distribution of rains is more important for millet then the total amount of precipitation. The lower limit of precipitation ranges from 100 to 150 mm / growing period. The temperature optimum ranges from 22 to 28 °C during the growing cycle (70 to 105 days, depending on millet species). Some varieties mature in less than 90 days from planting. Very important for the cultivation of pearl millet are soil physical properties. The fertility of a soil is determined by both its physical properties and its nutrient resources, but the efficient utilization of the nutrients depends on the suitability of the soil as a rooting medium. This important precondition is mostly fulfilled by the soils of the old aeolian dunes, which generally have depths over 90 cm. A soil depth greater than 90 cm is propitious, a depth from 50 to 90 cm moderate and soil depth less than 50 cm results in limitations in pearl millet crop. From the precondition of soil depth we could draw the conclusion that only soils of the old dune complexes are useful as arable areas for pearl millet cultivation. Therefore, various landscape profiles and soil transects were taken in the range of the dune bodies to map the soils and their physical properties. Moreover, in a region in which annual rainfall is generally less, and often much less, than potential evapotranspiration, water supply is frequently the factor that most limits crop growth, and the moisture storage characteristics of a soil are an important factor in its productivity. The highly seasonal nature of the rainfall means that periods of water excess may be expected even in the drier areas, and the risks of serious loss of soil and nutrients through erosion are high. An understanding, therefore, of the physical properties of the soils, particularly with regard to water control, is essential for the effective utilization and conservation of the soil resources of the entire region.

Methods

The methods of the study are divided into two main parts. First, field surveys have turned out to evaluate the different soil units and to derive robust parameters for the remote sensing analysis afterwards. During the field surveys, soils are mapped and soil moisture is measured by tensiometers. In the preliminary studies the preconditions or basic requirements for potential pearl millet locations in the study area are worked out. Afterwards, methods to detect possible sites for pearl millet growth via remote sensing data are discussed.

Typical Soil Sites Along a Transect and Soil Condition Data

Ten soil profiles are dug along the landscape transect of Saouga and analyzed in the laboratory of the University of Mannheim. The entire transect (see Fig. 3) is divided into three parts: the dune complex (S1 to S3), the river complex (S4 to S5) and the complex of the glacis and pediments (S6 to S10). One soil profile of the dune complex is described in detail. The profile of the location Saouga 3 (S3, dune complex) and the laboratory results (chemical analysis) are represented in detail (Fig. 4; Table 1). The soil at location S3 belongs to the soil unit of arenosols (Qf after the FAO classification) and is very typical as a potential site for millet cultivation on the dune complex. The second soil complex describes the soil units within the sphere of influence of the River Gorouol. These soils belong to the unit of fluvisols. Millet cultivation relating to traditional farming methods is not possible on these soils because tools are few and simple, the most important being the cutlass and the short-handled hoe. The third soil complex covers a laterite plateau between the river Gorouol and an inselberg in the north. All soils of this complex are usually fairly shallow and have a sharp contact to the stony underground. Most profiles are less than 60 cm deep. The soils at the locations S6 to S10 are classified to the soil units of luvisols and cambisols, which have strong limitations relating to the cultivation of pearl millet. To sum up the results, it can be said that detection of potential sites for pearl millet growth via remote sensing has to focus on sandy soils of the old dune complexes. Although topsoils are usually sandy, the sand is often largely within the fine sand fraction (0.20–0.02 mm). It is this feature, combined with the present of some silt, that causes soil surfaces to have a widespread tendency to form a crust after rain. In the extreme, the soils develop into laterites, in which clay and iron pans may occur quite close to the surface, seriously limiting water movement and crop rooting. There are also large areas in which the soils contain considerable quantities of ironstone fragments and ferruginous concretions, derived from the erosion and reworking of ironstone outcrops and ferruginous subsoils.

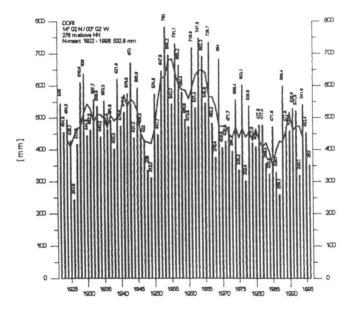

Fig. 2. Long-term precipitation at DORI (*line* = 5-year running mean)

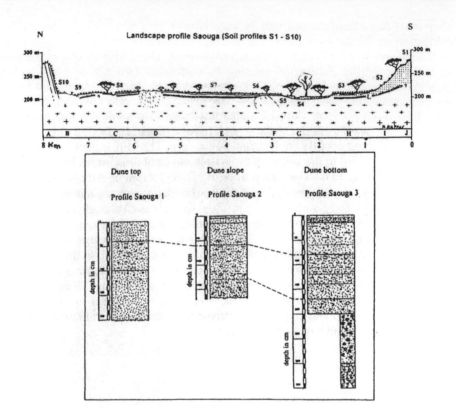

Fig. 3. Landscape transect Saouga with ten soil profiles and a catena from top to bottom of a dune

Location Saouga (S3): Ferralic Arenosol (FAO–UNESCO)

Entisols (oxic quartzipsamments) (US soil taxonomy)
Sols ferrugineux tropicaux peu lessivés peu différenciés (sur sables éoliens pauvres en argile et limon-erg recent) (Classification des sols francaises)

Location: 1 km northeast of Saouga, altitude: 220 m, physiography: glacis of an aeolian dune
Drainage: Impeded at depth
Parent material: aeolian sand, generally poor in organic carbon, nitrogen, exchangeable bases and P_2O_5, pH = 6, slightly acid
Vegetation: Cultivation of *Pennisetum glaucum, Acacia albida, Combretum micranthum*

Evaluation of Potential Land Use Sites in Dry Areas of Burkina Faso

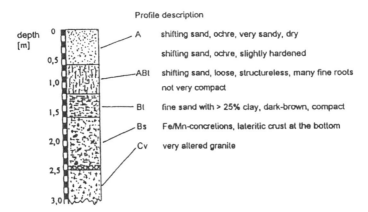

Fig. 4. Soil profile Saouga 3 in detail

Table 1. Laboratory results (chemical analysis) for all investigated soils of the Saouga transect

Sample	pH (CaCl$_3$)	EC Saltc. %	N$_2$ ‰	C %	*1.72 Humus %	Fe dith. ‰	Al dith. ‰	Mn dith. ‰	Si dith. ‰
Saouga 1	4.31	0.0033	0	0.07	0.12	2.999	0.51	0.038	0.134
Saouga 2	4.79	0.0044	0.072	0.08	0.14	3.781	0.609	0.036	0.144
Saouga 3	4.99	0.071	0.1	0.18	0.31	7.125	0.732	0.179	0.372
Saouga 4/1	4.89	0.0081	0	0.26	0.44	12.64	1.302	0.042	1.171
Saouga 4/2	4.64	0.0139	0	0.38	0.66	13.72	1.524	0.35	1.005
Saouga 5	8.07	0.0438	0	0.18	0.3	22.55	1.799	1.832	1.177
Saouga 6	6.44	0.0177	0	0.31	0.54	23.33	2.158	0.404	0.934
Saouga 7/1	6.74	0.0231	0.059	0.27	0.47	23.89	2.071	0.807	1.13
Saouga 7/2	7.9	0.0501	0	0.28	0.48	20.08	1.811	0.816	0.868
Saouga 8	6.05	0.0265	0.194	0.23	0.4	12.13	1.523	0.548	0.6
Saouga 9	7.19	0.0323	0	0.25	0.43	11.9	1.092	0.363	0.637
Saouga 10	6.69	0.0151	0.12	0.15	0.26	12.37	1.212	0.304	0.654

Effective C-exchange capacity in mval / 100g soil; exchange with NH$_4$Cl, soils with pH over 6

Sample	K mval	Na mval	Mg mval	Ca mval	Total acidity	pH	KAK sum mval	Base saturation %
Saouga 1	0.015	0.005	0.068	0.137	0.288	4.39	0.513	43.86
Saouga 2	0.009	0.005	0.362	0.434	0.188	4.79	0.998	81.16
Saouga 3	0.05	0.017	1.126	1.692	0.113	4.99	2.998	96.23
Saouga 4	0.414	0.131	4.578	7.285	0.125	4.89	12.533	99.02
Saouga 5	0.258	0.081	4.105	7.45	0.25	4.64	12.144	97.94

Measurement of Soil Moisture

The soil water tension of the soil types is measured by moisture tensiometers at different depths (surface moisture at 20 cm depth, 50, 100, 150 and 200 cm depth). Soil water tensions were measured between 0 and -900 hPa at all experimental sites (dune, river, glacis complex). The soil profile of the S3 experimental field was typical for a sandy soil of the dune complex commonly used for millet production in the study area. Millet roots formed mainly in the upper 40 cm, but roots were found down to a depth of 180 cm. Results of the field-determined infiltration (double-ring infiltration) indicated that the soil can absorb rainfall intensities up to 250 mm h^{-1} without any runoff. The water contents averaged 0.18 cm^3 cm^{-3} and 0.06 cm^3 cm^{-3} for -100 and -900 hPa. The soil moisture retention characteristics showed a water content of 0.40 cm^3 cm^{-3} at 0 hPa and water content decreased rapidly to values of 0.16 and 0.05 cm^3 cm^{-3} at -100 and -900 hPa. There was an increase in water-holding capacity from the top to the bottom of the soil profile. During the crop cycle the sandy soils of the dune complex, in contrast to the other sites, show the best seasonal water balance. In order to use the investigations of water balance on a regional scale, for example to classify potential production areas, it would be useful to combine them with remote sensing data. Field work and soil water measurements verify that the sandy soils of the old dune complexes should be the favoured areas for pearl millet planting. The next question is now how to detect such places with the help of remote sensing data.

The Use of Remote Sensing Data

In the current project we developed a different approach (see Fig. 6) based on indices values, simple linear combinations of bands of known thematic significance.

$$\text{B-IB} = \sqrt{\frac{(XS - 1^2 + XS - 2^2 + XS - 3^2)}{3}} \quad [\text{B-IB} = \text{brightness index}]$$

B-IC = (256 XS-2) / (XS1 + XS-2) [B-IC = colour index]
B-IR = XS-2^2 / XS-1^3 [B-IR = redness index]
B-ICFn = 2 * XS-3 - XS-2 - XS-1 / XS-2 - XS-1 [B-ICFn = form index]

The combination of the different indices (soil based indices) leads to a spatial differentiation of soil properties. As an example the cotent of iron in the soil surface can be monitored by the calculation of the redness index. In a combination with the colour index other colour effects that belong not to iron can be diminished. As a simple result you get a relative precise map of the iron content in the upper soil horizon. With the help of further combinations (overlay technics in a GIS) and the consideration of field experience and laboratory results a knowledge based interpretation of the soil surface and its underlying properties (shallow – not shallow, iron content, potentail moisture) can be derived. In future hyperspectral technics will help to characterize the reflection behaviour of the different soil types and surfaces in order to separate potential soil units for agricultural use.

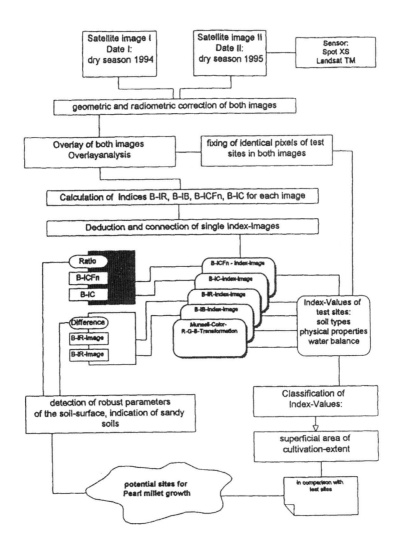

Fig. 5. Flowchart to derive potential landuse sites esp. for Pearl millet growth via soil indices

References

Fechter J (1993) The simulation of pearl millet growth under the environment conditions of Southwest Niger, West Africa. Hohenheimer Bodenkundl Hefte Nr 10, Stuttgart

Kappas M (1994) Fernerkundung – nah gebracht. Dümmler Verlag, Bonn

Kappas M, Wandelt B (1996) Vegetations- und bodenkundliche Arbeiten im Sahel Burkina Fasos. Mannheimer Geogr Arb, No 15, Mannheim

Kappas M (1997) Detection of soil and vegetation pattern in a Sahelian landscape of Burkina Faso with the help of remote sensing. In: Geowissenschaftliche Untersuchungen in Afrika III, Würzburger Geogr Arb, pp 1–26

Remark of the editors:
The originals of the figures were not available from the author.

Degradation of the Vegetation in the Central Kyzylkum Desert (Uzbekistan)

Lyuba Kapustina

Keywords. Anthropogenic influence, crop of the pastures, GIS, map

Abstract. The extent of degradation in the vegetation of the Kyzylkum Desert was determined by topography and GIS. Topography is an important variable in many ecological processes. GIS provides a number of methods for analyzing topography. The ability to analyze digital topographic data has significantly advanced ecological modelling (Johnston 1998). Geobotanic investigation, large-scale mapping and GIS make it possible to find conventionally climax associations and their anthopogenic modification in the Central Part of the Kyzylkum Desert. Zonal types were formed on the Old Xerophilous and Old Mediterrenean bases in the Paleogene and Neogene. The main types are Turanian semishrub desert type, Turanian psammophyton type, Irano–Turanian psammosavanna type and halophyton turanicum type on the different type of soils. Before the 20th century, climax associations of the first and second types were predominant in the plant cover. At present, associations of the third type predominate on sandy desert soils. Natural renewal of the fodder plants in intensively used pastures practically does not occur and weeds and unpalatable species predominate. Ninety percent of the degraded vegetation was determined by spatial analysis. Average dry mass of the pasture plants is about 105 kg ha^{-1} centner/hectare on grey-brown soils and 55 kg ha^{-1} on sandy desert soils.

Material and Methods

Investigation of present-day vegetation in the Central Part of the Kyzylkum Desert was carried out according to traditional floristic, geobotanical and ecological methods (Hill et al. 1996). Ecological profiles were taken on the plains and the low mountain slopes during field geobotanical investigations. Modifications of the vegetation were compared with relief changes, salinization and mechanical composition of the soils.

Crops of the pastures were determined on the typical plots (4 x 50 m) by the hay method for the forage grass and edible part shrub species. Different species of the shrubs and trees were counted on these plots. Five of the hay grounds (0.5 x 2 m) were selected for each pasture type. Average dry mass of the plants and average protein content in dry mass of the plants was defined according to the

266

Methodical Instructions for Geobotanical Investigations of Pastures in Uzbekistan (1980).

Vegetation polygons were interpreted on cosmic photographs of 100 m resolution after field investigation. The total information on vegetation was transferred to topographical maps (1: 200 000). Mapped vegetation data were entered into GIS with a digitizing table and registered to a common coordinate system in the Ecology Center, Inc. (Montana, USA). ArcView 3.0 programme was used to develop the maps: Vegetation for Central Part of the Kyzylkum Desert, Average dry mass of the plants and Average protein content in dry mass of the plants.

Vegetation attributes were described by structure, dynamics and ecological conditions of community distribution. More than 200 polygons were detailed for 19 vegetation types on the different soil types.

Results

Composition of the plant associations and their anthropogenic modification largely depends on the physical and chemical properties of the soils. In recent times, two group types – extremely arid and semiarid of the temperate floras foracenotype – an be distinguished in the desert vegetation of Uzbekistan (Kamelin 1979). The first group of extremely arid types includes climax associations and their anthropogenic modifications on grey-brown soil.

The Turanian semishrub desert type was formed on the basis of Old Mediterranean and Old Xerophilous floras in the Neogene. Groups of oligothermal and methothermal euxerophilous semishrubs prevail (Kamelin 1979). Climax associations are formed by *Artemisia diffusa* H.Krasch, *A. turanica* H. Krasch., *Salsola orientalis, S. arbuscula* Pall. with *Carex pahystylis* Gay. *and Poa bulbosa* L. (elements of the semisavanna group). Xeric semishrubland on grey-brown loam soils [*Artemisia diffusa* communities with *Artemisia turanica, Salsola orientalis, S. arbuscula* and *Aellenia subaphylla* (CAM.)Aellen and xeric pelitic semishrubland on grey-brown clay soils; *Artemisia turanica* communities with *Salsola orientalis, Salsola arbuscula, Artemisia diffusa, Carex pachystylis* and *Poa bulbosa*] predominate on the south mountain plains. Xeric psammophytic semishrubland on sandiest grey-brown soils [*Artemisia diffusa* communities with *Salsola arbuscula, Ferula foetida* Rgl., *Calligonum leucocladum* (Schrenk) Bge., *Ceratoides latens, Haloxylon aphyllum* (Minkw.) Iljin] and xeric shrubland on the loam grey-brown soils (*Salsola arbuscula* communities with *Artemisia diffusa, A. turanica, Carex pachystylis* and *Salsola orientalis*) are spread on the north and west mountain plains. *Artemisia diffusa* communities and *Salsola arbusculiformis* Drob communities with *Artemisia terrae-albae* H.Krasch, *Rhamnus sintenisii, Convolvulus fruticosus* Pall and *Atraphaxis spinosa* L. predominate on the calciumphitic slopes of low desert mountains, *Salsola orientalis* communities with *Anabasis brachiata* F. et M, *A. truncata* (Shrenk) Bge. , *Nanophyton erinaceum* (Pall) Bge. and *Artemisia turanica* on the saline soils of the mountain plains and

Degradation of the Vegetation in the Central Kyzylkum Desert 267

rocks. Dry mass content of the forage plants varied form 65 to 130 kg ha^{-1} for the xeric semishrub communities on grey-brown soils. Protein content in dry mass of the forage plants varied from 4.5 to 7.5 kg ha^{-1}.

Natural and anthropogenic successions on sandy soils are connected with the process of overgrowing (Kurochkina 1978). Climax and subclimax stages are represented by communities of the Turanian psammophyton type, syngenetic stages by sparse communities of psammodendron Irano–Turanicum type on loosely fixed sandy soils.

Turanian psammophyton type was formed on the basis of the Old Mediterranean flora in the Neogene. In the Pleistocene–Holocene the flora was enriched by different species. There are groups of oligothermal and less often, mesothermal euxerophilous psammophilous trees and shrubs (Kamelin 1979). *Salsola arbuscula, Calligonum microcarpum* Borszcz. and *Haloxylon persicum* Bge. with *Calligonum leucocladum , C. setosum* Litv., *Haloxylon aphyllum, Ferula foetida, Dorema sabulosum* Litv. form climax and serial associations on the sandiest and sandy desert soils. Mobility of sand dunes depends on the structure of plant communities. Synusiae of *Carex physoides* is usual in these shrubly-psammophilous associations on fixed sandy soils. Dry mass content of the forage plants varied from 90 to 30 kg ha^{-1}; average protein content is about 4 kg ha^{-1}.

Irano–Turanian psammosavanna type was formed from the Old Xerophilous flora in the Paleogene–Neogene. Later, some communities became relict and were enriched by Turanian psammophyton species. There are groups of mesothermal xerophilous and psammophilous trees, shrubs, grasses (Kamelin 1979). *Ammodendron conollyi* Bge. communities with *Convolvulus korolkovii* Rgl., *Salsola richteri* Karel., *Peganum harmala* L.; *Stipagrostis pennata* communities with *Calligonum setosum, Acanthophyllum borszczowii* Litv. and *Salsola richteri* predominate on erosion sandy desert soils. Dry mass content of the forage plants is 30–40 kg ha^{-1}; average protein content in dry mass of the forage plants is about 1kg ha^{-1}.

Halophyton Turanicum type was formed on the basis of Old Xerophilous and Old Mediterranean floras in the Paleogene–Neogene. There are groups of methothermal euxerophilous halophilous trees, semishrubs, perennials and, rarely, succulent semishrubs (Kamelin 1979). *Halohylon aphyllum* communities with *Girgensohnia oppositiflora* (Pall) Fenzl, *Salsola praecox* Litv., *Artemisia diffusa, Calligonum erinaceum* Borszcz. are spread on saline sandy desert soils. Salties pelitic semishrubsland [*Anabasis salsa* (CAM) Benth. *communities with Salsola orientalis, Ceratocarpus utriculosus* Bluk., *Anabasis aphylla* L., *Halocnemum strobilaceum* (Pall) MB., *Limonium suffruticosum* (L) Ktze., *Kalidium caspicum* (L.) Ung–Sternb] are spread on saline grey-brown soils and solonchaks. Dry mass content of the forage plants is 50–60 kg ha^{-1}; average protein content in dry mass is about 3.5 kg ha^{-1}.

The analysis of modern communities detected about 90% degraded vegetation in the Central Kyzylkum Desert. Overgrazing and cutting down of trees and shrubs are the main reasons for the degradation of plant associations: 7000 semishrubs of different species (*Convolvulus hammadae, Artemisia diffusa* etc.)

are cut for weaving silkworm (*Bombyx mori*) cocoons every year; 400 tons of *Artemisia* species are cut for fodder every year for state farms. The anthropogenic influence is the main reason for the change in vegetation structure – semishrub communities of semiarid type alternate with serial groups of Irano–Turanian psammosavanna type. Sparse communities of *Peganum harmala* with *Ceratocarpus utriculosus, Ferula foetida, Kochia iranica* Litv, *Convolvulus hamadae* V.Petr. communities with *Iris falcifolia* Bge. are spread on degraded loam and sandiest grey-brown soils, and *Iris songorica* Schrenk communities with *Peganum harmala* on degraded sandy desert soils. The anthropogenic dynamics of the vegetation is a cause for the improverishment of the pastures. The average crop of the pastures was 125 kg ha^{-1} and average protein content was 12.5–8.5 kg ha^{-1} between 1960 and 1979. At present, the average dry mass of the forage plants is 75–40 kg ha^{-1} and average protein content is 2.5–3 kg ha^{-1} on sandy desert soils.

References

Hill J, Sommer S, Mehl W, Megier J (1996) Use of Earth observation satellite data for land degradation mapping and monitoring in Mediterranean ecosystems: towards a satellite-observatory. Environm Monitoring and Assessment 37 (1–3):143–158

Johnston C (1998) Geographic information systems in Ecology. Blackwell Science, Minnesota, USA, 239pp

Kamelin R (1979) Kuchistanskij okrug of Central Asia mountains. Nauka, Leningrad (in Russian), 116 pp

Kurochkina L (1978) Psammophyllous vegetation of Kazakhstan desert. Alma–Ata (in Russian), 271 pp

Methodical Instructions for geobotanical investigations of the pastures in Uzbekistan (1980) FAN, Tashkent (in Russian), 170 pp

Modern Geomorphological Processes on the Kazakhstanian Coast of the Caspian Sea and Problems of Desertification

Farida Akijanova

Keywords. Transgression, modern tectonic, natural-anthropogenic processes

Abstract. The modern processes of desertification of the Caspian Sea territory have resulted in a complex influence on both external (climate, sea transgression, human activity) and internal factors (modern tectonic movement, rock structure and the level of their stability towards destruction and transportation, geomorphological and hydrogeological conditions, the structure of regional biocomponents), most effective of which is economic human activity. Irrational mastering of the semi-desert and desert regions of the Kazakhstanian zone of the Caspian Sea, along with the global and regional climatic changes, have led to significant and sometimes irreversible degradation of the many environmental components.

The modern transgression of the Caspian Sea essentially affected the intensity of the desertification process, and also aggravated the ecological and social economical problems of the region.

The data are made into a map of the modern natural-anthropogenic processes of the Kazakhstanian coasts of the Caspian Sea at a scale 1:500 000.

Analysis of the area development and degree of intensity of modern natural -anthropogenic processes permits a regional classification on the degree of destabilization in ecosystems under coastal management in view of modern transgression by the sea.

Introduction

The modern processes of desertification of the Caspian Sea territory have resulted in a complex influence on both external (climate, sea transgression, human activity) and internal factors (modern tectonical movement, rock structure and the level of their stability towards destruction and transportation, geomorphological and hydrogeological conditions, the structure of regional biocomponents), most effective of which was economic human activity. Irrational mastering of the semidesert and desert regions of the Kazakhstan zone of the Caspian Sea, along with the global and regional climatic changes, have led to significant and sometimes irreversible degradation of many environmental components.

Main Factors of Modern Relief Formation

Among the factors influencing modern relief formation are all the natural environment components such as recent and contemporary tectonic movements, lithology and morphology of the coast, hydrometeorological conditions, specialization and level of economic development.

General characteristics of the Caspian Sea relief in the Kazakhstan sector show large geotectonic structures. The Quaternary accumulative marine, alluvial-deltaic and low eolian plains of the northern coast and the adjacent shallow shelf belong to the Pricaspian Basin, which is the southeast periphery of the Pre–Paleozoic Russian platform. The structure is complicated by plicative and discontinuous dislocations which were active during the Pliocene Quaternary (recent) period. The modern valley of the River Ural is confined to a large submeridional fault. Its activity during the Quaternary period is proved by the raised surface of the Khvalyn marine terrace on the left shore; 2–3 m more than the right one. The characteristic feature of the Pricaspian basin tectonic structure is the salt dome distribution. Lack of Neogene Quaternary deposits on the domes as well as their manifestation in modern relief are indicative of tectonic activity during the neotectonic stage. A strip of bar-like uplifts on the shelf (North Caspian) and on the coast (South Emba) is confined to the southern edge of the Pricaspian basin.

In the south along the deep fracture, the Pricaspian basin is bordered by the Epigercinian Turan plate, represented by structures of the North Usturt syneclise, the Mangyshlak–Usturt system of dislocations and the South Mangyshlak monocline. In modern relief these structures are in conformity with a rather raised structural denudation plateau on relatively lithified Neogene substrate, framed by chinks (steep slopes) with abrasive and abrasive-accumulative shores. The Bozaschi Peninsula is an exception as, according to its relief, it is close to the Pricaspian lowland, thus emphasizing its location between two large structures. Mud volcanoes connected by their roots with faults are indicative of neotectonic activity at the northern periphery of the North Bozaschi dome and the southern slope of the Kultuk trough.

Analysis of data indicates that a large part of the region, including the Pricaspian northern part, has risen during the past 30–50 years at 0.2–2 mm a^{-1}. Within the Mangyshlak–Usturt system of dislocations the elevations are considerably larger 1.6 to 3.8 mm a^{-1}. Maximal velocities of vertical movements (> 4 mm a^{-1}) are observed in the northern part of the Tub–Karagan Peninsula and the southern region of the Bozaschi Peninsula. Rather small contemporary lowering (-0.3 to -0.6 mm a^{-1}) is observed on the northeast Caspian coast in the arch of the South–Emba uplift.

Modern processes of coastal morphogenesis are closely interconnected with water surges, depending on the background level of the sea, wind direction and

Modern Geomorphological Processes on the Kazakhstanian Coast

intensity, and the slope and relief of the shore. According to the maximal height of water surges, the Kazakhstan coast is divided into 15 areas (Sydykov et al. 1995). Water surges of 1 m in height are usual, caused by a wind velocity of 10–15 m s^{-1} during 12–24 h. Maximal surges (2.5 m) occur on the eastern low coast, beginning from the mouth of the River Emba to the Komsomolets bay. Vast areas of the near-shore plains are flooded and the width of the immersed territory is 15–30 km from the background water line.

Due to water rise and water remaining in the lowering of the relief, the groundwater level is also increasing considerably, affecting modern relief formation. At present the great majority of recently dry sors have become damp sites, including such vast areas as Mertvyi Kultuk, Kaidak and Bolshoi Sor. They are filled with water as a result of water surges and rise in groundwater level.

In the valleys of the rivers Ural, Emba, Sagiz and in a number of ephemeral water streams on the Mangyshlak Peninsula, modern relief formation accelerates during the spring flood: 70% of the River Ural's annual discharge inflows during this period. The average flood tide for many years near the village of Topoly is 5.3 m (maximum 9.8), near the town of Atyrau it seldom exceeds 2.3 m (maximum 3.8 m). Processes of lateral erosion intensify the accumulation of suspended and involved deposits on flood plains and interbedded areas, forming river bed accumulation ramparts. Water discharge of the Emba and Sagiz Rivers, drying out in summer, is by an order lower than that of the River Ural. The average spring flood level over many years does not exceed 1–2 m, fluvial plains are flooded only in spring; in summer the reaches are silted up. Economic activity promotes relief formation processes: increasing the technogenic load caused a rise in groundwater level, salinization and gully formation over large areas. Hydraulic engineering and meliorative constructions caused significant reduction of sediment transport of the Volga and Ural Rivers, intensifying the abrasion processes on the delta banks and their rapid regression. Redeposited, changed and technogenic rocks and relief forms within the oil fields significantly deteriorated the ecological state of the region. On the territory of old oil fields, surface subsidence with sor formation occurs. Seismic activity induced by change of pressure in oil-bearing beds in deep horizons due to withdrawal of great amounts of hydrocarbonic raw material and underground water is also of current concern.

Coastal Relief

Studying the geomorphological structure and dynamics of the coastal zone, we have analyzed the works of Leontev (1964, 1969), Leontev et al. (1977), Aristarkova (1970), Ignatov et al (1992), Nurmambetov (1991), Akiyanova et al. (1994). Using remote sensing methods, aerovisual and natural observations allowed the authors to map the geomorphological structure of the Kazakhstan coast as well as the characteristic features of modern-relief forming processes at a scale of 1: 500000.

Among the groups of coastal accumulative relief, marine plains of different age predominate. The early Khvalyn plain is well developed on the Bozaschi Peninsula; it is composed mainly of clayey, loamy, sandy loamy deposits with a thickness of up to 12 m, dating back to the first half of the Late Pleistocene. The surface of the plain is on 10–20 m above sea level, and is poorly reworked by exogenous processes. Only at the areas of former sand ridges have small massives of hummock-cellular sands with relative uplifts from 1 to 6–8 m formed as a result of eolian reworking. On the Mangyshlak coast fragments of the Early Khvalyn plain extend as a narrow strip (up to 2 km) and unite at 48 m above sea level. It is composed of sands of hammered hempseed husk with gravel shingle. The surface of the terrace is dissected by a thick network of ravine; erosion and land-slides are widespread on the bench.

The late Khvalyn plain is wider and spreads to the 0 m mark. On the Bozaschi Peninsula at the same level there are weakly pronounced coastal bars of this transgression. The surface of the plain in the district between the rivers Volga and Ural, mainly composed of sands up to 5 m thick, was considerably reworked by eolian processes with hummock-cellular and ridge form formation. In the areas adjacent to the deltas of the Emba and Sagiz Rivers, the Late Khvalyn deposits are loamy, the plain surface is complicated by numerous sors up to 5–7 m in depth, which inherit the primary roughness of the sea bed. Along the Mangyshlak coast, the Late Khvalyn plain was preserved in the form of narrow strips with indistinct elements. The most widespread on the Kazakhstan coast is the marine New-Caspian plain dividing into two generations: early and late. The Early New Caspian plain surface is sloping; it is complicated by sors with vast liman lowerings. It is composed of sands, sandy loams and clays. The external border is distinct, rising up to the mark of 22 m above sea level: on the north coast it passes along the Naryn sands frame, on the east along the Karakum Sands and on Bozaschi Peninsula along the edge of the sandy massives where it is emphasized by coastal bars up to 3 M high.

The Late New Caspian plain is composed of silty clays, fine sands with numerous well-preserved malacofauna of 2.5 m thickness. Its flat surface, situated up to −25 or −26 m from the modern coastal line, has insignificant gradients (0.0002–0.0009). On the former islands and sandy banks there are small uplifts overgrown by *Salicornia* up to 1 m high. On the edge of the terrace (-25, -26 m below sea level) gently sloping shore banks and abrasive scarps up to 0.5 m composed of sandy loams and fine shelly sands are preserved.

The Golocene alluvial-deltaic accumulative plains are exceptional at the beach between the mouths of the Rivers Volga, Ural and Emba. They are composed of fine sands, aleurites and clays. The deltaic plains have a slightly convex profile and a slight slope towards the sea. The surface is cut by streams and mort lakes to a depth of 5 m. At the interriver-bed areas of the River Volga Delta, there are massifs of rewinnowed and consolidated hummock sands. The major part of the deltaic plains, according to hypsometry, corresponds to a high fluvial plain except for uplifts of tectonic genesis at the salt dome outbursts (for example Kamennyi, Kusanbai and Chernorechenskii in the River Ural Delta) or complex formations (Baer knolls).

Modern Geomorphological Processes on the Kazakhstanian Coast 273

The Mangyshlak coastal relief is mainly denudational. It is represented by flat surfaces covered by a nearly horizontal plate of Neogene limestones overlapped by thin Quaternary alluvium. Coastal benches at the Mangyshlak plateau are dissected by a branched network of karst-erosional gullies up to 10 m in depth and more than 100 m in length. In gully mouths, opened towards the sea, proluvial material accumulates, often flowing together forming trains. At present, debris cones and trains are abraded. In the coastal benches (scarps), composed of limestones and marls, karst and abrasion processes are active, forming shelters, niches and caves. According to Potapova (1980), in the upper parts of benches there are many caves 3–8 m in length, while at the benches basement they are up to 50 m in length and from 3 to 20 m in diameter.

Type of the Coasts

On the accumulative coast of the Pricaspian lowlands "dried" coasts are most widespread (-800 km, total length 1100 km). During the last regression (1930 –1970) the dry land area increased by tens of meters per year. Dried surface leveling by water surges, subsequent reworking by eolian and erosional processes, as well as anthropogenic influence, slightly changed the appearance of the territory. At present, passive inundation of the dried shoaly shores occur (slope gradient is 0.0005%). At the adjacent territories, processes of rise in groundwater level, swamping and salinization are in progress due to the groundwater table increase. Data obtained from aerovisual observations are indicative of passivity in the relief-transforming processes in zones of shallow flooding. Only at the interfluvial area of the Rivers Volga and Ural does marked washout of the eolian forms occur; they have become archipelagos of small islands.

Reconstruction of the coastal zone becomes morphologically marked when slopes of the dried shores increase up to 0.0005%. The formation of a bar along the coast 0.5–1.2 m in height, up to 30 m in width and approximately 6 km in length which separates a small lagoon on the west coast of the Bozaschi Peninsula may serve as an example. Deltaic coasts are reconstructed more actively. During the latest regression, Kazakhstan territory on the Volga Delta (Kigach Channel and others) increased annually up to 100–250 m and the River Ural Delta during 1929–1978 by 30 to 40 km. The increase in sea level caused flooding of lowlands in the deltas of the Volga and the Ural, abrasion of near-delta islands and washout of bars. The sea is encroaching on the River Emba deltaic plain, which was continental before. Spring flood and water surges cause backwater, contributing to the accumulation of alluvial-deltaic deposits.

Contemporary transgression has had a great effect on abrasion, abrasion -accumulative and even accumulative Mangyshlak coasts, changing in the coastal and submerged part of the slope. Washout of bars, beaches, beach barriers and modern low marine terraces takes place, as well as formation of new accumulative forms. On the north and south coasts of the Tub–Karagan Peninsula on the shores of Kazakh Bay previously, extinct cliffs are reappearing. During the past 20 years the extent of abrasive sections has increased from 8 to 13%. The research undertaken will allow a forecast assessment of the development of coastal processes for well-founded engineering protection of economic and natural objects.

The complex estimation of desertification includes mapping negative displays of natural-antropogeneous processes and the degree to which they influence a modern landscape. The basic method during the study of this process was the interpretation of medium-scale space snapshots from different years, the analysis of the field materials and specific researches on the condition of a natural environment (geological-geomorphological, neotectonical, hydrological landscape). The data on specialization and the degrees of economic development have allowed us to map the modern natural-anthropogenic processes of the Kazakhstanian coasts of the Caspian Sea at a scale of 1:500 000.

Analysis of the area development and degree of intensity of modern natural -anthropogenic processes makes it possible to carry out regional classification of the degree of destabilization of the coastal ecosystems, and is necessary for the development of an optimum management strategy in view of the modern transgression by the sea.

References

Akiyanova F, Nurmambetov EI, Potapova GM (1994) Modem relief of Kazakhstan's coast and shelf of the Caspian Sea. Collected volume Scientific-technical progress and ecology in West Kazakhstan, Atyrau, pp 21–23

Aristarkova LB (1970) Geomorphology of Podural plateau and Pricaspian lowland. J Geol of the USSR 21,2:283–293

Ignatov EI, K.aplin PA, Lukyanova SA, Soloveva GD (1992) Influence of temporary transgression of the Caspian Sea on dynamics of its coasts. J Geomorphology Nl: 12–21

Leont'ev OK (1964) Role of the sea level fluctuations in Caspian coast formation. Nauka, Moscow, pp 122–130

Leont'ev OK (1969) Relief of the coast. Morphology and dynamics of the coast. The Caspian Sea. Publishers House of Moscow University, Moskow, pp 17–44

Leont'ev OK, Maev EM, Rychagov GL (1977) Geomorphology of the Caspian Sea coasts and bed. Publishers House of Moscow University, Moscow, 208pp

Nurmambetov EI (1991) Pricaspian accumulative plain. Relief of Kazakhstan, part 2, Science, Alam–Ata, pp 44–155

Potapova GM (1980) Some features of the development and distribution of carst on the forms of Southern Mang Yshlak. Science, Alma–Ata, pp 153–156

Sydykov MS, Golubzov VV, Kuandykov BM (1995) The Caspian Sea and its coastal zone. Olke, Almaty, 211 pp

Anthropogenic Transformation of Desert Ecosystems in Mongolia

Ekaterina I. Rachkovskaya

Keywords. Vegetation, semidesert, exremely arid desert, desertification, Gobi

Abstract. Some aspects of the differentiation of Gobi desert ecosystems caused by climate (sectoral position of the territory on Eurasian continent, latitudinal -longitudinal changes) are reviewed. Diversity of dominant species of plant communities is noted in different regions. Ratios between areas of the different soil-lithological groups of ecosystems are presented. Main factors of anthropogenic transformation of ecosystems have been revealed and main criteria for the assessment of modern conditions of the main ecosystem components (soils and vegetation) have been developed. Ratios between ecosystems subject to different degrees of anthropogenic transformation have been calculated on the basis of ecosystem maps at a 5° scale of transformation. It is shown that anthropogenic transformation of semideserts is greater than that of deserts. Recommendations have been proposed for conservation of the ecosystems of the Gobi desert.

A differential approach is essential for the organization of a land use system in the Mongolian deserts. The natural diversity of ecosystems should be taken into account as a reason for the different sensitivity of ecosystems to natural and anthropogenic desertification.

An ecosystem map of Mongolia, scale 1:1 000 000, was made by a research team with our participation in 1995. The map reflects the pattern of spatial distribution of natural ecosystems in relation to climate, relief and soil-lithological conditions. It provides information on expert assessment of the modern state of each ecosystem contour in connection with economic activity.

According to the longitudinal pattern of vegetation distribution, the Mongolian Gobi desert is situated in the East Siberian–Central Asian sector of Eurasia (Volkova 1997) and occupies approximately one third of the Mongolian territory. On a planet scale, climate continentality has maximum expression in this region (long cold winter and short warm summer, substantial daily, seasonal and yearly fluctuation of temperature). The Gobi deserts are characterized by an uneven distribution of precipitation and a short biologically active period. Maximum vegetation growth occurs in summer (in some years in late summer) and is explained by the monsoon pattern of precipitation.

The set of climatic factors, first of all the balance of heat and humidity, changes in the Gobi from north to south. In this connection, change in zonal type of desert is observed at a relatively short distance (Beresneva and Rachkovskaya 1978; Table 1).

Table 1. Zonal types of ecosystems

Main zonal types of ecosystems	Climate	Soils	Coverage by plant communities (%)
Small bunchgrass desert	r=113 t=2830 R/Lr=7	Brown desert soils	15 (12–16)
Semishrub and shrub desert with grasses	r=105 t=2810 R/Lr=9.4	Pale-brown desert soils	10–12
Semishrub and shrub desert	r=75 t=2770 R/Lr=13	Grey-brown desert soils	7–8 (1–10)
Extra arid deserts	r=20 t=3500 R/Lr=20	Extraarid soils	0

r precipitation (in mm); *t* sum of temperatures above 10 °C of soil surface and air (in °C); *R/Lr* radiation index of dryness; expresses relationship between radiation balance and sum of temperatures required for evaporation of annual precipitation

Most of the Gobi belongs to the cold-temperate semideserts and deserts. The climate is cold and sharp continental. The annual sum of temperatures above 10 °C is very low, between 2600 to 3000 °C. Winters are cold with a small amount of snow. The dryness index is 7–13 (i.e. two times higher than in the North Turanian deserts, including Kazakhstan).

The distinctive characteristic of semidesert ecosystems is the codominance in their plant communities of steppe and desert plant species. A particular northern zonal belt of semideserts – desert grasslands – is distinguished in the Gobi (called by some authors desert steppes). Plant communities are formed by small-bunch feather grasses (*Stipa gobica* Roshev., *S. glareosa* P. Smirn.), geophytes (*Allium polyrrhizum* Turcz. ex Regel) and desert dwarf semishrubs (*Anabasis brevifolia* C.A. Mey., *Salsola passerina* Bunge).

To the south they are replaced by grass-dwarf semishrub and grass-dwarf shrub deserts (semideserts). The main type of community structure is the dominance of dwarf semishrubs and dwarf shrubs. Grasses and geophytes form the dominant sinusia. Some of the dominant species are listed in Table 2.

Dwarf semishrub and shrub deserts (true deserts) are distinguished by the dominance of Central–Asian dwarf semishrubs, dwarf shrubs and shrubs in their plant communities.

Warm-temperate deserts are characterized by warmer and drier climate. The annual sum of temperatures above 10 °C is 3500 through 5000 °C with average annual temperature two times higher than in cold-temperate deserts.

Anthropogenic Transformation of Desert Ecosystems in Mongolia 277

Table 2. Dominant species of Gobi vegetation

	Entire Gobi	Western part of Gobi	Eastern part of Gobi (to the east of 100° latitude E)	Dzungaria (depression Barun–Hurai)
S	*Amygdalus pedunculata* *Caragana leucophloea* *Ephedra przewalskii* *Haloxylon ammodendron* *Nitraria sibiricai* *N. sphaerocarpa* *Zygophyllum xanthoxylon*	*Caragana bungei* (Depression of Great Lakes and Valley of Great Lakes)	*Ammopiptanthus mongolicus* *Amygdalus mongolica* *Caragana korshinskii* *Caragana brachypoda* *Potaninia mongolica*	
M	*Ajania fruticulosa* *Anabasis brevifolia* *Artemisia xerophytica* *A. xanthochroa sphaerocephala* *Ceratoides papposa* *Kalidium gracile* *K. foliatum* *Reamuria songorica* *Salsola arbuscula* *Sympegma regelii*	*Chenopodium frutescens* *Artemisia klementzae* (Depression of Great Lakes) *Iljinia regelii*	*Ajania trifida* *Brachentemum gobicum* *Salsola laricifolia* *Salsola passerina*	*Anabasis elatior* *A. truncata* *A. salsa* *A. aphylla* *Artemisia gracilescens* *A. shrenkiana* *A. sublessingiana* *A. terrae alba* *Atroplex cana* *Halocnemum strobilaceum* *Nanophyton erinaceum*
H	*Stipa glareosa* *S. gobica* *Allium polyrhizum*		*Stipa breviflora* *Psammochloa villosa*	*Stipa orientalis*

Life forms: S shrubs, M semishrubs, H herbs

Precipitation is very low (less than 100 mm, mainly 50 mm and lower) with maximum in summer. Winters are dry without snow; the dryness index exceeds 20. Warm-temperate deserts in Mongolia are represented by extremely arid deserts (Rachkovskaya 1977). Within this zone watersheds (mesoplacors) have no higher plants and are divided, according to deposits into crashed stony, stony and stony -puffed-crashed stony (gypsiferous). Plant communities exist in depressions, basically along temporary dry waterways. Slopes of hills and low mountains also have no vegetation, it is accumulated only in gullies and gorges. Active phytocoenose formers along temporary waterways are the Gobi species *Ephedra przewalskii, Sympegma regelii, Reaumuria songorica, Nitraria sphaerocarpa* and the West Gobi species, *Iljinia regelii.* A particular type of extremely arid soil is characteristic to watersheds of this region (Evstifeev 1980). The distinguishing property of this soil is the high content of easily soluble salts, carbonates and gypsum, irrespective of soil formation and underlying rocks. The presence of blue-green algae and a short-term activity of microorganisms in the soils should be noted for these soils.

In Mongolia, extremely arid deserts are distributed in West Gobi, and in China – predominantly in Kashgaria and Bei-Shan. Fragments of extremely arid deserts are found in the lower parts of the Dzhungarian depression (China) and the Ili depression (Kazakhstan).

Gobi deserts are the stronghold of unique Central Asian desert flora and vegetation communities of original composition and structure (Rachkovskaya and Volkova 1977; Hilbig 1990).

The basis of the vegetation cover is desert species developed in the Neogene, representatives of genera *Salsola, Anabasis, Sympegma* and *Iljinia*. Vegetation communities are also formed by the ancient (Paleogene) xerophytic species *Zygophyllum xanthoxylon, Ephedra przewalskii, nitraria shaerocarpa, Reaumuria soongorica* and by the living fossils (Cretaceous–Paleogene) *Ammopipthanthus mongolicus, Potaninia mongolica* and *Gymnocarpus przewalskii* (Rachkovskaya 1993).

The ecosystems of the Gobi desert have the following characteristic features: absence of complicated multiple-storey communities, presence of the summer -autumn sinusia of annuals and absence of the spring-early summer sinusia of annuals typical for the North Turanian deserts and low diversity of dominants and their combinations when forming communities on different substrates.

Our study (Ecosystems of Mongolia 1995) showed that stony and fine earth -crashed stony deserts predominate in the Gobi (64%), sand ecosystems occupy 12%, gypsum deserts approximately 10%, halophytic deserts 10% and semihydromorphic and hydromorphic 4%.

The extreme ecological conditions of a sharp continental climate cause the particular vulnerability of the Gobi desert ecosystems, which is expressed by the deterioration of self-regulation processes and a sharp reaction to anthropogenic influences.

From an economic point of view, the Gobi desert is one of the main grazing rangelands. The vegetation of the Gobi deserts is the main, and often the only, fodder resource for pasturable stock-raising with year-round grazing. A great number of desert plant species are of high nutritional value for livestock.

The main factors of anthropogenic transformation of deserts are as follows: agricultural (pasturable stock-raising, irrigated agriculture), industrial or related to construction (mines, quarries, settlements) and cutting plants for fuel, transport (uncontrolled network of roads without surfacing).

The main traditional type of anthropogenic influence is grazing. The following indicators were used to assess grazing impact on plant communities: change in species composition, change in habitus of plants, abundance of digressionally active and ruderal species; in relation to soil: condition of soil surface, degree of water erosion and deflation, salt and humus content. For the set of indicators, a five level scale was developed to assess anthropogenic transformation of

Anthropogenic Transformation of Desert Ecosystems in Mongolia

ecosystems subject to grazing: undisturbed or very slightly disturbed, slightly disturbed, medium disturbed, strongly disturbed and very strongly disturbed.

Results of expert assessment and analysis of the ecosystem map revealed different degrees of anthropogenic transformation of rangelands in different subzones. Undisturbed, nearly climax ecosystems occupy only 9% of the area of desert grasslands and are located predominantly in mountains, hills and saline depressions. On the plains, undisturbed ecosystems make up less than 1% of the area. In this subzone 25% of the area is subject to permanent grazing of medium pressure. Ecosystems of this area have decreased productivity and biological diversity. An area of 3% suffers a strong degree of degradation. In the subzone of steppe deserts 6% of the area is occupied by nearly climax ecosystems, 68% is under permanent grazing at low pressure, 25% is represented by substantially degraded pastures. A completely different situation occurs in the zone of true deserts. Almost half of the area (45%) is not touched by grazing and 55% is slightly transformed by grazing. Extremely arid deserts are represented mainly by nearly climax ecosystems, which make up 91% of their area. The reason is that pastures of these deserts have low productivity. It is more rational to use ecosystems of extremely arid deserts as grazing lands for wild herbivores. The practice of irrational grazing has led to overgrazing of some areas (around settlements) and insufficient grazing at others. An area of 6–7% in semideserts and 2–3% in deserts is subject to a strong and very strong degree of anthropogenic transformation. It should be noted that the new economic policy of the state is changing these tendencies for the better.

Analysis of the quantitative distribution of rangelands by degree of disturbance has revealed that substantial degradation of rangelands occurs in semideserts. True deserts and extremely arid deserts in particular are in quite a good state. However, it should be remembered that disturbance of these ecosystems would lead to their slower restoration if compared with semideserts.

In additon to grazing, there are other forms of human influence on desert ecosystems which are usually of local character. Trees of *Haloxylon ammodendron* are cut for fuel everywhere. The clearing of particular gallery forests along dry waterways is noticeable. The area of woodlands of *Ulmus pumila* L. is reduced in the eastern and woodlands of *Populus diversifolia* Shrenk. in the western part of the Gobi. Long-term studies implemented in the Gobi allow the following recommendations: organization of the monitoring of state of rangelands and assessment of degree of their degradation, distribution of small farms, organization of livestock migration, prohibition of summer and winter grazing at the same area, withdrawal of degraded rangeland from use, establishment of fixed routes for livestock migration.

A particular land use system should be developed for the unique oasis ecosystems with rich meadow and woodland vegetation. Negative processes of desiccation, salt accumulation and erosion should be avoided here. All water sources in deserts should not be used for irrigation; part of the natural oases should be kept as watering places for wild animals, which are the national wealth of the country.

In order to increase the productivity of rangelands, the road network must be reduced, together with the construction of roads with solid surfacing. One of the conservation measures for desert ecosystems is the organization of protected areas (National Parks, reserves).

The ecosystems of the Gobi are unique biological resources of the planet, calling for the organization of rational land use based on the assessment of their modern state.

References

Beresneva IA, Rachkovskaya EI (1978) Factors of zonality in the south part of Mongolia. Probl Desert Dev 1:19–29

Ecosystems of Mongolia (1995) Map of 1:1,000,000 scale, 15 pp and Ecosystems of Mongolia: distribution and modern state. Nauka, Moscow, 218 pp

Evstifeev YuG (1980) Extremely arid soils of Gobi. Probl Desert Dev 2:20–30

Hilbig W (1990) Erforschung Biologischer Ressourcen der Mongolischen Volksrepublik. Martin–Luther–Universitat, Halle–Wittenberg, Wissenschaftliche Beiträge, Halle, Saale, 145 pp

Rachkovskaya EI (1977) Extremely arid types of deserts in Central Asia. In: Problems of ecology, geobotany, botanical geography and floristics. Nauka, Leningrad, pp 99–109

Rachkovskaya EI (1993) Vegetation of Mongolian Gobi deserts. Nauka, Sankt–Petersburg, 132 pp

Rachkovskaya EI, Volkova EA (1977) Vegetation of Trans–Altai Gobi. In: Vegetation and animal world of Mongolia. Nauka, Leningrad, pp 46–74

Volkova EA (1997) System of zonal-sector distribution of vegetation on Eurasian continent. Bot J 8:18–34

Assessment of the Modern State of Sand–Desert Vegetation in Kazakhstan

Gulbarshin K. Bizhanova

Keywords. Anthropogenic transformation, degradation, mapping, vegetation cover, vegetation dynamics, pasture use

Abstract. This chapter notes that the main leading factors disturbing the vegetation cover of Kazakhstan's sand deserts cause different degrees of anthropogenic transformation. Methods for mapping anthropogenic successions and forecasting vegetation dynamics are discussed. A series of maps has been made to reflect the modern state and extent of anthropogenic transformation of the psammophytic vegetation: maps of modern vegetation, potential vegetation, anthropogenic transformation of vegetation, pastures, intensity of pasture use, desertification, and anthropogenic dynamics of vegetation.

There are more than 50 sand massifs in Kazakhstan (Kurochkina 1978). Nowadays, in connection with increasing human activity, damaging processes occur in the vegetation communities of desert areas. The leading factors of anthropogenic disturbance of vegetation cover in deserts are agricultural land use (grazing, clearing of shrubs and undershrubs, hay making) and technogenic influences (mining, construction activity, transport). These anthropogenic factors cause a different degree of transformation of vegetation cover. To study the modern state and degree of anthropogenic transformation of psammophytic vegetation of Moiynkums, Lesser and Greater Barsuki, the following series of maps have been made: of modern (actual) vegetation, of potential vegetation, of anthropogenic transformation of vegetation, of pastures, of intensity of pasture use, of desertification, and of anthropogenic dynamics of vegetation (Bizhanova and Kurochkina 1989; Bizhanova 1998).

The map of present-day vegetation for this territory is made in different scales (1:25 000, 1:300 000, 1:500 000). The method developed for making the *Map of Vegetation of Kazakhstan and Middle Asia* (1995) was adapted. First, zonal and subzonal positions were taken into consideration; much attention was given to the determination of interrelationships between the distribution of plant communities and soil-ground edaphic conditions. The following factors, significant for the distribution of vegetation, were carefully considered: mechanical composition of soils (firmness of sand) and salinity and moisture conditions. A series of plant communities, aggregates and combinations of series were used as mapping units. A psammoseries combines communities of successional range, which are represented by stages of sequential successions caused by eolian processes on ecologically homogeneous territories. Aggregates of series combine series of different successional ranges and are represented by a regularly repeated alternation of certain sets of series connected with the forms of mesorelief. Combinations are used to show on the map the heterogeneity of vegetation cover composed of regularly repeated types of communities belonging to genetically different territories.

The map of present-day vegetation is the basic inventory map reflecting the ecological situation of the territory, first the pattern of vegetation distribution caused by peculiarities of relief and edaphic conditions. Soil differences in soil firmness (mellow, dusty-sandy, firm sands and others) correspond to relief elements such as tops, slopes of different expositions and depressions. Different edaphic variants of vegetation correspond to these soil differences: eupsammophytic vegetation on dusty-sandy hilly sands, hyperpsammophytic and hydropsammophytic on hilly-ridge mobile eolian sands, hemypsammophytic and pelipsammophytic on overgrown hummocky sands and rolling sand plain with firm sand soils. Eupsammophytic variants of vegetation predominate among others.

The map of potential vegetation was based on literature and cartographic sources as well as research material of the author on progressive successions of vegetation. The map shows mainly the distribution of the group of associations for the prevailing plant formations in the mapping area. Besides each subdivision of the legend the conventional climax vegetation has been determined and changes in vegetation cover to the present have been shown. It can be assumed that conventional climax vegetation for the sands is composed of species of the following genera *Haloxylon, Calligonum, Artemisia, Agropyron and Krascheninnikovia* (Cherepanov 1995).

The map of anthropogenic transformation of vegetation reflects the presence of active weed species, as indicators of pasture digression. The type of weed infestation is shown for each sand massif. The main types of weed infestation are *Artemisia arenaria* DC, *A. scoparia* Waldst. et Kit, *Leymus racemosus* (Lam.) Tzvel, *Stipagrostis pennata* (Trin.) et Winter, *Eremosparton aphyllum* (Pall.) Fisch et C.A. Mey, *Ceratocarpus utriculosus* Bluk. and others. Successional series and type of weed infestation are presented in the legend to the maps of anthropogenic transformation for each conventional climax type of vegetation. For example, the typical climax community with *Calligonum aphyllum* (Pall.) Guerke, *Agropyron fragile* (Roth) P. Candardy is transformed into a community with *C. aphyllum, Ammodendron bifolium* (Pall.) Yakovl., *Atraphaxis spinosa* L., *Artemisia tomentella* Trautv., *A. arenaria, A. fragile* and then into a community with *A. bifolium, A. spinosa, C. aphyllum, A. arenaria, A. tomentella*. Types of weed infestation are *Artemisia arenaria, Eremurus inderiensis* (Stev.) Regel.

The map of pastures shows the distribution of pasture types. By pasture type is meant the union of identical or close (by composition or fodder quality) to other associations or groups of associations. Ecological (type of habitat) and phytocoenotic approaches were used in determining the pasture types. Types of pastures have been determined for sand massifs; the characterization of their vegetation, habitats, seasons of development and mean productivity has been

Assessment of the Modern State of Sand-Desert Vegetation in Kazakhstan 283

made. Modifications of these pastures have been revealed according to infesting weed species. The basis of pastures of the sand massifs is made of shrub pastures and the more vulnerable *Agropyron fragile, Artemisia terrae-albae* Krasch., and *Krascheninnikovia ceratoides* (L.) Gueldenst. pastures.

Table 1. Legend to the map of anthropogenic dynamics of vegetation of Lesser Barsuki

Past (1900)	Present	Future (after 2000)	
		Strong pasture use	Improved conditions and conservation
1	2	3	4
Eupsammophytic and hemipsammophytic vegetation			
1. Calligonum aphyllum, Agropyron fragile, Atraphaxis spinosa, Krascheninnikovia ceratoides	*1.; 2. Ammodendron bifolium, Calligonum aphyllum, C.murex, Krascheninnikovia ceratoides (Artemisia arenaria)*	*1. Artemisia arenaria, Ammodendron bifolium* with *Eremosparton aphyllum, Leymus racemosus, Iris tenuifolia, Alhagi pseudoalhagi*	Under controlled grazing-restoration of *Calligonum aphyllum, Agropyron fragile, Krascheninnikovia ceratoides* pasures
2. Calligonum aphyllum, Artemisia terrae-albae	*3. Calligonum aphyllum, Artemisia tomentella, A.arenaria*	*2. Sparced Calligonum aphyllum (Artemisia arenaria, A.scoparia, Euphorbia sequieriana)*	If *Artemisia terrae-albae* are sown the sagebrush pasture are restored
Hyperpsammophytic and hydropsammophytic vegetation			
3.Artemisia tomentella, Agropyron fragile, Calligonum aphyllum, C.murex, Phragmites australis, Salix caspica, S.rosmarinifolia	*4. Artemisia tomentella (Eremosparton aphyllum, Leymus racemosus, Stipagrostis pennata), Salix caspica, S.rosmarinifolia*	*3. Artemisia tomentella, Eremosparton aphyllum, Leymus racemosus* on disturbed sands	If sands are fixed and sown with fodder plants the *Artemisia tomentella* pastures and *Agropyron fragile* pastures can be restored
	5. Artemisia arenaria (Euphorbia sequieriana, Asperula danilevskiana, Chondrilla ambiqua), Phragmites australis, Salix caspica, S.rosmarinifolia	*4. Calligonum murex, C.aphyllum, Artemisia arenaria* with *Eremosparton aphyllum, Stipagrostis pennata, Chondrilla ambiqua* on disturbed sands	*Agropyron fragile, Artemisia tomentella*

Table 1 continued.

Past (1900)	Present	Future (after 2000)	
		Hemipsammophytic vegetation	
4. *Artemisia terrae -albae, Krascheninnikovia ceratoides*	6. *Artemisia arenaria, A.terrae -albae, A.scoparia, Krascheninnikovia ceratoides*	5. *Artemisia arenaria, A.terrae -albae, A.scoparia, Koeleria glauca, Festuca sulcata, Agropyron fragile, Ceratocarpus utriculosus*	If pastures are sown with *Artemisia terrae-albae, Krascheninnikovia ceratoides, Agropyron fragile* and subject to controlled grazing, the pastures are restored
		Pelitopsammophytic vegetation	
5. *Agropyron fragile, Krascheninnikovia ceratoides*	7. *Agropyron fragile, Artemisia arenaria, A.terrae-albae, Krascheninnikovia ceratoides*	6. *Artemisia arenaria, Agropyron fragile, Ceratocarpus utriculosus, Krascheninnikovia ceratoides*	If grazing and hay making are controlled and pastures are sown with fodder plants, *Agropyron fragile, Artemisia terrae -albae, Krascheninnikovia ceratoides*, pastures are restored

Under long-term unsystematic grazing, the deflation and vegetation change affect more pastures of sand massifs. The map of intensity of pasture use reflects degree (weak, moderate, strong) and character of their use (grazing, hay making, cutting woody vegetation for fuel). The vegetation of sands according to degree of economic exploitation can be assigned to moderately used pastures, since state farms practice seasonal use of pastures.

Degradation of natural ecosystems leads to processes of desertification caused by overload of pastures by livestock. The degree of desertification (close to climax, slight, moderate, strong, very strong) of the psammophytic vegetation is shown on the map of desertification which was based on the basis of the map of modern vegetation. From this map we notice that the largest area is classified as a moderate degree of desertification of dwarf shrub vegetation; this is explained by multiseasonal grazing accompanied by the significant influence of drought.

A cartographic model of anthropogenic dynamics of vegetation was made for each sand massif. This model and the legend to it present temporal and spatial successions in the past, present and future as well as the assumed distribution of vegetation by the beginning of the this century (2000), nowadays, and prognosis of future changes under two variants of pasture use: under increasing anthropogenic pressure and under conditions of restoration and conservation. Additionally, the map shows factors of succession (aggravation of unregulated grazing, clearing of *Haloxylon* and shrubs etc.), factors of improvement (planting of *Haloxylon*, sowing fodder species etc.) and nature of influence (disturbing, regulating, restoring) (Fig. 1 and Table 1 give the example of the desert massif Lesser Barsuki).

These maps allow an assessment of the modern state of vegetation of Kazakhstan's desert sands and can be used for objective and prompt registration of ecological potential and biological resources of the territory.

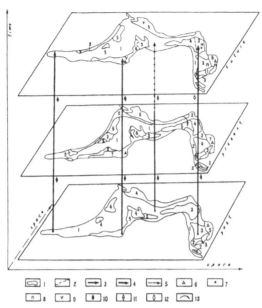

Fig. 1. Map of anthropogenic dynamics of vegetation of Lesser Barsuki. Boundaries: 1 contour; 2 prognosticated boundary. Succession factors: 3 assumed; 4 current; 5 finishing; 6 hay making; 7 conservation, protection in reserves. Factors of improvement: 8 sowing and planting of sand fixing plants; 9 sowing of fodder plants. Character of influence: 10 disturbing; 11 regulating; 12 restoring; 13 extrapolation

References

Bizhanova GK (1998) Anthropogenic transformation of vegetation of sand deserts in Kazakhstan. Dissertation for the degree of Doctor of Biological Sciences. Almaty, 56 pp (in Russian)

Bizhanova GK, Kurochkina LY (1989) Anthropogenic successions of Moiynkum pastures and their mapping. Nauka, Alma–Ata, 164 pp (in Russian)

Cherepanov SK (1995) Vascular plants of Russia and neighboring countries. Mir Semya, Saint Petersburg, 991 pp (in Russian)

Kurochkina LY (1978) Psammophytic vegetation of the deserts of Kazakhstan. Nauka, Alma-Ata, 271pp (in Russian)

Vegetation map of Kazakhstan and Middle Asia (within desert zone) (1995). Euro-Asia association, Moscow (three sheets)

Part V:

Reclamation

Water-Harvesting Efficiency in Arid and Semiarid Areas

Aaron Yair

Keywords. Runoff generation, rocky surfaces, soil covered surfaces, flow discontinuities, climatic change.

Abstract. Water-harvesting techniques were, and are, often applied in arid areas. Cultivated fields are irrigated by collecting overland flow from adjoining hillslopes. The purpose of the present chapter is to analyze water-harvesting efficiency in the northern Negev Desert of Israel, where annual rainfall varies from 280 mm in the north to 70 mm in the south. The hypothesis advanced is that runoff water-harvesting efficiency is primarily controlled by surface properties rather than by the absolute amounts of storm and annual rain amounts, the controlling factor for runoff generation being the extent of rocky versus soil -covered surfaces. Rocky surfaces respond quickly to rainfall, due to their low water absorption capacity. At the same time, soil covered surfaces, with a high porosity and high water absorption capacity, represent efficient sinks. As extensive rocky surfaces are more widespread in arid than in semiarid areas, where soil cover is more extensive, it appears that rocky arid areas are more suitable for runoff water-harvesting than climatically wetter semiarid areas. Hydrological data collected at two instrumented watersheds, located one in an arid rocky area, and the second in a semiarid soil-covered area, support the hypothesis. The implications of data obtained for runoff water-harvesting under changing climatic conditions are analyzed. A drier climatic regime can be expected to improve runoff water-harvesting perspectives in a semiarid area, while reducing such perspectives in rocky arid areas. An opposite trend for both areas is assumed during a transition to wetter climatic conditions.

Introduction

Ancient agricultural systems in arid areas are based on ingenious water-harvesting techniques. Water concentration, from a large contributing area into a limited collecting area, was the only way to supply the water needs of crops in areas where rainfall is limited in its amount and frequency and is highly variable both within the rainy season and from one year to the next. Such installations are known from deserts in the Middle East (Aharoni et al. 1960; Evenari et al. 1971; Kedar 1975; Shanan and Schick 1980) and North Africa. Cultivated fields occupied the flat valley bottoms irrigated by collecting overland flow from adjoining hill slopes. It is quite evident that successful and efficient water-harvesting practices rely highly on the frequency and magnitude of runoff generation. Both variables are closely controlled by rainfall properties and surface conditions prevailing in a given area.

Arid areas belong to the morphogenetic zone of the globe where physical weathering processes predominate over chemical processes. The resulting landscape is often rugged, with a limited soil cover. Hillslopes are usually subdivided into two distinct units. The upper slope sections are characterized by extensive bedrock outcrops and a patchy, shallow soil cover. The lower slope section has a contiguous soil cover composed of a mixture of gravels and finer particles. The fine-grained material is mainly of eolian origin. The extent of the rocky slope section increases with increasing aridity, whereas the extent of soil and vegetation cover increases towards the more humid areas. An interesting question is: what are the relative roles of rainfall and surface properties on the frequency and magnitude of surface runoff? Long-term hydrological monitoring, conducted at the Sede Boqer Experimental Watershed, located in the Negev Highlands (Fig. 1), clearly shows that runoff generation is faster on relatively impervious, bare, rocky surfaces than over soil-covered surfaces, with a high porosity and a high water-absorbing capacity (Yair 1992, 1999). A higher flow frequency and a higher magnitude of runoff should therefore be expected in arid areas where rocky areas prevail, than in more humid areas where soil cover is more extensive. On the other hand, it may well be that the fast response of rocky areas to rainfall can be offset in more humid environments by the higher frequency of rain events, the higher storm rain amounts, higher annual rainfall and smaller deviations from the long-term average. In other terms, we need to know which of these two main factors, rainfall properties or surface properties, plays the major role in runoff generation and potential water-harvesting efficiency.

Aim of Present Work

The purpose of the present chapter is to analyze runoff water-harvesting efficiency in the northern Negev area. Average annual rainfall decreases from 280 mm in the north to 70 mm in the south. The hypothesis advanced is that runoff water-harvesting efficiency is primarily controlled by surface properties rather than by the absolute amounts of storm and annual rainfall, the controlling factor for runoff generation being the extent of rocky versus soil-covered surfaces. As rocky surfaces are more widespread in the southern part of the area, whereas soil-covered surfaces are more extensive in its northern part, a higher water-harvesting efficiency can be expected in the former than in latter area. This is despite the fact that the northern area is climatically wetter than the southern.

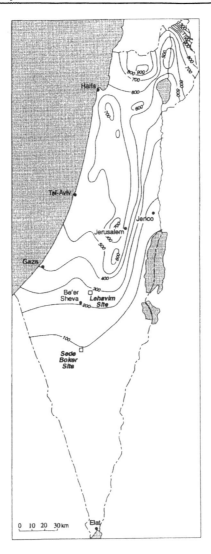

Fig. 1. Location map

Watershed Hydrology

Data presented here were collected in two instrumented first order drainage basins located in the northern Negev (Fig. 1). The Sede Boqer Experimental Watershed, located in the Negev Highlands, represents the arid environment. Average annual rainfall is 93 mm. The Lehavim Experimental Watershed, located north of the

Beersheva Depression, represents the semiarid environment. Average annual rainfall is 280 mm. The rainy season is limited to the winter months from September to May, and is usually shorter in the south. Potential annual evaporation is on the order of 2000–2600 mm (Rosenan and Gilad 1985).

The Sede Boqer Research Site

Site Description

The Sede Boqer instrumented watershed, carved in a limestone terrain, represents a first-order drainage basin that extends over an area of 2.1 ha. In terms of hydrological response units it is subdivided into five units (Fig. 2).

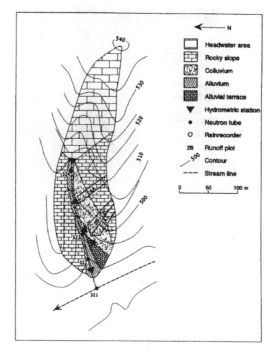

Fig. 2. The Sede Boqer instrumented watershed

1. The headwater area. This unit extends over 0.8 ha. Very extensive rocky outcrops, almost devoid of any soil and vegetation cover, form most of the surface.
2. Rocky-colluvial valley side slopes. This unit extends over 0.81 ha with slopes ranging from 55 to 72 m in length. The upper part is made of massive limestone and displays a stepped topography with soil strips at the base of the slopes. Bare

rock outcrops form 60–80% of the surface. The massive limestone is underlain by thinly bedded and densely jointed limestone. Most of the latter formation is covered by a gravelly colluvial mantle, which thickens quickly downslope.

3. Rocky valley side slopes. These slopes are much shorter than the opposite ones. They are carved in the thinly bedded limestone and lack a colluvial mantle at their base. The rock weathers into gravels and cobbles. Soil cover is shallow and patchy.

4. Alluvial reach. The alluvial fill extends from the rocky headwater area down to the mouth of the valley. It is 50 cm thick and mainly composed of fine-grained material. It covers 0.2 ha.

5. Alluvial terrace. This low terrace separates part of the colluvial mantle from the present-day active channel.

Instrumentation

The experimental watershed is equipped with devices for automatic and simultaneous measurements of rainfall and runoff (Fig. 2). The location of hillslope and channel hydrometric stations is based on the geomorphic-hydrologic response units described above. Hillslope runoff is measured at two plots. Plot 4A drains a whole slope whose upper part is rocky and lower part is colluvial. Plot 2 is subdivided into three subplots. Subplot 2C drains the rocky slope section; plot 2B the colluvial slope section, and plot 2A the adjoining slope from top to bottom. This design allows us to study the specific hydrological response of colluvial and rocky surfaces to rainfall, as well as the response of a combined rocky-colluvial slope. Channel runoff is measured at two stations. The upper one is located at the transition from the rocky headwater area into the alluvial reach, the second close to the mouth of the drainage basin.

Results

Long-term hydrological and environmental data, collected at the Sede Boqer site, were published in several publications (Yair 1994, 1999; Yair and Danin 1980; Yair and Lavee 1985; Yair and Shachak 1987). Only data relevant to the problem of water-harvesting will be presented here. Available data indicate that the threshold level of daily rainfall necessary to generate runoff in the rocky areas is as low as 2–4 mm. This threshold is greater for the stony colluvial soils (4–8 mm) and even greater for the lower channel (6–15 mm). The characteristic hydrological response of hillslopes, whose upper part is rocky and lower part colluvial, is displayed in Fig. 3. Three sprinkling experiments were carried out on 3 consecutive days over the whole hillslope. Flow lines were mapped using different dies. For the first run, conducted under dry surface conditions, runoff commenced within the upper rocky slope section 4 min after the onset of rainfall, and was continuous thereafter until the end of the run. All runoff generated within the rocky area infiltrated into the soil on reaching the upper colluvial slope section. A similar trend was observed on the second day. It was only on the third run, performed under very wet conditions and a very high rain intensity of 60 mm h^{-1}, that an integrated flow developed, after 10 min, along the whole slope. Long-term runoff data recorded at the Sede Boqer site show that a continuous flow along a slope 60 m long can be expected to occur in only 5% of the rainstorms in the area (Yair and Lavee 1985).

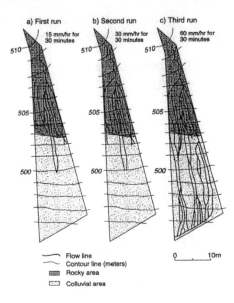

Fig. 3. Flow lines mapped during sprikling experiments

Similar discontinuities in the runoff process occur along the channel. Figure 4 presents the response of the upper and lower channel to an extreme rainfall event. Whereas runoff at the upper channel was quite continuous during the storm, it was quite discontinuous at the lower channel, indicating important losses into the alluvial fill. The phenomenon of flow discontinuity on passing from rocky into soil-covered surfaces (colluvium or alluvium) is due to the difference in their infiltration and absorbing capacity. This phenomenon is, however, greatly enhanced by the rainfall properties in the study area. An important characteristic is that most storms actually consist of several separate showers (Fig. 4), often of short duration and low rain intensity, each totaling a few millimetres. Consecutive rainshowers are separated by time intervals ranging from a few minutes to more than an hour, during which time the soil can drain and dry. The hydrological response to such rainstorms is that runoff events are brief, resulting in flow discontinuities upon reaching areas whose infiltration rate exceeds rain intensities.

Three distinct locations of flow discontinuities, and hence water concentration, were identified at the Sede Boqer site: the first the transition between the upper rocky and the lower colluvial slope, the second the transition from the rocky headwater area into the alluvial reach, and the third the confluence from the first order stream with the main channel.

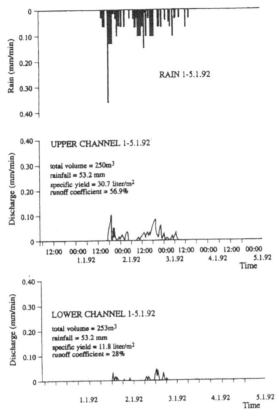

Fig. 4. Sede Boqer site. Storm of January 1992

Implications for Water-harvesting

The identification of the sites where runoff water concentration occurs under natural rainfall conditions led to the idea that the positive effect of water concentration can be further increased by constructing earth dams where trees could be planted. The earth dams were constructed at the following locations: (1) Small furrows (called minicatchments) were dug at the interface rocky-colluvial slope. The furrows drain an area of 200–250 m^2 and can collect a water volume of 1500–2000 l; one tree was planted in each minicatchment. (2) A small earthen dam was constructed across the channel, at a short distance downstream of the interface rocky-alluvial channel. A group of four to six trees can be planted behind the dam. (3) In a similar way, an earthen dam was constructed at the outlet of small watersheds, allowing for the plantation of seven to ten trees. Runoff water -harvesting efficiency at the above sites is presented in Table 1. The period considered (1982–1994) is characteristic of the high annual variability in rainfall and in runoff to be expected in an arid environment. Data obtained clearly show the important role of rocky surfaces as runoff-contributing areas. Minicatchments and upper channel dams collect substantial water volumes on wet and average years; lesser amounts are collected on dry years. When runoff volumes collected are expressed in millimetres (per unit area) the figures are extremely high. In an area where average annual rainfall is 93 mm, the average annual water input into the minicatchments is equivalent to 2000 mm, with 3000–5000 mm in the wettest years (1990,1991) and 200 mm in dry years (1992,1993). The latter figure is three times higher than the annual rainfall amount recorded. The lower channel dams collect substantial water amounts in wet years, but very small amounts in average years and no runoff input on dry years. Runoff input per unit area is always much smaller than in the minicatchments and upper channel dams.

Table 1. Sede Boqer Site, runoff water-harvesting efficiency at sites of flow discontinuity

| Rainfall (a) | Annual rainfall (mm) | Runoff volume collected (M3) | | |
| | | Location of water-harvesting site | | |
		Minicatchment	Upper channel	Lower channel
1982–83	136.0	6.75		
1983–84	72.0	1.45		
1984–85	79.0	1.40	No	No
1985–86	138.0	9.82		
1986–87	64.3	1.83	Data	Data
1987–88	96.6	3.03		
1988–89	95.6	1.38	10.25	2.27
1989–90	73.3	0.97	6.25	1.22
1990–91	139.4	12.90	74.07	63.27
1991–92	163.9	10.2	70.03	38.74
1992–93	66.8	0.42	0.46	0.00
1993–94	63.2	0.56	0.46	0.00
Average	99.0	4.26	26.92	10.55

The Lehavim Research Site

Site Description

The instrumented watershed, a first-order drainage basin, covers an area of 9 ha. Hillslopes, carved in thinly bedded chalky limestone, are 70–120 m long. Slopes are steep. The thinly bedded bedrock forms small rock steps and weathers into gravels and cobbles. The soil is a typical regolith, stony, with fine-grained material of eolian origin. Soil cover is more extensive than at the Sede Boqer Site and rocky surfaces more discontinuous. Bedrock outcrops are limited to small rocky steps, more common at the upper slope sections. A colluvial mantle covers the lower slope sections. The combination of a more extensive soil cover and a higher average annual rainfall results in a more extensive vegetal shrub cover than at Sede Boqer.

Instrumentation

Hillslope runoff was measured with plots that vary in their dimensions (Fig. 5).

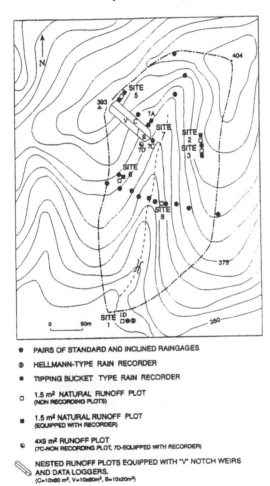

Fig. 5. The Lehavim research site (after Kossovsky 1994)

1. Small plots. They drain an area of 1.5 m². They were installed on opposite east- and west-facing slopes. They represent various microenvironments, such as vegetated plots, non vegetated plots, colluvial and rocky plots.
2. Medium plots. 4 x 8 m². These plots include vegetated and non-vegetated surfaces.
3. Large plots. 200–800 m². These are nested plots, constructed on the east-facing slope, where the soil and vegetal cover are less extensive than on the opposite slope. One subplot drains the lower colluvial slope. The second, above it, the rocky slope and the third covers the whole slope.
4. Channel runoff. Channel runoff was measured close to the watershed's outlet with a Parshall flume, equipped with a stage recorder.
5. Rainfall. Rainfall distribution was measured with small orifice rain-gage collectors and a Hellman–type rain recorder.

Results

Runoff data collection was limited to two rainy years (1990 and 1991). Data relevant to the purpose of this chapter are displayed in Figs. 6 and 7. Figure 6 presents runoff coefficients obtained at representative small, medium and a large plot.

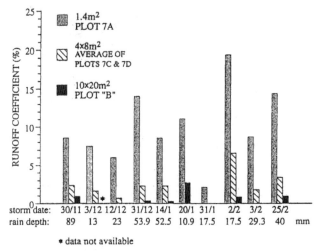

Fig. 6. Lehavim site: runoff coefficients in 1991–1992 (after Kossovsky 1994)

Runoff frequency and especially runoff magnitude decrease with increasing plot area. Losses are quite significant on passing from the small to the medium plot, indicating the importance of losses within very short distances of 4–8 m. Figure 7 displays an extreme rainstorm of 105 mm, with an estimated recurrence interval of 75–100 years. The storm occurred at the end of the rainy season and lasted for 36 h, with a high temporal variability in rain intensity. Peak rain intensities of 85–93 mm h^{-1} were recorded.

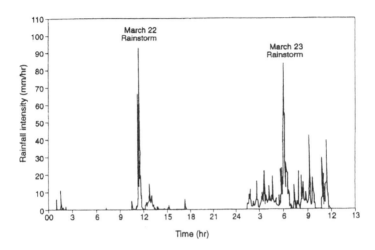

Fig. 7. Lehavim site, rain storm of march 1991 (after Kossovsky 1994)

However, during most of the storm, rain intensities were below 10–15 mm h^{-1}. Higher rain intensities lasted only a few minutes and were separated by time intervals varying in the range of 15 min to several hours. The response of the watershed to this extreme storm is presented in Table 2. As could be expected, the highest runoff coefficients were obtained at the small plots (20–37%). A sharp decrease in runoff coefficient was recorded on passing from small to medium -sized plots. Runoff coefficients for plots draining a slope length of 8 m were limited to 6–7%, and for a plot draining a slope length of 20 m only 1.5%. The most significant result is that no runoff was recorded at the channel hydrometric station. Data obtained are explained by the combination of the following factors:

1. Gravelly surfaces tend to increase infiltration by limiting surface-sealing process connected to the impact of raindrops on bare soils, devoid of gravels (McIntyre 1958; Grant and Struchtemeyer 1959; Jung 1960).
2. Shrubs, with gravel mounds at their upper part, represent efficient sinks that limit flow continuity.
3. Rain intensities in frontal rains are on the whole quite low. Intensities in excess of infiltration rates are short; and when they occur they usually last for a few minutes, a duration not long enough to allow a continuous flow.
4. Rainfall is highly intermittent. Each rainstorm consists of numerous separate rainshowers, of a short duration, each totaling a few millimetres. Under such conditions the concentration time (the time required for a continuous flow along a whole slope) is much longer than the duration of most, and probably all, effective rainshowers.

Discussion

The discussion will focus on the suitability of arid and semiarid areas for water-harvesting. It is mainly based on the hydrological response of the Lehavim and Sede Boqer watersheds to the extreme storm of March 1991. This storm covered the whole of the northern Negev desert. Storm rain amount and peak rain intensity, recorded at the Sede Boqer site, were approximately half those recorded at Lehavim (Table 2). Despite this important difference, runoff coefficients were much higher at Sede Boqer than at Lehavim. They were on the order of 50–65% for rocky areas; 25% for colluvial areas. Runoff was recorded at the lower channel, with a 37% runoff coefficient.

Table 2. Storm of March 1991, rainfall and runoff at Sede Boqer and Lehavim sites

Rainfall and runoff	Lehavim site	Sede Boqer Site
Rain amount (mm)	105.3	45.7
Peak rain intensity (mm h^{-1})	80–90	34.2
Runoff coefficients	Small plot (1.5 m^2): 20–37%	Rocky areas (200–8000 m^2): 51–65%
	Medium plot (4x8 m^2): 6–7%	Rocky-Colluvial Hillslope (360–1400 m^2): 25%
	Large plot (10X20 m^2): 1.5%	See above
	Channel Runoff (9 ha) : No Runoff	Lower Channel (2.1 ha): 37%

The basic difference between the two watersheds is connected with the different ratio of rocky versus soil-covered surfaces. Bare rock outcrops form 60–80% of the headwater area and upper part of slopes at Sede Boqer. These surfaces are continuous allowing, due to their very low infiltration and water absorption rate, for a high degree of flow continuity. An opposite situation prevails at Lehavim. Bare rocky outcrops are limited and patchy. Any runoff that develops over such surfaces is immediately absorbed by gravelly, and vegetated soil patches, separating two adjoining rocky surfaces. The soil-covered surfaces can therefore be regarded as efficient sinks, able to absorb any runoff generated upslope, even on extreme rain events. The extremely high infiltration losses within hillslopes at the Lehavim site are best indicated by the storm of March 1991, when no channel runoff had been recorded at the outlet of the watershed.

Conclusions

The main conclusion derived from data presented is that runoff water-harvesting efficiency is more affected and controlled by surface properties than by storm or annual rainfall amounts. Rocky arid areas, due to their very low water absorption capacity, low infiltration rate and low threshold for runoff generation, represent the most responsive surface units to rainfall. Such areas allow some water -harvesting even on bad rainfall years. On the other hand, climatically wetter semiarid areas, characterized by a more extensive soil and vegetation cover, are able to absorb most rainwater at most storms. Runoff frequency and magnitude are limited, thus preventing the positive effect of water concentration from a large contributing area into a small collecting area, without which sustainable agricultural practices would not be possible in a desert environment.

Hydrological data presented in this chapter bear some importance in view of an expected climatic change related to the global warming process. Following are some of the possible implications:

1. Semi-Arid Areas.
The transition to drier and warmer climate could have some positive effects in semiarid areas. Drier and warmer conditions reduce moisture available to support plants. The resulting reduction in vegetation density may lower infiltration rates thus increasing the frequency and magnitude of runoff and soil erosion. The stripping of the soil cover from the hillslopes would increase the extent of runoff -contributing bedrock areas. By uncovering the underlying bedrock, whose runoff efficiency is high, more water is made available for the adjoining soil-covered areas, permitting increased productivity. An opposite process is to be expected in the case of a wetter and cooler climate. An increase in soil moisture will increase the soil and vegetation cover, increase infiltration losses and limit the frequency and magnitude of runoff, thus reducing the positive effect of water concentration.
2. Arid Areas.
Trends opposite to those described for semiarid areas are to be expected in arid areas. Under drier and warmer conditions, a decrease in the frequency of runoff will take place, thus reducing runoff water collection. Analysis of the areal distribution of some ancient agricultural systems in the Negev Desert (Yair 1983) shows that such installations are limited to rocky areas where average annual rainfall exceeds 70 mm. With annual rainfall below 70 mm, sustainable agriculture becomes problematic, practically impossible, mainly due to the uncertainty connected with the increase in the deviation from the average annual rainfall, with a parallel increase in the number of consecutive dry years. On the other hand, a climatic transition to wetter and cooler conditions, with a lower variability in annual rainfall, would be most beneficial to rocky arid areas. An increase in both the frequency and magnitude of runoff is to be expected, providing more water to cultivated fields.

Acknowledgements. The study conducted at the Sede Boqer research site was supported by the Arid Ecosystems Research Centre of the Hebrew University. I am grateful to Mr. E. Sachs for his help in data collection and processing. The study at the Lehavim site was supported by a research grant of the Binational USA–Israel Agricultural Research Program, in the frame of an MSc thesis by Mr. A. Kossovsky. Thanks are also due to Mrs. M. Kid.ron of the Department of Geography for drawing the illustrations.

References

Aharoni Y, Evenari M, Shanan L, Tadmor NH (1960) The ancient desert agriculture of the Negev. An Israelite settlement at Ramat Matred. Isr Explor J 10:23–36

Evenari M, Shanan L, Tadmor NH (1971) The Negev. The challenge of a desert. Harvard University Press, Cambridge, Massachusetts, 345 pp

Grant JW, Struchtemeyer RA (1959) Influence of a coarse fraction in two main potato soils on infiltration, runoff and erosion. Soil Sci Am Proc 23:391–394

Jung L (1960). The influence of the stone cover on runoff and erosion on slate soil. IAHS Publ. 53:143–153

Kedar Y (1975) Water and Soil from the Negev: some ancient achievements in the central Negev. Geogr J 123:179–187

Kossovsky A (1994) Generation of runoff in first order drainage basins in a semiarid region, Lahav Hills, Negev, Israel. Msc Thesis, The Hebrew University, Institute of Earth Sciences, Jerusalem, 124 pp, (in Hebrew)

McIntyre TA (1958) Permeability measurements of soil crusts formed by raindrop impact. Soil Sci 85:185–189

Rosenan N, Gilad M (1985) Atlas of Israel, meteorological data, sheet IV/2, Carta, Jerusalem

Shanan L, Schick AP (1980) A hydrological model for the Negev Desert highlands: effects of infiltration, runoff and ancient agriculture. Hydrol Sci Bull 25 (3):269–282

Yair A (1983) Hillslope hydrology, water-harvesting and areal distribution of some ancient agricultural systems in the northern Negev desert. J Arid Environ 6:283–301

Yair A (1992) The control of headwater area on channel runoff in a small arid watershed. In: Parsons T, Abrahams A (eds) Overland flow, chap 3. University College Press, London, pp 53–68

Yair A (1994) The ambiguous impact of climate change at a desert fringe: Northern Negev, Israel. In: Millington AC, Pye K (eds) Environmental change in drylands: biogeographical and geomorphological perspectives, chap 11. John Wiley, New York, pp 199–227

Yair A (1999) Spatial variability in the runoff generated in small arid watersheds: Implications for water-harvesting. In: Hoekstra TM, Shachak M (eds) Arid lands management toward ecological sustainability, chap 14. University of Illinois Press, Chicago, pp 212–222

Yair A, Danin A (1980) Spatial variations in vegetation as related to the soil moisture regime over an arid limestone hillside, northern Negev, Israel. Oecologia (Berl) 47:83–88

Yair A, Lavee H (1985) Runoff generation in arid and semiarid areas. In: Anderson MG, Burt TP (eds) Hydrological forecasting. John Wiley, New York, pp 183–220

Yair A, Shachak M (1987) Studies in watershed ecology of an arid area. In: Berkofsky L, Wurtele MG (eds) Progress in desert research, chap 10. Rowman & Littlefield, Totawa, New Jersey, pp 145–193

The Effect of Landscape Structure on Primary Productivity in Source–Sink Systems

Karin Nadrowski and Gottfried Jetschke

Keywords. Shrubland, patchiness, runoff, productivity, simulation model, sink limitation

Abstract. The objective of this study was to understand how landscape structure influences primary productivity in arid shrublands. We propose that analyzing spatial arrangement of shrub patches can help to restore degraded landscapes by indicating limiting factors to landscape productivity.

Based on field data from a shrub land in the Northern Negev Desert, Israel, a computer model was developed to simulate the process of surface runoff generation and redistribution by two functional units: shrub patches acting as sinks and a matrix of crusted soil acting as source. Response variables are landscape productivity and water leakage. The model shows two different kinds of ecosystem behaviour: one where water availability limits productivity and one where productivity is not affected by water availability. Spatial arrangement of patches affects landscape productivity only in a water-limited system. We propose that analysis of spatial arrangement of shrub patches in arid shrublands can indicate whether the system is limited by water availability.

Introduction

An understanding of the limiting factors for productivity is important for the restoration of desertified ecosystems. Arid shrublands can be the result of limited water availability (Shmida 1985), but they can also be the result of human overexploitation (Orshan 1986; Schlesinger *et al.* 1990, 1996; Graetz 1991). We propose that the analysis of landscape structure may indicate whether a system is limited by water availability, because the spatial arrangement of shrub patches in these systems should be naturally optimized to intercept as much surface runoff water as possible.

Arid shrublands have a conspicuous landscape structure, consisting of "fertile islands" around shrubs, interspersed in a matrix of bare or crusted soil with only a few shorter plants (Noy–Meir 1985). There are functional source–sink relationships between these two contrasting landscape units: crusted area are sources of surface runoff water, whereas shrub patches act as sinks for it (Garcia –Moya and McKell 1970; Shachak and Pickett 1997).

Landscape structure has a major influence on the redistribution of resources within the system (Cerda 1997; Evenari *et al.* 1982) and may hence influence primary productivity. If there are too few sinks to absorb the resources generated, the system will leak resources. Shachak (Shachak *et al.* 1998) calls these ecosystems sink limited.

Not only the amount of shrub patches, but also their arrangement within the system influences the redistribution of resources. The aim of this study is to look at the response of a functional source–sink system to the spatial arrangement in water- and sink-limited ecosystems. This was accomplished by incorporating data on biomass production and landscape structure from a shrub land into a model simulating a source–sink system.

Material and Methods

The Research Site

Data for model calibration come from Sayeret Shaked Park in the Northern Negev, Israel. The region receives an average annual precipitation of 200 mm (Shachak et al. 1998). In this area, as in most of the Negev, most of the vegetation production is determined by herbaceous plants. The landscape mosaic is characterized by a matrix of soil crust covered with a microphytic community, and small patches of dwarf bushes, mainly *Noaea mucronata, Atractylis comosa* and *Thymelaea hirsuta*. Patch sizes range from 0.005 to 3 m^2 (Shachak et al. 1998). The microphytic matrix is composed of cyanobacteria, bacteria, algae, bryophytes and lichens.

Model Description

The simulation model was developed within the RAMSES modelling environment (Fischlin 1998). The model simulates rain falling on a patchy landscape followed by the successive generation and redistribution of surface runoff water. It calculates water leaking from and biomass produced within the landscape.

Patches and matrix have differing capacities to absorb water: patches 250 mm, matrix 40 mm. Depending on the balance between the input of precipitation and the output of water being taken by patches or matrix, water does or does not remain for surface runoff generation.

To be able to quantify spatial arrangement of patches along only one axis, model landscapes were constructed based on data from a real landscape, a 5 x 10 m plot (see Fig. 1a) arbitrarily chosen from a larger map drawn during fieldwork. While working with the same patch cover of 27%, the number of patches was reduced from 103 to 98, to arrange them regularly within the landscape (Fig. 1c). All patches were given the same circular form and the same size.

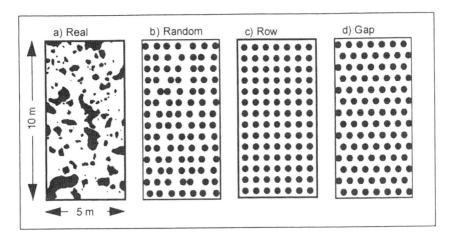

Fig. 1. a The original map and **b–d** the resulting model landscapes. Patch cover in all maps is 27%, patch number was slightly reduced for the model landscapes. Only the model landscapes are used in the simulation runs, because the spatial arrangement can be quantified using only one parameter. **b** The landscape Random was used for simulation varying amounts of rainfall, while **c** the landscape Row and **d** Gap represent both extremes of the six maps used for varying spatial arrangement

Samples taken from the herbaceous vegetation during field work confirmed the notion that biomass production on the soil crust is much lower than within shrub patches (Noy–Meir 1985; Boeken and Shachak 1994; Shachak and Pickett 1997). Mean productivity per area of 18 shrub patches and adjoining matrix area was 17 g m^{-2} for the matrix and 167 g m^{-2} for the patches. Simple extrapolation leads to a landscape productivity of 56.7 g m^{-2} for the chosen plot.

Matrix productivity per area is assumed to be constant, while patch productivity per area increases linearly with the amount of water accumulated by the patch. It reaches a saturation point at the maximum storage capacity of water by the patch.

Results

Water Availability

Landscape productivity (Fig. 2) of the landscape Random (Fig. 1b) increases with rainfall until it reaches a saturation point at 180 mm precipitation with 56.4 g m^{-2}. Landscape water leakage (Fig. 2) increases with rainfall. The slope of the function increases up to 180 mm rainfall, it remains constant at 1 l m^{-2} per mm rainfall for values greater than 180 mm. Since this is exactly the water provided by one additional mm precipitation, it means that for precipitation values exceeding 180 mm, all additional water leaks from the landscape.

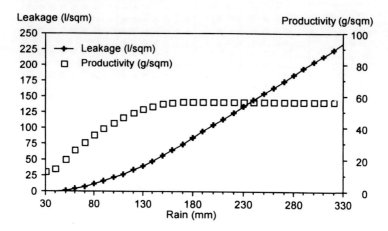

Fig. 2. Simulation results for landscape productivity and leakage as a function of increasing precipitation. The underlying landscape was the landscape Random (Fig. 1b)

Spatial Arrangement

In the case of 100-mm rainfall (Fig. 3a), biomass production increases and water leakage decreases with changing spatial arrangement from Row to Gap (cf. Fig. 1 c, d). At 300 mm rainfall (Fig. 3b), biomass and water leakage have much higher values than at 100 mm, but remain approximately constant for different landscape types.

Conclusion

There are two types of ecosystem behaviour, one where productivity is increased by additional water and one where all water added leaks from the system without affecting productivity. Systematic variation of patch spatial arrangement affects landscape productivity only in the water-limited type of system. Since productivity increases monotonously from row to gap position, this axis represents an axis of optimality for spatial arrangement. The landscape Gap (Fig. 1) is better suited to intercept resources redistributed by the process of surface runoff water generation than other arrangements.

Systematic variation of patch spatial arrangement in systems not limited by water does not affect landscape productivity. This is because the matrix area above each patch provides enough or more water to reach the capacity of the patch to store water. An optimal arrangement of patches is not needed in these kinds of systems. If the spatial arrangement of patches in an arid shrubland (with low rainfall) is random, the conclusion might be drawn that this ecosystem is not in equilibrium and still limited by the absence of sink units intercepting resource flow.

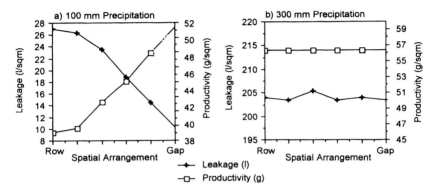

Fig. 3. a.,b. Simulation results for varying spatial arrangements at **a** 100 mm and **b** 300 mm precipitation. To vary spatial arrangement, six differing landscapes were constructed by moving patches along the x-axis of the landscape. Figure 1c and d show the two extremes of these landscapes

Discussion

Water availability is the main factor determining maximum vegetation cover and primary productivity in arid ecosystems (Walter and Breckle 1983; Shmida 1985). A patchy distribution of vegetation is the system's response to limited water availability by concentrating water from the matrix into the patches; but maximum cover and productivity is only reached in equilibrium systems.

Disturbance, for example by human overexploitation, can convert full cover to patchy vegetation (Schlesinger et al. 1990, 1996) and can further reduce patch cover (Shachak et al. 1998). These degraded systems may represent only transient states with unknown local stability and may be converted to higher cover systems by management action (Shachak et al. 1998). This is not the case in equilibrium systems. Discriminating water-limited equilibrium systems from disturbed transient systems will make management efforts more efficient.

The most optimal spatial arrangement to intercept runoff water is banded vegetation (Aguiar and Sala 1999). Since the shrubland observed does not show a banded vegetation pattern, this indicates that there are already other organizing factors for patch arrangement apart from surface runoff water; but this study shows that even not banded vegetation can show optimized spatial arrangement.

Optimized distribution of patches only affects the redistribution of water within a source–sink system if water is limited. It indicates that an optimized spatial arrangement may be the result of limiting factors in the establishment, growth or mortality phase of the patches, while a less optimized spatial arrangement may indicate that there are no limiting factors (or disturbances).

We suggest that analyzing spatial arrangement of patches in respect to the direction of surface runoff water flow can help to decide whether a system already has reached maximum cover of vegetation or not.

Acknowledgements. We thank Stefan Halle, Dirk Strauch and Karin Nadrowski (sen.) for reviewing the manuscript and the Savannization team of the Mitrani Group for Desert Ecology for providing research fascilities and fruitful discussions. Special thanks to Moshe Shachak, Bert Boeken and Yarden Oren for supervising the fieldwork. Financial support was provided by the German Academic Exchange Program and the Jewish National Fund.

References

Aguiar MR, Sala OE (1999) Patch structure, dynamics and implications for the functioning of arid ecosystems. Trends Ecol Evol 14:273–277

Boeken B, Shachak M (1994) Desert plant communities in human-made patches – implications for management. Ecol Appl 4:702–716

Cerda A (1997) The effect of patchy distribution of *Stipa tenacissima* L. on runoff and erosion. J Arid Environ 36:37–51

Evenari M, Shanan L, Tadmor N (1982) The Negev. The challenge of a desert. Harvard University Press, Cambridge, Massachusetts, 437 pp

Fischlin A (1998) Ramses. 2.2. ETH Zürich. http://www.ito.umnw.ethz.ch/SysEcol, Zürich

Garcia–Moya E, McKell CM (1970) Contribution of shrubs to the nitrogen economy of a desert-wash plant community. Ecology 51:81–88

Graetz RD (1991) Desertification: a tale of two feedbacks. In: Medina E, Schindler DW, Schulze E–D, Walker BH, Mooney HA (eds) Ecosystem experiments. John Wiley (SCOPE 45), New York, pp 59–88

Noy–Meir I (1985) Desert ecosystem structure and function. In: Evenari M, Noy–Meir I, Goodall DW (eds) Ecosystems of the world, vol 12A, Hot deserts and arid shrublands. Elsevier, Amsterdam, pp 93–103

Orshan G (1986) The deserts of the Middle East. In: Evenari M, Noy–Meir I, Goodall DW (eds) Ecosystems of the world, vol. 12B, Hot deserts and arid shrublands. Elsevier, Amsterdam, pp 1–28

Schlesinger WH, Reynolds JF, Cunningham GL, Huenneke LF, Jarrell WM, Virginia RA, Whitford WG (1990) Biological feedbacks in global desertification. Science 247:1043 –1048

Schlesinger WH, Raikes JA, Hartley AE, Cross AF (1996) On the spatial pattern of soil nutrients in desert ecosystems. Ecology 77:364–374

The Effects of Landscape Structure on Primary Productivity in Source-Sink Systems   309

Shachak M, Pickett STA (1997) Linking ecological understanding and application: Patchiness in a dryland system. In: Pickett STA, Ostfeld RS, Shachak M, Likens GF (eds) The ecological basis of conservation. Chapman and Hall, New York

Shachak M, Sachs M, Moshe I (1998) Ecosystem management of desertified shrubland in Israel. Ecosystems 1:475–483

Shmida A (1985) Biogeography of the desert flora. In: Evenari M, Noy–Meir I, Goodall DW (eds) Ecosystems of the world, vol. 12A, Hot deserts and arid shrublands. Elsevier , Amsterdam, pp 23–77

Walter H, Breckle S–W (1983) Ökologische Grundlagen aus globaler Sicht. Gustav Fischer, Stuttgart

Sedimentary Environments in the Desiccated Aral Sea Floor: Vegetation Recovery and Prospects for Reclamation

A. Yair

Introduction

Desert shrublands, all over the globe, are characterized by a patchy distribution of plants. Biomass and productivity are limited to small fertile sites. The distance between the patches, and the extent of the fertile patches, are controlled by the specific local edaphic conditions. In a non-saline environment the spatial distribution of the fertile spots is determined by spatial and temporal changes in the water regime, related to spatial differences in water input and water infiltration depth. Under a given climatic regime the differences in the water regime are basically controlled by surface properties. A compacted topsoil surface would limit infiltration depth and enhance runoff generation. Runoff generated at compacted areas is absorbed, on its way downslope, by riparian areas with high infiltration rates, where infiltration at a depth beyond that allowed by the direct rainfall occurs, creating local fertile spots. Patchiness increases tremendously in areas where, as is the case of the Aral region, differences in salinity and subsurface water drainage accompany the spatial non-uniform water regime.

From the said above it clearly appears that the detailed knowledge of the extremely intricate character of the coastal and marine microenvironments, uncovered by the desiccation of the sea floor, is required in any attempt to understand the recovery processes of the natural vegetation as well as the prospects for land reclamation. This knowledge covers the spatial variability in the composition of the sediments and the evolution of the shallow and saline groundwater following the retreat of the seashore.

Aim of Present Study

The aim of the present study is twofold:
- To describe and analyze the factors that control the evolution of the drainage conditions and soil properties in the desiccated area. As a result of this analysis several basic landscape units with specific edaphic conditions will be described and defined.

Sedimentary Environments in the Desiccated Aral Sea Floor 311

- To discuss possible ways of improvement of the water regime in some of the landscape units defined, and propose an approach for phytoamelioration experiments based on an understanding of the local edaphic conditions.

Sedimentary History of the Aral Sea

Sedimentary History Prior to Recent Desiccation Phase

The recent geological history of the Aral Sea was reviewed by Letolle and Mainguet (1993). The Aral depression developed during the late Pliocene, in response to regional tectonic activity. Quaternary sediments, deposited in the Aral Sea, vary in thickness from 20 to 140 m. Sediments deposited in the last 20,000 years have a thickness of 12–15 m. It is these young sediments that are being exposed at the surface, following the shrinking process of the Aral Sea, since 1960. The area of the Aral Sea was never constant. Aladin (1998) has shown the changes in the Aral Sea area that took place since the Holocene period. The changes in the extent of the Aral Sea are explained by several factors: Tectonic movements, that changed the course of the rivers feeding the Aral Sea and climatic fluctuations that affected the discharge of the rivers and evaporation rates from the water surface. Water input into the Aral Sea was also influenced by irrigation schemes, during periods of flourishing human activity, along the Amu- and Syrdarya rivers and their respective delta areas. Coastal and lacustrine sediments are found some 10 km east of the 1960 shoreline. Aladin (1998) identified eight distinct coastlines at different elevations. The highest shoreline is at an elevation of 72.73 m a.s.l. some twenty meters higher than the 1960 shoreline. The lowest is below the present coastline, at the elevation of 31 m. At each regression or transgression phase coastal dunes and shallow lagoons develop along the coast. This evolution leads to the development of parallel to sub-parallel belts of sandy ridges separated by lagoons. This structure is very well expressed in Figure 3 derived from a hydro-geological map of the north-eastern part of the Aral Sea, at a scale of 1:500,000 (Poriadin and Podolny, 1989). Sandy ridges, trending roughly N–S, are represented here by the light blue color. Whereas the coastal ridges are made of well-sorted sand with shell fragments; sediments in the lagoon depressions vary in their particle size composition from coarse sand to clay, with a predominance of the silty-clayey fractions. Sediments deposited in the lagoons, and on the sea bottom, are horizontal to subhorizontal, and well compacted. The thickness of the sedimentary layers deposited in the latter environments may vary greatly both in space and time, from few centimeters to meters, depending on the composition and amounts of sediment brought by the rivers, the distance from the shoreline, sea currents and local topography. An example of the highly variable sedimentary context is shown in Table 1 that describes the sedimentary sequence in a soil pit dug at the western part of the Bayan transect (Map reference: N 46 00 241; E 61 00 052)

312 A. Yair

Table 1. A characteristic sedimentary sequence in the dried out sea bed

Depth (in cm)	Description of sedimentary layer
0–40	Structureless dry fine sand
40–60	Horizontal sand, moist
60–70	Hydromorphic loam
70–85	Layered medium sand, reddish, moist
85–105	Hydromorphic loam, grey-blue
›105	Coarse sand, hydromorphic

Sedimentary Environments and Evolution of Surface Properties in Dried-Out Areas

The following factors and processes will affect surface evolution following sea recession.

Particle size composition

The development of the sea bottom material is largely determined by its texture. In sandy material infiltration rates are high and capillary water movement limited. Salt leaching will be fast and development of salt crusts inhibited. The depth to the saline groundwater will be determined by the thickness of the sand deposit. The thicker the sand the lower is groundwater level, eliminating the risk of surface salinization processes. Sandy deposits in the desiccated areas are characterized by a non-uniform topography. Old coastal dunes and offshore sand bars have the shape of linear ridges, roughly parallel to the coast line; while sand deposited in lagoons and in depressions below sea level will be horizontal, and well sorted. In addition to the coastal and marine sandy sediments, uncovered by sea recession, new sandy areas developed in some parts of the extensive desiccated areas. Windblown sand led to the formation of active barkhan fields. Needless to say the salinity of the young shifting dunes, unconnected to the saline groundwater, is much lower than that of sand deposited in marine or lagoon environments.

A different water regime governs fine-grained silty to clayey material. Due to the high water holding capacity of fine textured soils, coupled in saline environments with clay dispersion processes, these soils will always be deposited horizontally, or sub-horizontally, and highly compacted, limiting water infiltration and soil aeration. If groundwater is close to the surface water capillary rise is active, leading to the development of salt crusts. Such areas and their soils are known as "Solonchak".

Fine grained layers deposited by flooding in the delta areas, are less saline. In the absence of a shallow and saline groundwater, salt crusts will not develop, or will only form to a minor degree. Clay crusting will take place leading to the formation of flat "Takyr" surfaces. Due to clay dispersion and compaction infiltration is low. They are often covered with a shallow water layer at the end of the snowmelt season.

Stratigraphy and Structure

These terms cover two different aspects. The term stratigraphy refers to the vertical sequence of sedimentary units, deposited one on top of the other, as well as to the thickness of the sedimentary units. The structure refers to the depositional environment, whether it was a closed or an open one. Table 1 clearly shows that the number of possible stratigraphic combinations and sequences may be extremely high. Following is an analysis of three basic sequences:

Sandy Layer on Top of a Silty-Clayey Layer

Such a sequence is characterized by a high infiltration rate of the sandy topsoil layer, underlain by the compacted and impervious fine textured layer, often hydromorphic and highly saline. The effect of such a sequence on soil salinity and vegetation recovery will depend mainly on the thickness of the upper sandy layer. A sand layer some 40-cm is thick enough to improve salt leaching from the sand, while reducing capillary rise of salts from the underlying saline unit. As salt leaching in a sandy substratum is very fast tamarix trees, adapted to high groundwater salinity, get established quickly. Such a situation was clearly observed in the Chevchenko bay where *Tamarix* trees developed on top of parallel, low recessional sandy ridges (Colour Plate 13). An increase in sand thickness has three beneficial effects. Salinity is decreased, due to a greater salt leaching depth and fast lateral drainage of the saline marine waters towards the shrinking lake. More importantly, due to the combination of the two processes described above, the depth to the saline groundwater increases. Such a sequence allows for the development of a thick soil layer with reduced salinity. In addition, the deep infiltration of rainwater and snowmelt waters leads to the creation of a thin, perched, freshwater layer on top of the saline material. The establishment of species sensitive to salinity becomes possible. *Tamarix* trees are replaced here by Saxaul trees (*Haloxylon aphyllum*), a variety of shrub and annual species. A pit dug in a Saxaul forest (Colour Plate 14) shows a local groundwater level below one meter. Similar relatively good edaphic conditions characterize the new sand fields that developed on top of the desiccated sea bottom following seawater recession.

Silty-Clayey Layer on Top of a Sandy Substratum

Such a sequence represents the most unfavorable conditions for soil improvement following sea recession. A fine textured layer 40–50 cm thick is thick enough to prevent salt leaching. The dispersed and compacted material may be regarded as completely impervious, eliminating any possibility of deep-water infiltration and salt leaching. Due to the high water holding capacity of clays lateral drainage of the saline marine waters is negligible, and groundwater is at a shallow depth. The fine textured material remains saline and saturated at a depth of 20–30 cm (Colour

Plate 15). Drying of the top 20 cm leads to the development of "solonchak" soils with saline crusts at the surface (Colour Plate 16). The high salinity coupled with the anaerobic conditions inhibits the establishment of almost any vegetal cover. The prospects of soil amelioration by natural processes following the exposure of the clayey material is negligible, especially when the thickness of the fine textured layer exceeds 1 m, and saline groundwater is at a shallow depth.

The prospects for soil amelioration do however exist if the silty-clayey layer is thin, up to some 20 cm, and underlain by a thick sandy unit. The fast lateral and vertical drainage of the saline marine water in the sandy unit inhibits capillary rise, reducing thus the salt source for the fine textured surface layer. Upon drying the latter layer is prawn to deflation, loosing part of its salt content. This process is enhanced by salt leaching along the dense network of cracks caused by shrinking of the fine textured layer when dry. The processes described above lead to a gradual desalinization of the topsoil layer, and to the transformation of a surface with originally "Solonchak" properties into a surface with "Takyr" properties.

Complex Stratigraphic Sequences

The two cases analyzed above represent simple stratigraphic sequences. As shown in Table 1 the sedimentary reality is far more complicated with several sandy units that vary in their thickness; underlain each by a fine-grained unit whose thickness also varies. In general, the higher number and thickness of silty-clayey layers the lower are the prospects for soil amelioration. Each silty-clayey layer acts as a barrier for deep infiltration and lateral drainage of the saline marine waters, preserving thus the initial hydrophobic and saline conditions. In other terms a perched saline water table is preserved on each of the fine-grained units following the recession of the shoreline, and groundwater table remains close to the surface.

Spatial and Temporal Evolution of Sedimentation Processes

To the vertical variations in the sedimentary sequences one has to add the lateral variability in the sedimentary units. This variability is primarily controlled by the location of the shoreline. As coarse particles settle more quickly than fine-grained particles, an increase in the area and thickness of fine-grained deposits are to be expected from the old to the more recent sea shorelines. This trend is accompanied by a parallel increase in salinity, caused by evaporation and salt concentration in the diminishing volume of the shrinking sea. Initial salinity in 1960 was ~ 10 g/l. It is now four times higher (Letolle and Mainguet, 1993). The extent of "Solonchak" surfaces can therefore be expected to increase with decreasing surface area. The edaphic significance of this situation is that the chances for environmental improvement, by natural processes, decrease significantly from the older to the more recent shoreline, unless new sand deposits cover the inner exposed lake bottom.

However the spatial changes in particle size and salinity from the old to the present shoreline are far from being gradual. They are controlled by the non -uniform topography of the sea bottom, affected by tectonic activity, and by the sequence of recession and transgression phases outlined in paragraph 3.1. Each of these phases resulted in the development of sandy coastal ridges, sand bars and lagoon depressions. Once submerged the lagoons may be regarded as closed or semi-closed systems where fine-grained deposits and salt accumulate and preserved after sea recession. An example of such closed systems can be identified in the Aralsk map, at a scale of 1:50,000. Marshy areas, with perched groundwater levels are found at elevations higher than that of the present shoreline, pointing at the lack of vertical and lateral movement of the saline groundwater.

Vegetation Recovery

The succession dynamics starts with very few fast invading pioneer species, but after a few years a mosaic of various plant communities develops, according to diverse substrate conditions (see Wucherer et al., in this volume)

Prospects for Phytoreclamation

General Approach

The sedimentary history of the Aral Sea area, characterized by frequent fluctuations of sea level, resulted in a complicated mosaic of environmental and edaphic conditions. This initial puzzle was affected later on by sand deposits in some parts of the desiccated sea bottom. This evolution, coupled with the high sensitivity of desert plants to slight spatial and temporal changes in salinity and water regime, determines the guidelines for a realistic phytoreclamation management policy. The following principles are hereby advanced:

- Reclamation measures cannot be based on a single principle. They should be closely adapted to the local environmental conditions. In other terms the recovery of the natural vegetation, and the prospects for reclamation, are not only a function of time since sea recession, but rather a function of the local sedimentary sequence.
- The prospects for reclamation are highly conditioned by the particle size and thickness of the sedimentary units, which determine soil salinity and soil water regime.
- Sandy areas, characterized by a low salinity, good infiltration, good water preservation and low groundwater levels, represent the most favorable environment for the natural recovery of the vegetal cover as well as for phytoreclamation. At the other end of the spectrum we find the "Solonchak" surfaces characterized by a very high salinity, limited infiltration and shallow saline groundwater. These areas are devoid of any vegetation and represent the

main source of airborne dust and salt transported towards the cultivated lands. They represent the most problematic areas for reclamation, being in addition the main cause for the deterioration of adjoining arable lands. Nothing can better prove the difference in the reclamation potential of sandy and clayey solonchak soils than the experiment conducted by a team of Kazak scientists (Meirman 1999). According to this work plantation was conducted in a clayey solonchak area where soil started to desalinize. However, out of 17 indigenous halophyte species planted only three species (*Suaeda, Salsola* and *Micropeplis*) survived after one year. The experiment was more successful in an area with a sandy topsoil layer, overlying a clayey substratum. Six out of nine species planted survived here, with *Haloxylon, Salsola* and *Climacoptera* showing a good vitality.

Practical Recommendations

The results of the experiment presented above should serve as the basis for reclamation policy in the area. They lead to the conclusion that solonchak areas are very resilient to change in their edaphic conditions. Planting without changing the environmental conditions is not the appropriate way to improve the situation. Reclamation of such areas must therefore be based on the alteration of their hydrological regime, if not all over the surface, at least locally.

Sand deposition over arable lands is very often regarded as a major cause for land degradation and desertification. This approach is not relevant in the case of sand deposition over silty-clayey, saline, compacted and hydrophobic areas, such as solonchaks, underlain by shallow saline groundwater. Sand deposition over such surfaces has only a beneficial effect. As stated earlier, a topsoil sandy layer has five beneficial effects: it improves infiltration depth, lowers the level of the saline groundwater, increases salt leaching, reduces capillary rise and increases water storage by the creation of a perched freshwater lens on top of an impervious saline substratum. Any reclamation policy should therefor attempt to artificially increase sand cover on top of saline solonchak surfaces. Following are two suggestions for field experiments. Experiments proposed are based on the variety of sedimentary sequences that occur in the desiccated sea bottom described earlier. As the natural recovery of vegetation is facilitated and relatively rapid in sandy areas the recommendation for reclamation will focus on the saline problematic areas.

Silty-Clayey Layer Underlain by Thick Sand

A saline and compacted silty-clayey layer some 40-cm thick is thick enough to eliminate deep infiltration into the underlaying sand, preserving thus the properties of a solonchak. By removing the topsoil clayey layer the sand is uncovered, allowing for the beginning of a positive feedback in terms of salinity and water regime. The material removed may be used in order to build a sloping surface. As this material is impervious runoff may develop on the sloping surface; increasing water input into the area where the sand had been uncovered. This technique of

Sedimentary Environments in the Desiccated Aral Sea Floor 317

water harvesting is often applied in arid and semi arid areas. Concentrating water from a large impervious contributing area into a small collecting area, where infiltration is high, will increase locally infiltration depth. Water storage will be beyond that allowed by the direct annual precipitation, resulting in the creation of a "fertile island". These fertile sites may be planted with trees or shrubs, starting a positive feedback. Sand will accumulate around the shrubs, and the area submitted to deflation processes of dust and salt will be reduced.

Thick Silty-Clayey Layer with Shallow Saline Groundwater

In areas where the topsoil fine-textured layer is thick, beyond 1-m, and groundwater is close to the surface the application of the previous method is not realistic. In such conditions it is proposed to dig pits or trenches, 1.5 m deep and 1 m wide, and to fill them with sand from a nearby sandy area. In the absence of a deep sand layer groundwater will remain close to the surface preventing the plantation of many indigenous species. Such sites would be appropriate for the plantation of *Tamarix* trees adapted to high salinity and shallow groundwater level. This approach is based on field observations displayed in Figure 4 where *Tamarix* trees developed on a thin sandy bar overlying a clayey saline unit with groundwater at a depth of some 80-cm. The same method may be applied in areas with a high vertical variability in the sedimentary sequence such as that described in Table 1.

Acknowledgements. This research is funded by the German Federal Ministry for Education and Research (BMBF grant 0339714).

References

Aladin NV (1998) Some palaeolimnological reconstruction and history of the Aral Sea Basin and its catchment. In: Ecological Research and Monitoring of the Aral Sea deltas. Unesco, Paris, pp 3–12
Dimeyeva LA (1999) The ways of conservation and restoration of vegetal cover in the Aral Sea coast. In: Breckle S–W, Scheffer A, Veste M, Wucherer W (eds) Ecological problems of sustainable land-use in deserts. Bielefelder Ökologische Beiträge 15:13
Letolle R, Mainguet M (1993) Aral. Springer, France, 357 pp
Meirman G (1999) Phytoreclamation on the dry floor of the Aral Sea area. In: Breckle S –W, Scheffer A, Veste M, Wucherer W (eds) Ecological problems of sustainable land -use in deserts. Bielefelder Ökologische Beiträge 15:25
Poriadin VI, Podolny OV (1998) Hydrogeological map of the Aral sea area. National Centre for Geodesy and Cartography, Uzbekistan
Wucherer W, Breckle S–W (1999) Vegetation dynamics on the dry sea floor of the Aral Sea. In: Breckle S–W, Scheffer A, Veste M, Wucherer W (eds) Ecological problems of sustainable land-use in deserts. Bielefelder Ökologische Beiträge 15:41

Seeding Experiments on the Dry Aral Sea Floor for Phytomelioration

G.T. Meirman, L. Dimeyeva, K. Dzhamantykov, W. Wucherer and S.-W. Breckle

Keywords. Halophytes, desertification, soil formation, vegetation cover, salt regime, eolian processes

Abstract. The dry sea floor of the Aral Sea at present is a huge open salt flat. The enhancement of vegetation cover by means of phytomelioration is a realistic way for stabilization of the dry sea floor surface. This will support the natural processes of vegetative and generative propagation and the creation of seed banks for natural dissemination. Two experimental plots were established for identification of perspective plants for phytoreclamation of saline soils. Seeding experiments were started on two kinds of different ecological environments. Seeds of perennial and annual halophytes: *Halocnemum strobilaceum, Haloxylon aphyllum, Halostachys caspica, Climacoptera aralensis*, etc. were sowed. Experiment has shown that sandy soils have more favourable conditions for implementation of phytomeliorative measures than clay soils; the aridity of the first vegetation period plays a major role on the establishment of seedlings and saplings; species from local flora are more effective for phytoreclamation.

Introduction

The problems of the Aral Sea and the Aral Sea basin have no analogous features in size and intensity elsewhere (Glazovskii 1990; Walter and Breckle 1994; Letolle and Mainguet 1996; Klötzli 1997; Breckle et al. 1998). The centre of natural and anthropogenical influence, and thus of the Aral Sea crisis, is the territory of the dry sea floor.

In the first years after water retreatment, marsh lands are formed with shallow underground water table. This land is inhabited by *Salicornia europaea, Suaeda acuminata, S. crassifolia* and then by *Climacoptera* and *Petrosimonia* species and other annual salt-tolerant herbs. The process of intensive evaporation and increasing salinization can be observed everywhere. The further phase of salt crust formation is determined by the lithology of the sea floor deposits. A light lithology of surface horizons leads to the formation of sandy soils. Where heavy lithology and salinization processes prevail, nepkha landscapes are formed; 7–10 years after sea floor exposure and ongoing salinization or partial desalinization, the soils are inhabited by perennial halophytes, e.g. *Halocnemum strobilaceum, Halostachys caspica, Kalidium caspicum* and *Haloxylon aphyllum.* Observation of natural colonization on dry areas leads to the conclusion that phytomeliorative measures would be more rationally implemented where soil is undergoing the stage of primary soil desalinization. The process of desalinization on the dry sea floor can continue for about 7–10 years after drying. This period of soil formation is the most dangerous factor for soils of light texture. Once the saltcrust on the sandy upper stratum is blown off by winds, these areas serve as an arena for sand and salt storms. This enhances desalinization, but also affects the wider environment very severely. In contrast to light soil lithology, the processes of desalinization on heavy-textured substrates of the sea floor deposits are extremely slow. Phytomeliorative measures, again, can help to expedite and improve conditions of natural vegetation development and to replenish the poor seed bank of the sea floor substrate. The primary purpose of phytomelioration certainly must be to originate puddles on exposed bottom substrates with good diverse seed banks, to ameliorate shrubby vegetation and, in the long run, possibly to expedite the agricultural use of parts of the new territory.

Our exploration of a territory of the eastern part of the dry floor of the Aral Sea began from the primary sea shore to the Uzunkayr sandy barkhan region. That territory belongs to areas which fell dry in the 1960s–1970s, and in the 1980s were mostly covered with dwarf semishrubs, semishrubs, bushes and small tree plants: *Halocnemum strobilaceum, Kalidium caspicum, Halostachys caspica, Tamarix* –species and *Haloxylon aphyllum.*

Field Experiments

Our two field experiments were made with different kinds of halophytes from the local flora on typical saline soils close to the former Kaskakulan island in 1997 –1998.

Experimental plot no. 1 is located 18 km from the original coastline to the southwest of Kaskakulan island on a crust solonchak without vegetation. It is a flat plain in the desiccated sea bottom of the 1970s. The solonchak is surrounded by blown sands with halocnemum-saxaul vegetation (*Haloxylon aphyllum, Halocnemum strobilaceum, Atriplex fominii, Climacoptera aralensis*). The soil is saline. Salt content at the stratum is 0-2 cm 24.5% (Cl^-: 10.4%, SO_4^{2-}: 5.1%, $Na^+ + K^+$: 7.8%). Due to the strong evaporation, the capillary upwards movement of moisture brought large amounts of salt to the upper horizon of the soil. As a result of this, the effusion regime phenomenon, where the main parts of the salts are concentrated in an upper salty crust, is very prominent. The soil layer of between 2 and 50 cm depth is made up of medium loam with thin sublayers of packed clay. These layers are rather wet and the moisture content increases with depth. Apparently, the movement from underground water takes place. The groundwater table at the start of the year of the experiment was found at a depth of 3 m. The groundwater had a mineral content of 48 g l^{-1}. Seeds of 17 species of halophytes were sowed there in November, 1997 (Table 1).

Table 1. Halophyte phytomelioration on crusty saline clay soil of the Aral Sea dry floor

Plant / date	Quantity of plants on recorded areas			Plant height (cm)	
	22.04.98	01.06.98	23.08.98	01.06.98	23.08.98
Halogeton glomeratus	27	27	27	4.0	9.0
Suaeda acuminata	77	62	51	1.0	5.0
Climacoptera aralensis	35	32	11	1.0	2.0
Atriplex fominii	43	30	10	1.0	8.0
Atriplex tatarica	125	89	0	1.2	-
Salsola nitraria	23	23	0	0.5	-
Petrosimonia brachiata	39	37	0	1.5	-
Kalidium caspicum	14	6	0	1.0	-
Climacoptera lanata	17	16	0	1.3	-
Limonium gmelinii	-	-	0	0	-
Karelinia caspia	-	-	-	-	-
Pseudosophora alopecuroides	-	-	-	-	-
Halostachys caspica	-	-	-	-	-
Haloxylon aphyllum	-	-	-	-	-
Suaeda microphylla	-	-	-	-	-
Salsola australis	-	-	-	-	-
Halocnemum strobilaceum	-	-	-	-	-

Experimental plot no. 2 is located 11.6 km from the original coast on a puffic solonchak without vegetation. It is a flat plain in the desiccated sea bottom of the 1970s. The solonchak is surrounded by saxaul vegetation (*Haloxylon aphyllum, Salsola australis, Climacoptera aralensis, Halogeton glomeratus*). The profile of this sandy saline soil is typically layered with a prevalence of light layers: 27 cm sand and clay, then 137 cm light clay intermitted with sandy clay and in the lower strata 137–157 cm and 157–177 cm sand clay turns into light clay. The salt regime of the given soil is also effusional but with an apparent prevalence of sulphate salinization. This type of salinization corresponds to saline soils, where formation of salt crusts is less pronounced than in clay soils, though even there was a noticeable deposition of salts on the surface horizons: 0–2 cm, 2.63%; 2–7 cm, 2.92%; 7–17 cm, 1.46%. In the lower horizons of the soil profile from 17 to 137 cm, the quantity of salts decreases to about 0.43–0.19%, in deeper horizons the salt content sharply increases again. Special gravity throughout the profile section fluctuates within 2.73–3.90 g cm^{-3}. Seeds of nine species were sowed in November, 1997 (Table 2).

Experiment 1 took place on a soil characterized by properties and tendency towards heavy lithology. The soil of the plot for experiment 2 is a typical soil of light lithology. These two different soils were chosen with the purpose of creating different conditions for plants of the local flora. We wanted to trace their growth and development under the arid climate with long drought periods and under more remote stages of ecosystem formation. During the period of the experiments four expeditionary trips were organized with the purpose of observing the status of the seedlings and checking growth conditions. In three stages (April 23, June 1, August 23) calculations of plant growth and measurements of shoot length on

Seeding Experiments on the Dry Aral Sea Floor for Phytomelioration 321

specifically allotted test areas (15 x 50 cm) in each repetition were made. The Vitality of all seedlings was evaluated on September 10.

Table 2. Halophyte phytomelioration on saline sandy soil of Aral Sea dry sea floor

Plant / date	Quantity of plants in recorded areas			Plant height (cm)	
	22.04.98	01.06.98	23.08.98	01.06.98	23.08.98
Haloxylon aphyllum	21	20	17	5.4	7.0
Salsola nitraria	24	23	8	1.5	4.5
Petrosimonia brachiata	17	16	3	2.2	3.3
Climacoptera lanata	16	16	11	3.3	6.7
Climacoptera aralensis	65	64	41	4.7	8.3
Salsola australis	20	13	1	4.4	6.5
Kalidium caspicum	9	4	0	1.0	-
Suaeda microphylla	17	17	0	0.2	-
Halocnemum strobilaceum	0	0	0	-	-

First Results and Conclusions

The test results of possible cultivation of halophytes for experiment 1 showed that eight species (47%): *Limonium gmelini, Karelinia caspia, Pseudosophora alopecuroides, Halostachys caspica, Haloxylon aphyllum, Suaeda microphylla and Halocnemum strobilaceum* produced no shoots at all. The halophytes *Atriplex fominii, A.tatarica, Salsola nitraria, S.australis, Petrosimonia brachiata, Kalidium caspicum* and *Climacoptera lanata* produced shoots, but after about 40–50 days they died when the plants were about 0.50–1.58 cm (see Table 1). Thus, only four species (23.5%) were still alive by the end of the vegetation period: *Halogeton glomeratus, Suaeda acuminata, Climacoptera aralensis* and *Atriplex fominii*. By a five-point marking system, their vitality under these extraarid conditions was estimated at 3–4, 2, 2 and 2 points, respectively. In experiment 2 *Halocnemum strobilaceum* gave no shoots at all, and *Kalidium caspicum* and *Suaeda microphylla* died after June 1 at a negligible height (0.2–0.9 cm). Six species (67%) were preserved and developed more or less satisfactorily. An especially good perspective was exhibited by *Haloxylon aphyllum, Climacoptera aralensis* and *C. lanata* (see Table 2). From these species, about 60–80% of the shoots were preserved by the end of the vegetation period. The vitality of the seedlings at the end of the vegetative stage was as follows: *Haloxylon aphyllum* 4, *Climacoptera aralensis* 2, *C. lanata* 2, *Salsola nitraria* 2, *S. australis* 2 and *Petrosimonia brachiata* 1.

Thus, the experiments show that the environmental conditions of plot 1 were more problematic for phytoreclamation. Only a few species could grow and establish (23.5%). Germination of many seedlings in spring resulted from desalinization of the surface after the snow melt. Dying off of seedlings started in June, caused by salinization of the soil surface and blowing out of roots by strong winds. Ecological conditions in plot 2 were more favourable. Sand cover allowed not only germination but 67% development of seedlings through the vegetative stage. Most of the plants achieved good vitality, and annuals produced seeds.

The following conclusions can be made on the results of experiments 1 and 2 carried out in the area of Kaskakulan island:

- On sandy soils of the 1970s there are more favourable conditions for implementation of phytomeliorative measures than in clay soils.
- The aridity of the first vegetation period plays a major role in the establishment and thus the survival rate of seedlings and saplings.
- Species of local flora are more effective for phytoreclamation.
- On the dry sea floor of the Aral Sea the development of soils and formation processes is continuing; a final balance between herbaceous cover and shrubs has not been reached. Soil formation and vegetation changes are rapid.

Acknowledgements. This research is funded by the German Federal Ministry for Education and Research (BMBF grant 0339714).

References

Breckle S–W, Agachanjanz OE, Wucherer W (1998) Der Aralsee: Geoökologische Probleme. Naturwiss Rundsch 51:347–355

Giese E (1997) Die ökologische Krise der Aralseeregion. Geogr Rundsch 49:293–299

Glazovskii NF (1990) Aral'skii krizis. Prichiny vozniknoveniya i puti vykhoda. Moskva (in Russian)

Klötzli S (1997) Umweltzerstörung und Politik in Zentralasien – eine ökoregionale Systemuntersuchung. Eur Hochschulschriften, Reihe IV, Geographie, vol 17, 251pp

Letolle R, Mainguet M (1996) Der Aralsee: eine ökologische Katastrophe. Springer, Berlin Heidelberg New York, 517pp

Walter H, Breckle S–W (1994) Ökologie der Erde. Band 3: Spezielle Ökologie der gemäßigten und arktischen Zonen Euro–Nordasiens. 2. Aufl. by Breckle S–W, Agachanjanz OE. Fischer, Stuttgart, 726pp

Rehabilitation of Areas of Irrigation Now Derelicted Because of Strong Salinization in Ecologically Critical Zones of Priaralia

Alexsei Rau

Keywords. Toxic salts, leaching, rice

Abstract. The main purposes of the project are to study the chemical composition of soils in areas of irrigation which were withdrawn from agricultural production because of strong salinization, and to develop a technology for planting salt -tolerant cultures on such lands. The main tasks of this year: (1) to wash out strongly salinized soils in the areas which were withdrawn from agricultural rotation in Kazakhstan, Priaralia, up to the threshold level of salt toxicity in the 1 -m-layer of soil; (2) to plant rice, lucerne (alfalfa), melilot and saflor on the lands that were ponded in 1997 with a water norm of 9700 m^3 ha^{-1}.

An experimental site was established on the left bank of the Syrdarya River, within the Kzyl–Ordinskiy area of irrigation, where the area of lands withdrawn from agricultural rotation exceeds 30 000 ha.

Structure and Soil Texture on the Experimental Site

Before ponding, the strongly salinized soils of the experimental site were classified as solonchak-like soils, with weak salinity up to a depth of 80 cm and strongly salinized from a depth of 1 m. A groundwater table at a depth of 3–3.5 m showed the mineral content of the groundwater about 7.0 g l^{-1} and higher. Under ponding, the topsoil horizon became salinized as a result of a rise in the groundwater table and capillary water to the soil surface. After evaporation from the soil surface, salts were accumulated in the topsoil horizon, and the soils should now be classified as solonchaks.

Soil samples for chemical and mechanical analysis were obtained from the strongly salinized soils of barren lands in the Kzyl–Ordinskaya oblast. Within the soil profile, three horizons were defined with subhorizons A^1, A^2, B^1, B^2, B^3, C^1 and C^2.

A^1: puffic solonchak, depth 0–5 cm. On the soil surface a white-coloured salt crust, 1–1.5 cm depth with dry soil at the plant roots. Vegetation cover: scanty salt-resistant plants, occasionally *Halostachys* (Karabarak) bushes, light-grey -coloured.

A^2: wet soil, depth from 5–33 cm, loam, light-brown-coloured soil. Traces of decomposed roots. Structureless compact layer, homogeneous colour. Sharp transition to the B^1 layer.

B^1: dark-coloured; transit to B^2 by fibres. Inclusion of white carbonates. Depth 33–48 cm; moderate, heavy loams. Soil is wet, granular, colour is not uniform, with blue-grey tint.

B^2: depth 48–53 cm, light-brown-green-bluish-coloured, compact, wet, heavy loam and clay. Plant fragments are absent.

B^3: depth 53–77 cm, with sparse inclusions of carbonates, homogeneous colour, heavy loam. Definite transit to C^1 layer.

C^1: chocolate-coloured, compact, wet, heavy clay. Depth from 77 to 135 cm, colour is homogeneous. Transition to C^2 is definite.

C^2: green-bluish-coloured, wet, water releases after compression. Moderate loam to loamy sand. Depth of the layer 135–185 cm. Groundwater lies at a depth of 2.8 m.

The composition of soils of the strongly salinized barren lands is loam and clay that were found within all soil profiles. Within the soil profiles to a depth of 1.2 m heavy loam and heavy clay dominate (Table 1). Aquifer lies below 2.5 m, where groundwater is under pressure and rises up to 0.5 m.

Table 1. Mechanical composition of soils

Depth	W (%)	Fraction content in dry soil						Soils
		Sand			Silt		Clay	
		1–0.25	0.25 –0.05	0.05 –0.01	0.01 –0.005	0.005 –0.001	< 0.001	
0–5	2.2	1.90	47.17	13.29	9.33	14.40	13.91	Loam
5–10	2.6	17.46	39.83	17.45	6.08	6.78	12.40	Light loam
10–20	0.8	0.76	50.54	20.53	10.39	12.77	5.01	Light loam
20–40	2.0	1.84	22.98	28.16	16.49	14.37	16.16	Heavy loam
40–60	1.0	0.35	18.72	59.84	7.88	7.31	5.90	Light loam
60–80	2.2	0.97	12.73	40.12	14.32	17.63	14.23	Heavy loam
80–100	2.4	1.99	4.15	30.45	18.98	22.83	21.60	Heavy clay
100–120	2.0	1.10	12.04	22.98	21.18	21.27	21.45	Heavy clay
120–140	1.8	1.09	13.33	14.62	19.80	29.61	21.55	Heavy clay
140–160	1.0	2.75	29.05	13.82	10.30	19.68	24.40	Medium clay
160–180	3.2	16.04	42.60	9.13	3.39	10.29	18.55	Loam

W, hygroscopic water

The ground on the experimental site is layered in structure, and is very varied in composition, lithology and water-physical and chemical properties. This stratification changes the filterability of the soils in vertical and horizontal dimensions, and strongly influences the interdrain distances, which had not been considered in the course of constructing the irrigation systems.

In Table 2 data on porosity, volume and specific weight and water exchange in the soil horizons under leaching by 9700 m^3 ha^{-1}, are given. The lowest water capacity indicates a volume of soil water which remains after surface ponding and free seepage downward.

Table 2. Water-physical properties of soils and soil water exchange under washing, norm 970 l m^{-3}

Sl (cm)	Vw (gm^{-3})	Sw (gm^{-3})	P (%)	W (%)	H (%)	C (%)	D (%)	S (l m^{-3})	L (l m^{-3})	E
0–20	1.40	2.59	46	4.6	32.2	32.4	9.8	74.5	591.8	7.9
20–40	1.48	2.69	45	4.7	28.5	34.8	6.7	72.4	477.5	6.6
40–60	1.50	2.69	44	4.6	30.4	39.7	1.2	70.9	309.5	4.4
60–100	1.52	2.74	45	4.4	30.7	40.3	1.7	146.2	-	-
100–160	1.47	2.61	44	4.0	31.3	37.6	3.5	216.0	-	-

Sl, soil layer; *Vw*, Volume weight; *Sw*, specific weight; *P*, porosity; *W*, hygroscopic water; *H*, volume humidity before ponding; *C*, volume lowest water capacity; *D*, volume of soil water drainage; *S*, volume of salt-dissolved water; *L*, volume of water leached through the soil; *E*, water exchange in soil horizon

Volume weight was determined under field conditions by digging from the soil layer an intact soil sample with a volume of 50 cm^3. Soil was obtained from pits that had reached a groundwater layer. Samples were obtained from all soil horizons.

The use of the strongly salinized solonchak soils for cultivation will be possible only after removal of the highly soluble toxic salts from the topsoil horizon to make an allowable threshhold at which planting is possible.

In 1998 experiments on ponding of strongly salinized lands were carried out by using water norms of 4000, 8000 and 12 000 m^3 ha^{-1}.

Desalinization of soils begins from the topsoil horizon, and the soil salt content in the lower layers decreases along with an increase in the washing norm and with the volume of total seepage. 40-cm water layer (norm of washing water is 4000 m^3 ha^{-1}) leaches the 20-cm soil layer to the allowable threshold level of salt toxicity. Within the soil profile, an accumulation of salts in the lowest horizons of 1.2–1.6 m is observed (water norm being 4000 m^3 ha^{-1}).

At washing water norms of 8000 and 12 000 m^3 ha^{-1}, desalinization of the whole soil profile 0–1.6 m is observed, but the allowable salt toxicity level may be reached only after leaching with a norm of 12 000 $m^3 ha^{-1}$. The relation between the leaching norm and the depth of soil desalinization is shown in Fig. 1.

Leaching was carried out with a 7–10 cm water layer on the soil surface. The period of leaching was 2 days with a water volume of 4000 m^3 ha^{-1}, 3–4 days with 8000 m^3 ha^{-1}, and 5–6 days with 12 000 m^3 ha^{-1}.

Rice cultivation forms a groundwater pressure regime on the fields, and the degree of salt loss depends on the soil salt content and the rate of filtration. On well-drained lands, where the redistribution of salts over the soil profile exceeds their removal, filter water provokes overwatering, waterlogging and salinization of the areas in the depressions. On weakly drained lands, where redistribution of salts over the soil profile is less than their removal with filter water, a process of salt removal from the soil does not occur, but diffusive translocation of salts over the soil profile and their accumulation in the layer of heavy mechanical composition takes place. For soils in the rice-planting areas in Priaralia an optimum volume of filtration runoff from the ponded rice field is 4–8 mm day^{-1} (Fig. 2).

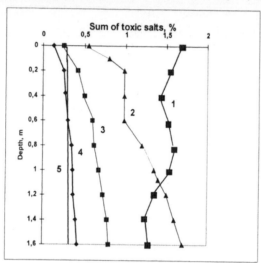

Fig. 1. Content of toxic salts in soil profile. *1* Before leaching; *2,3,4* after leaching with water volume 4000, 8000, 12000 m^3 ha^{-1}, correspondingly; *5* allowable level of salt toxicity for rice, lucerne, melilot planting

A high concentration of salts in the soil solution has a toxic effect on cultivated plants. Especially harmful are the salts composed of carbonate and chloride Na, which cause high osmotic pressure in the soil solution, so that water from the soil is not available to the root system of vegetation. On strongly salinized lands of Priaralia with a salt content in dry residual of more than 2% in the tillable layer, seeds of lucerne, melilot and saflor sowed in May did not sprout. Rice, after ponding with an 8–10 cm water layer, gives seedlings, but total yield depends on the soil salt content: at a soil salinity of 2.340% the rice harvest is 1.5 t ha^{-1}; at 1.857% 2 t ha^{-1}; at 1.567% 2.9 t ha^{-1} (Table 3).

Rice cultivation on salinized lands will be possible after removal of salt from the ploughed layer with filtration water. At a filtration rate of 10 mm day^{-1} in 10 days the salt content in the 0–30-cm layer of soil decreases by two to three times. This provides rice swelling and growth.

The relation between the rice harvest, mineral content of the groundwater and volume of seepage from the soil surface on the rice field is shown in Fig. 3.

A high salt content in soil and water decreases rice productivity (1) due to the seedlings being lost and emptiness of ears. Rice productivity (2) decreases at the rate of water seepage from the soil surface less than 4 mm day^{-1} and at a rate higher than 8 mm day^{-1}. At low water seepage, less than 4 mm day^{-1}, the decrease in rice productivity is determined by an insufficient washout of toxic salts from the topsoil horizon with downfiltration. At a high rate of water filtration from the soil surface (more than 8 mm day^{-1}), both harmful salts and plant nutrients are washed out.

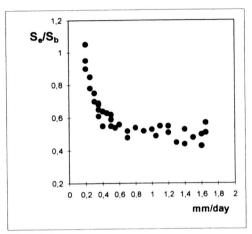

Fig. 2. Dependence of the coefficient of seasonal accumulation of salts on the volume of filtration from a ponded rice field

Table 3. Relation between the rice harvest and the soil salt content

Salt content in soil before rice seeding (%)	Including		Rice productivity, (t ha^{-1})	Salt content in soil after rice harvesting (%)
	Cl$^-$	HCO$_3$		
2.340	0.231	0.062	1.5	1.529
1.857	0.283	0.067	2.0	1.209
1.567	0.233	0.071	2.9	0.826

Conclusions

Soils of lands earlier irrigated, but currently derelict because of strong salinization, require leaching for reclamation.

In 1998 experiments on leaching of salinized lands were carried out, with water volumes of 4000, 8000 and 12 000 m^3 ha^{-1}. The water norm for leaching strongly salinized soils up to the admissible level of toxic salt threshold amounts was 12000 m^3 ha^{-1}. Desalinization of a 1-m layer of soil by a water norm of 8000 m^3 ha^{-1} does not attain an allowable level of salt toxicity.

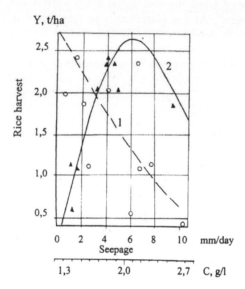

Fig. 3. Relation between the rice harvest, mineral content in the groundwater (1) and rate of seepage (2); C, mineral content in the groundwater

Rice cultivation, after leaching of strongly salinized lands, requires an optimal volume of filtration from a rice field 4–8 mm day^{-1}, which will prevent seasonal accumulation of salts.

With a decrease in water seepage from the soil surface of a rice field to a rate less than 4 mm day^{-1}, rice productivity declines because of insufficient salt washout from the root layer; with an increase of water seepage to more than 8 mm day^{-1} the rice productivity declines because of the washout of plant nutrients.

On strongly salinized lands which were leached in 1997 with a water volume of 9700 m^3 ha^{-1}, in the spring of 1998 experiments on rice, lucerne, melilot and saflor planting were carried out. Seeds of lucerne, melilot and saflor did not germinate. Rice sprouted after ponding with 8–10 cm water layer, but productivity depends on the degree of soil salinity: at a salinity level of 2.340% (in dry residuals), rice productivity amounts 1.5 t ha^{-1}; at 1.857% 2 t ha^{-1}; at 1.567% 2.9 t ha^{-1}.

During autumn 1997 and spring 1998 on the areas which were washed out in summer 1997 with a water norm of 9700 m^3 ha^{-1}, a translocation of salts from low layers upward was observed owing to capillar evaporation from the soil surface.

The amount of salts in the topsoil horizon for this period increased from 1.243% to 1.57–2.34% (in dry residuals). As a result of secondary salinization, the quality of the soil shifted from mid-salinized to salinized, and of topsoil horizons towards strongly salinized, so that the seeds of lucerne, melilot and saflor did not germinate in spring 1998. For successful plant swelling and for a decrease in secondary salinization of topsoils, the soils should be developed and planted early as 1 month after leaching.

Desert Soil Recultivation and Monitoring of (Phyto–) Toxicity (DEREMOTOX): Pilot Project in Three Phases Lasting for 4 Years

G.K. Hartmann, J.U. Kügler, P. Belouschek, L. Weissflog, K.H. Weiler, H.Ch. Heydecke, G. Reisinger, G.S. Golitsyn, I. Granberg, N.P. Elansky, E.B. Gabunshina, V.V. Alekseev, E. Putz, G. Pfister, A. Steiner

Abstract. DEREMOTOX is a research and development (R&D) pilot project for the development of "smarter" technological modules especially for and in small ecosystems and for optimizing the also necessary monitoring and risk assessments. Mainly it deals with the low-risk further development of so-called prototype modules or processes. "Smarter" in this context means: resource -efficient, i.e. with higher resource productivity, long-lived, labour-intensive, low -cost mass production of simple (user-friendly), basically needed (technical) modules, adaptable for various local situations. It is proposed to combine/apply at least the following methods in the selected area for recultivation, the Kalmykian steppe:

• The (new), successfully tested soil-conditioning process proposed in (Belouschek and Kügler 1992) under the acronym SOREC (soil sealing and recultivation), combined with so-called brush walls or Benjes Hedges (BENHEDGE), and complemented with: Solar cooking and solar water steriliziation at the recultivation location and with the extraction of freshwater from atmospheric water vapour in arid regions.
• Modified and complemented phyto-toxicological investigations (PTI) as done in the context of the EU research project ECCA (see chapter 2). The PTI part is subdivided into three parts: (1) Analysis of local pollution pattern and its effects. (2) Investigation of pollution transport and deposition mechanisms. (3) Concept for a future pollution control and protection from pollution and other external hazards, e.g. through a greenhouse which should be simultaneously tested.

Introduction

"The significance of knowledge in the value-adding process is today just as great as that of the three classical production factors: labour, land, and capital. In addition, knowledge in unattached form, for example in patents, processes and freely available knowledge, increases the productivity of the other production factors, principally that of work. This enables businesses to reduce their personnel requirements and increase their return on capital" (Miegel 1997). This statement, made for national states like Germany, needs to be supplemented and modified with respect to countries and areas which have not participated in the so-called first wave industrial development and to those industrialized countries with fast –increasing, especially anthropogenically caused environmental problems. First of all, these classic three production factors must be supplemented by water and air. For all three existentials: air, water and land we have to care for a quality that allows (human) life and which is the precondition of any human activity and/or productivity. In an increasing number of regions where human beings live – often over a longer time span – this precondition is no longer fulfilled or the relevant quality standard has begun to deteriorate. Here we cannot think about using knowledge and capital to reduce work. We have to use labour and knowledge to efficiently restore and/or newly create – with as little capital as possible – the necessary preconditions for direct survival and for an onset of future market productivity and trade activities. We have to strive for an optimization of technologies, products, and processes, including monitoring, used within and vital for socio-economic-ecologic-technologic, cultural determined systems (see Figs. 1a, b), briefly called smarter system technology. Smarter in this context means: resource-efficient, i.e. with higher resource productivity, long-lived, labour -intensive, low-cost mass production of simple (user-friendly), basically needed technical modules, adaptable for various local situations. It is mainly based on:

- A synergetic combination of hardware and software from the relevant, but complementary low-tech and high-tech areas.
- Thorough testing, evaluation, and evaluation of the results, mainly based upon non-linear biocybernetic thinking, which essentially means minimization of energy and matter fluxes, in the system.

The less this is based on a friendly, interdisciplinary, intergenerational, intranational- and international-intercultultural comprehensive teamwork, and on diversity, the less efficient and successful it will be. Such teamwork, based not only on competition but also on symbiosis, can help to reduce the socio-economic -ecologically induced threats of large migrations of the poor and also the danger of ethnic cleansing. This finally implies that the less the classic management tasks can be complemented by so-called catalytic functions – denoted by G.K. Hartmann in analogy to a catalyst in chemistry – the less successful these projects will be. [The great importance of such a catalytic function for the success of a complex, international (scientific technological) project has been demonstrated by G.K. Hartmann in his role as the Principal Investigator (PI) of the MAS (Millimeter wave Atmospheric Sounder) successfully flown with the three NASA ATLAS Space Shuttle missions in 1992, 1993 and 1994 see also Web Side http://www.linmpi.mpg.de/english/projekte/mas/]

Desert Soil Recultivation and Monitoring of (Phyto-) Toxicity: Pilot Project

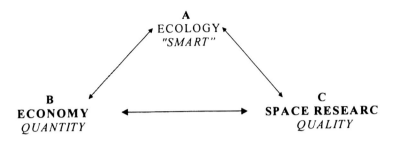

Fig. 1. a Modern production determining factors and production characteristics

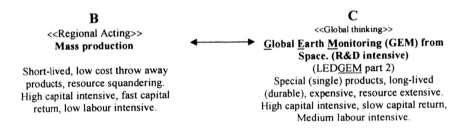

Fig. 1. b Example for Fig. 1. a; *LEDGEM: Local Ecovillage System Development and Global Environmental Monitoring (Hartmann 1998)

The presented DEREMOTOX pilot project proposal, a first step towards a later local eco-village system development (LED), emphasizes all these aspects. The envisaged activities for the Kalmykian steppe will be to a considerable extent complementary to those for countries with saturated markets, which are based on short-lived, low-cost mass productions. Here, we will first of all lay the foundations for low-cost mass production of basic technical goods with new smarter system technologies especially providing products with a longer life time. Excellent scientists and engineers capable of and motivated for the above -mentioned comprehensive teamwork – some with the additionally necessary catalytic abilities – will play a decisive role in approaching the goals of this project. The present DEREMOTOX team will meet this challenge very well. (Hartmann 1991; 1993; 1994; 1996a, b; 1997a, b; 1998; 2000a).

Remark About Science

Empirically based science contributes to a better way of seeing oneself in relation to the cosmos, complementary to the transcendence, enables technical and technological developments, and it is for the scientists a pretentious possibility of self-representation.

Remark About Resource Eefficiency

According to research carried out in 1994 in the USA (Allenby and Richards 1994) there was the following situation: 93% of the worked resources were never transformed into saleable products, 80% of all products were thrown away after first use, and 99% of the substances contained in the products became waste within 6 weeks. The basic problem lies in the fact that mass production for saturated markets depends on the sale of low-cost short-lived, i.e. throw-away products. Low costs result from mass production and they must be short-lived so that this mass production can continue.

DEREMOTOX, however, has a complementary aim, focusing on presently unsaturated or even barely existing markets – in order to create there a future viable living and market place. This also means in those locations an increase of personnel requirements, i.e. a decrease in the present unemployment.

The more "quick-dollar thinking"dominates the acting, the more we have to deal with:

- Less capital investment in the creation of new future market places.
- Less thorough synergetic combination of low-tech and high-tech hardware and software, a precondition for really significant technical innovations within the necessary R&D activities, not only for the sake of better competitive capabilities but also for the sake of a better symbiotic capabilities and thus for a more efficient and stable economic cooperation.
- Less striving for approaching a dynamic equilibrium between competition and symbiosis and for preserving diversity, an important precondition of evolutionary processes (Hartmann 2000a).

The Ecological Situtation in Large Areas in the West and Northwest of the Caspian Sea

This area with its steppes and semiarid regions constitutes a very complex ecosystem. In its Russian parts, precipitation between 200 and 260 mm per year

Desert Soil Recultivation and Monitoring of (Phyto-) Toxicity: Pilot Project 333

has been noted. High temperatures in summer, strong solar radiation, periodically occurring droughts and often hot and dry winds lead to an increase in ecological problems, especially for agricultural production. The extensive agricultural activities over many years in the past contributed very strongly to this pattern.

The worst situation exists at present in the Russian Republic of Kalmykia, in the southeast of the Russian European sector; 83% of its land, i.e. 7.6 million ha suffer from desertification. Per annum at present an increase of the desertification process of 40 000 to 50 000 ha has been observed in the Republic of Kalmykia and the region of Astrachan. It is anticipated by politicians and scientists that in 15 to 20 years this area will cover 1 million ha.

At the end of the 1950s the largest portion of this unique steppe was in fairly good condition and only 9% suffered from desertification. Today, we talk about more than 80%.

The following causes for the desertification have been investigated:

- overpasturing;
- oversalting;
- blown over (salted) sand from distant regions, e.g. from Middle Asia;
- occurrence of air-borne anthropogenic phytotoxic trichloroacetic acid (TCA);
- others to be determined.

Some Remarks About the Caspian Sea

The Caspian Sea lies east of the Caucasus Mountains and dominates the huge, flat expanses of western Central Asia. The elongated sea sprawls for nearly 1200 km from north to south, altough its average width is only 350 km. It covers an area of about 386 400 km^2 while its surface lies about 27 m below sea level. The maximum depth, towards the south, is 1025 m. The sea is bordered in the northeast by Kazakstan, in the southeast by Turkmenistan, in the south by Iran, in the southwest by Azerbaijan, and in the northwest by Russia. The Caspian Sea is the greatest salt lake in the world. The Caspian Sea is of exceptional scientific interest, because its history, particularly in respect to former fluctuations in both area and depth, offers clues to the complex geologic and climatic evolution of the region. Man-made changes, notably those resulting from the construction of dams, reservoirs and canals on the immense Volga River system (which drains into the Caspian Sea from the north), have had their own effect on the contemporary hydrological situation. The Caspian Sea is also of great importance in the transportation network of the region and in the production of oil and natural gas, and its sandy beaches are used increasingly for health and recreation resorts.

Taking into consideration the complex problems in the Kalmykian steppe and the Caspian region it becomes clear that a (linear) improvement of the presently used technologies alone will not be sufficient, i.e. we must consider the following technological activities if we want to significantly reduce the problems in this socio-economic-ecologic-system:

- improving the technology used at present;
- combining it with other not yet used, but available prototype like technologies;
- developing – with respect to hardware and software really new, essentially smarter technologies and systems, e.g. more resource-efficient, i.e. with higher resource productivity, and more user-friendly (simpler), needed for low cost mass production of basic technical modules. This means that, in the future, for our special case:

We will have to strive for a better balance between machine intelligence and machine labour and human intelligence and human labour (Hartmann 2000a) and we need to strive for a better synergetic combination of man and machine.

Remark

The pilot project LEDGEM, which is described in Hartmann 2000a (see also Fig. 1a, b), emphazises especially (2) and (3). It consists of two complementary parts, the combination of which needs, in addition to normal management, also human catalytic activities, the more the complexer the system gets and especially the more interdisciplinary, international and intercultural cooperation is needed. The DEREMOTOX concept is to a great extent derived from it.

Results of Biomonitoring

Through biomonitoring measurements at 13 different locations in Southern Russia within the EU project ECCA (Weißflog et al. 1999) the theoretically predicted occurrence of air-borne anthropogenic phytotoxic trichloroacetic acid (TCA) has been confirmed. The results are in agreement with the meteorological data of the Russian partners. The hypothesis claims that high volatile VCHs, (volatile chloro -hydrocarbons) coming from the Caspian Sea, are oxidized by atmospheric chemical processes to TCA and that TCA is deposited in the surrounding vegetation and leads there to phytotoxic effects. TCA and its derivates were used in the 1950s as herbicides for agricultural purposes and were banned in the 1960s thus that a direct anthropogenic input is not likely. It is assumed, however, that the measured TCA is produced in the atmosphere by hydroxylradically induced oxidation of C_2 chlorohydrocarbons. These substances are extensively used for cleaning purposes in the metal and textile industry and run with waste water from the industrial plants into the Rivers Wolga and Ural and into the Caspian Sea. However, also the oil industry contributes with its emissions to the total C_2 chlorohydrocarbon load of the Caspian Sea. Caused by strong evapouration processes, large amounts of VCHs compounds reach the atmosphere and are transported away from the Caspian Sea, e.g. towards the Kalmykian steppe.

It is therefore suggested that modified and complemented phyto-toxicological investigations (PTI) as done in the context of the ECCA, should be carried out here, too. The PTI part is subdivided into three parts:

1. Analysis of local pollution pattern and its effects.
2. Investigation of pollution transport and deposition mechanisms (Hartmann 2000b).
3. Concept for a future pollution control and protection from pollution and other external hazards, e.g. through a greenhouse which should be simultaneously tested.

* For references and more details see: Annual project report 1998 for the EU -Project: INCO-COPERNICUS No.: IC15 CT96-0106, Ecotoxicological Risk in the Caspian Catchment Area (ECCA). See also e-mail addresses at end of chapter.

Desert Soil Sealing and Recultivation (SOREC) and Erosion Protection

SOREC

A recultiviation of desert areas can only succeed if it is possible to create a sufficiently thick layer below the surface of the desert with soil temperatures of less than 20 °C being typical for some dm of root depth, to be made of the locally available soil plus suitable admixtures. Such a soil layer is characterized by the following properties:

- It must be capable of accumulating and retaining water and must be impermeable (water-tight) with respect to the deeper underground to prevent water losses in this direction.
- It must be rich in nutrients and suitable for plants to create roots to be held by the soil.

The basic idea concerning the solution for the recultivation of desert regions in order to prevent further desolation is that the above-mentioned, partly contradictory, specifications can be met by means of adding an extra soil layer under the surface of the present soil.

In a plantation and water-sealing layer with these features, the whole drainage can be done underground without significant evapouration processes in the root areas of the plants. Thus, water losses will be minimal. By means of an intermittent water circulation, a high salt content of the plant nutrient layer and the above protecting layer can be avoided. The nutrient and water-sealing layer must be designed according to the type of plantation process envisaged and the local soil and climatological conditions, which must be investigated in the planning phase A.

The soil-sealing or impermeabilization effect of water glass will play a major role in this re-cultivation process. Water glass is a melting product obtained from quartz sand and soda or quartz sand and potash. Its pH values range from 11 to 13 and, for a 1% solution, between 10 and 12. With decreasing pH value – 9 and smaller – the polymerization of silica acid species strongly increases and leads to insoluble polymeric silicate structures. The surface of these structures contains further so-called silanol groups that foster further condensation processes. Gels and precipitation products can also be produced by reaction of water glass solutions with metallic salt solutions, e.g. salts of alkaline earth metals etc. Furthermore, the gel formation will also be influenced by organic matter (Belouschek and Kügler 1992).

Erosion Protection

An erosion protection of the new recultivated area should be supplied, e.g. by creating so-called brush walls, also known as Benjes Hedges (Benjes 1998), from biomass that is normally used as firewood.

Extraction of Freshwater from Atmospheric Water Vapour

Many countries suffer from absence of freshwater, though its content in the atmosphere is significant. The use of natural processes allows large quantities of freshwater to be obtained without destroying the environment. Usually, atmospheric vapour content gives a possibility to extract 10–20 g of water m^{-3}. The use of atmospheric vapour has minimal effects on the environment, contrary to desalting sets. There are no exhausts and there is practically no energy supply needed. This resource must be reused. The condensate contains on 2–3 orders less toxic metals in comparison with sanitation norms. It contains practically no microorganisms and is well aerated. The cost of 1 m^3 of water may be 10–50 US cents, depending on the equipment characteristics and the atmospheric conditions. In Russia, until the 1880s, the town Feodosiya (in Krimea) obtained freshwater by this method. The water was supplied by gravity through potter pipes, which originated from artificial detritus heaps. During the heyday of the city – in the Middle Ages – there were up to 100 fountains, which provided water for 80 000 city dwellers. The system includes a heap condenser and well accumulator for the water. The purpose of the project is the creation of an experimental autonomous water-supply system, imitating the Feodosiyan-type condenser. It will have a volume of about 1/10 of the typical Feodosiyan condenser, i.e. approximately 200 m^3. The areas of the coasts of Cyprus, Greek and Israel, and the Caspian Sea, have a sufficient diurnal temperature course and sufficiently high absolute humidity. It is supposed that the system will give approximately 20 m^3 of freshwater per day. As a result of the project, the evaluation of system effectiveness for various seasons and various scales, and the check of theoretically calculated efficiency will be obtained (see also Shtmtnauer and Cereceda 1991; Alekseev and Beresin 1998). The estimated costs for the proposed experimental system are 200 000 –300 000 US $.

Solar Reflector Community Cooker and Water Sterilization

In order to avoid that the biomass needed for the brush walls is burned, e.g. for cooking purposes, a community solar cooker should be installed at the location where the recultivation takes place. It should also be modified/used for the sterilization of water perhaps also for solar distillation. For more information contact: K.H. Weiler, Solarprojekt Emden, Web Site: http://spot.fho -emden.de/hp/weiler/solar.html or http://spot.fho-emden.de/hp/weiler/links.htm, which is described in the following section.

More than 2 billion people worldwide have been using firewood for cooking. However, the northern industrialized countries, 20% of the earth's population, have been using too much of the planet's ecological capital, that is 85% of the global non-renewable fossile fuel consumption. Global climatic change, as well as deforestation or soil degradation, demand the development of renewable substitutes for the use of firewood, a sustainable treatment of the environment in both highly industrialized and poorly industrialized national states.

After some 100 000 refugees from Burma penetrated into the poor south of Bangladesh, and the rest of the former tropical rainforest were cut down, the new ecological training centre of Ukhia was planned by the Bangla German Sampreety (BGS), and with help from the Fachhochschule Ostfriesland (FHO), the Solar Project Emden, started at the beginning of 1995. This is a joint project between the two departments of Applied Natural Sciences and the Social Science at the FHO with co-author Prof. Walter Dissinger. After a mobile system (OSOW1) in Emden, the first solar reflector community kitchen in appropriate technology (OSOW2) was built in Dhaka at the end of that year and integrated into the Ukhia centre. A second one was carried to Tangail (OSOW3), in the north of Dhaka.

Now there are three different solar reflector kitchens in Emden, two in Bangladesh and another one in Nepal, where two reflectors were set into operation on Jan. 9, 1999 (OSOW6, OSOW7) in Itahari. There also biogas can be used alternatively in case of no or poor sunshine. OSOW4 is stationary at the Ecological Center Oekowerk, Emden. OSOW5 is at the FHO; its reflector carries 428 glass mirror facets, producing a very small focal spot at the entrance window of the oven. (There are now (1998) more than 150 solar reflector community kitchens (SRCC)).

An 8-m² parabolic reflector concentrates the sunlight on a small secondary reflector under a 50-l pot or a plate, thus producing up to 2400 W at the bottom of the pot. Automatic solar tracking by means of a pendulum clock-driven bicycle chain and gears keeps the focus on the solar oven. Thus, one can cook for about 80 people, which means a reduction of about 100 kg firewood day Operation after the same principle was proved from Emden (53° north) to the Equator (0°), provided the sun was shining.

The SRCC can also be used for water heating, sterilization, distillation and water vapour production. In some locations in Africa and India, the SRCCs use a storage tank during times of no cooking. Thus the water temperature is always high and the time needed for cooking can be reduced. In many cases the temperatures are above the boiling point so that sterilization is possible. Properly insulated tanks maintain the temperature even overnight. Furthermore, heat storage with stones or rocks is possible, i.e. a relevant device can be put into the focal area of the SRCC reflector and loaded with enough energy so that cooking is still possible after sun set. In India at the location of Mount Abu, Rajastan, a battery of SRCCs generates 600 kg of steam per day to feed a kitchen.

A Greenhouse for Comparison and Protection from Phytotoxicological Effects

It is also envisaged that parallel to the open air recultivation activities, a greenhouse with glas fibre structures should be installed to investigate what type and design have the greatest advantages in that area, especially compared to the open air agricultural activities, which very likely suffer from phytotoxicologic, atmogen deposits coming from the Caspian Sea and which need to be investigated, as described above.

Remarks on the Greenhouse

To produce vegetables, spices, herbs and flowers of high quality, and to produce these products independent of seasonal foil tunnels or greenhouses, is of advantage. Such protected production drastically reduces the cultivation risks (e.g. natural hazards, storm, heavy rain, hail, frost and anthropogenic ones like man-made phytotoxic air pollution) and the specific irrigation water demand. For any healthy plant production without or with only a minimum of pesticides, a sufficient technical standard of equipment is necessary, which furthermore needs to be adjusted to the local conditions. A low level of automation for irrigation and venting are important for success. Thus, the design of a resource-efficient, locally adjusted, full operational greenhouse, which uses primarily local available materials, requires some time, good know how and good teamwork (cooperation of all involved parties). The protecting structure can be used for rainwater collection which might be stored in a cistern or a pond. G. Reisinger[6] has designed and built such an experiemental greenhouse in Crete supported by the German Ministry of Research and Technology. It is also planned to build such an experimental greenhouse in the context of DEREMOTOX parallel to the outdoor recultivated and agriculturally used areas in order to be able to compare the cost efficiency of the two methods, outdoor versus indoor. The greenhouse might be covered with either glass or plastic foil.

A greenhouse constructed with Math-Web glassfibre elements has several advantages:

1. The material is of high strength but low weight. A Math-Web structure has only 25% of the weight of a comparable steel construction. This means that a greenhouse could be put together by persons with only average physical strength.

2. The Math-Web structures allow the construction and production of technical parts with exactly defined physical properties and high dimensional exactness and stability. Strength and load properties will be calculated absolutely precisely. Fastening points like bearings and bushings will be worked in during the production process, which makes the assembling with other parts easier. It is also possible to increase the strength of the product at specially loaded points.

3. Even after bending by overloading (wind pressure, high rainfall, snow) the Math-Web structure will return to its original shape and will be able to accept the full load again.

4. The structures need no maintenance or service. They are resistant to corrosion and chemical attack, for example by fertilizers.

5. A local production is possible without high investment. The production costs at present are comparable to steel and other materials. Math-Web structures could also be used for masts for communication purposes, wind power stations, as support or frames for solar energy systems, for small bridges and structures for houses and stables.

6. To work with Math-Web structures does not need expensive or sophisticated tools, and the low weight of the construction units leads to substantial cost savings by foundations: saving of material, time, wages and salaries and, again very important, no maintenance, e.g. painting, is needed.

Planning and Construction Phases

Phase A (first 12 months): Determination of the boundary conditions in a location to be selected in the Kalmykian steppe with respect to:
- political, legal, and infrastructural aspects,
- geographic and climatic aspects,
- available resources and local soil conditions,
- status of (phyto-) toxicity,
- rough cost estimates,
- fundraising.

Phase B (the following 6 months): Detailed end-to-end planning applying as far as possible the so-called MIPS concept (Lehner and Schmidt–Bleek 1999) and the

340 G. K. Hartmann et al.

priniciples of natural capitalism (Hawken et al. 1999). Assembling of the final DEREMOTOX team. Refined cost compilation for phase C and final fundraising. **Phase C** (last 30 months) construction and evaluation phase.

Remarks

- These methods are also required when we deal with preservation of still possible agricultural land use but which is endangered by erosion and pollution processes.
- Costs for phase A: to be determined
- MIPS: material input per service unit. Concept developed by F. Schmidt–Bleek. See Lehner and Schmidt–Bleek (1999).
- DEREMOTOX is a subproject of the already discussed pilot project Eco -Village System Development, which is Part 1 of the proposed pilot project LEDGEM (local eco-village system development and global environmental monitoring; G.K. Hartmann, 1998).
- G.K. Hartmann, Max–Planck–Institut für Aeronomie, Max–Planck Str. 2, D -37191 Katlenburg–Lindau, Germany, phone: +49 5556 979 336; fax: 240; e mail: ghartmann@linmpi.mpg.de

Conclusions

DEREMOTOX is a research and development (R&D) pilot project for the development of smarter technological modules especially for and in small ecosystems and for optimizing the also necessary monitoring and risk assessments. Mainly, it deals with low-risk further development of so-called prototype modules or processes. "Smarter" in this context means: resource -efficient, i.e. with higher resource productivity, long-lived, labour-intensive, low -cost mass production of simple user-friendly, basically needed technical modules, adaptable for various local situations. The envisaged DEREMOTOX activities for the Kalmykian steppe will be to a considerable extent complementary to those for countries with saturated markets, which are based on short-lived, low-cost mass productions. This also means in those locations an increased personnel requirement, i.e. a decrease in the present unemployment and the creation there of a future viable living and market place. The longer life time of the new products – mainly from already existing prototypes – will to a considerable extent be guaranteed by simple maintenance and/or service activities by man. Excellent scientists and engineers capable and motivated for a friendly, interdisciplinary, intergenerational, intranational- and international-intercultural comprehensive teamwork cooperation – some with the additionally necessary catalytic capabilities – will play a decisive role in approaching the goals of this project. The present DEREMOTOX team will very well meet these challenges and thus the risks for a succesful execution of this project are very small. Furthermore, the

Desert Soil Recultivation and Monitoring of (Phyto-) Toxicity: Pilot Project 341

chances to use these optimized technical modules in the context of future ecovillage system development are very promising and high; however, the aspects mentioned in the LEDGEM proposal (see Fig. 1a, b) should also be considered (Hartmann 2000a).

Acknowledgements. The first author, Gerd Hartmann, thanks the colleagues and friends from IEMA-UM in Mendoza Argentina, for the good cooperation and discussions of the IEMA project (proposal) Ecovilla, (Olivia et al. 1994), which supplied essential experiences for this proposed pilot study DEREMOTOX through his LEDGEM proposal, where Part 1 is concerned with a local ecovillage -system development (LED) and Part 2 with global environmental monitoring (GEM). He thanks in this context Dipl. Geogr. S. Engelmann for scientific technical documentation support and discussions on special geographic aspects of Part 1. He thanks Mr. G. Reisinger (Illertissen) and Mr. H. Ch. Heydecke (Math -Web International Inc., Eldingen) for their contributions for various possible greenhouse designs and constructions and the suggested test procedures. He finally thanks Prof. Dr. H. A. Fischer–Barnicol for detailed discussions on intercultural and ecumenic aspects of the LED concept. Finally the authors thank their institutions for supporting this proposal and the related activities.

An alphabetical list of producers, who contributed to the LEDGEM proposal, with new technologies, is obtainable from the authors.

References

Alekseev VV, Beresin MJ (1998) Freshwater from atmospheric vapour for arid regions. Renewable energy 3, pp 36–38

Allenby BR, Richards DJ (eds) (1994) The greening of the industrial ecosystem. National Academy of Engineering, Washington DC

Belouschek P, Kügler JU (1992) Wasserglasvergütete mineralische Dichtsysteme zur Abdeckung von Deponien und Altlasten unter dem Gesichtspunkt der aktiven Rißsicherung. In: Thome–Kozmiensky (ed), Abdichtung von Deponien und Altlasten, Verlag für Energie und Umwelttechnik, Berlin, pp 227 –247

Benjes H (1998) Die Vernetzung von Lebensräumen mit Benjeshecken (brush walls), 5th edn. Natur und Umwelt Verlag, Bonn

Hartmann GK (1992) International and intercultural cooperation. Text in English, German and Spanish in: Premio de Cooperación Cientifica Technologica International Dr. Luis Federico Leloir. (Edicion: SECYT, UM, MPAE). Further in: MPAE-L-66-91-24, 1991 and in: Universidad de Mendoza, Argentina (UM) UM 01-09-05-0669-0392

Hartmann GK (1993) MAS on ATLAS: an experience of "the in between" economy and ecology, MPAE-L-66-93-08, Max–Planck–Institut für Aeronomie, D-37191 Katlenburg -Lindau, Germany

Hartmann GK (1994) Responsibility with respect to fault, error and uncertainty occurring in the interfaces between man and its environment and man and machine, MPAE-L-66-94 -23, Max–Planck–Institut für Aeronomie, D–37191 Katlenburg–Lindau, Germany

Hartmann GK (1996a) Science responsibility and risk management: are we predetermining the results of research?, MPAE-L-015-96-17, see above

Hartmann GK (1996b) Data growth rate problems, In: Dieminger W, Hartmann GK, Leitinger R (eds) The upper atmosphere; data analysis and interpretation. Springer, Berlin Heidelberg New York, pp 956–993

Hartmann GK (1997a) More conserving utilization of our environment requires more claim of responsibility and more teamwork: a pleading for more cooperative learning and teaching, MPAE-L-015-97-03

Hartmann GK (1997b) Facts about data from the Earth's atmosphere, MPAE-L-015-97-24

Hartmann GK (1998) Space research between Russia and the USA. A chance for Europe?, MPAE-L-015-98-02

Hartmann GK (2000a) More caring utilization of the natural resources through interdisciplinary and intercultural scientific-technical cooperation. Paper for the Alexander von Humboldt Days, Concepción, Chile, July 14–16, 1999

Hartmann GK (2000b) The variabilities of H_2O fluxes in the Earth's atmosphere. Phys Chem Earth (C) 25, 3:189–194

Hawken P, Lovins A, Hunter Lovins L, (1999) Natural Capitalism. Creating the next industrial revolution; Little, Brown and Company, Boston New York London

Lehner F, Schmidt–Bleek F, (1999) Die Wachstumsmaschine. Der ökonomische Charme der Ökologie; Droemer Verlag, München

Miegel M (1997) The causes of unemployment in Germany and other countries from the first wave of industrialization. In: EXPO GmbH (ed) Zukunft der Arbeit, Nov 1997, pp 14–23

Oliva LA, Gelardi D, Esteves A, Meyer N, Martin F, Gantuz M (1994) Proyecto Ecovilla, report IEMA-UM, Ed. Idearium UM -02-03-05- 0696-0993, Universidad de Mendoza (UM), Mendoza, Argentina

Shtmtnauer RS, Cereceda P (1991) Fog-water collection in arid coastal location. AMBIO 20, 7:303–308

Weissflog L, Manz M, Popp P, Elansky N, Arabov A, Putz E, Schüürmann G (1999) Airborne Tricholoracetic acid and its deposition in the Catchement Area of the Caspian Sea. Environmental Pollution 104:359–364

Contributions to a Sustainable Management of the Indigenous Vegetation in the Foreland of Cele Oasis — A Project Report from the Taklamakan Desert

Michael Runge, Stefan Arndt, Helge Bruelheide, Andrea Foetzki, Dirk Gries, Jun Huang, Marianne Popp, Frank Thomas, Gang Wang and Ximing Zhang

Abstract. The ecological situation of oases at the southern border of the Taklamakan desert is shortly described, and the importance of a vegetation from indigenous species at the transition from the oases to the desert is emphasized. This vegetation serves as a shelter against sand drift and as a source of livestock feed as well as of fuel and construction material. Its destruction through overexploitation and other interventions during the last decades has considerably promoted sand drift and the deterioration of arable land. Therefore, a management of this protective vegetation is to be developed that leads to a sufficient regeneration and that ensures both its preservation and its use. A research project that is carried through jointly by Chinese and European scientists shall yield an ecological basis for this sustainable management.

Project Background

The Taklamakan Desert is the largest desert in China, covering about 337 000 km^2 of the westernmost Chinese province, the Xinjiang Uygur Autonomous Region (Yang 1991). The low precipitation in this desert is the result not only of its inner continental location but also, of its being enclosed by high mountain ranges that keep out rain-bearing clouds. These ranges are, above all, the Kun Lun Mountains in the south, maximally rising to 7712 m above sea level, and the Tien Shan in the north, where the highest peak extends to 7450 m. Together with the Pamir in the west, these ranges enclose the Tarim Basin, the largest part of which is occupied by the Taklamakan Desert (Walter and Box 1983).

Depending on the local climatic conditions, the mean annual precipitation varies between 31 and 51 mm and the annual potential evaporation between 1700 and 2800 mm (Xia et al. 1993). As a result of these very arid conditions, human settlements are restricted to the perimeter of the desert where rivers enter it from the adjoining mountains. The Taklamakan Desert is, thus, surrounded by an interrupted chain of river oases.

Irrigation of arable land in the Taklamakan oases involves the risk of soil salinization; but the oases at the southern border of the Taklamakan are particularly endangered by a further threat. Because of prevailing northwesterly winds and storms, these oases are exposed to severe sand drift and to a high risk that arable land and even settlements could be covered by blown sand (Coque et al. 1991; Xia et al. 1993; Zhang, this vol.). There is archaeological as well as historical evidence that people in this region of the old Silk Road have always been forced to cope with this threat (Xia et al. 1993), but during the past few decades, the situation has deteriorated even more. This is primarily the result of inappropriate use of the natural resources, produced by a rapid population growth. The Cele (Qira) Oasis can be considered as an example for describing the interacting processes. However, the description is valid in principle for all oases at the southern border of the Taklamakan Desert.

In the Cele Oasis, the water flux of the rivers differs strongly between seasons. During the winter the rivers carry very little and sometimes no water at all. According to a rough estimate, based on Xia et al.(1993), less than 3% of the annual supply reaches Cele from December to the end of February. From March to May this amount increases slowly to 9%. However, with the onset of snow melt and melting of glaciers in the Kun Lun, a peak water flood reaches Cele, and 77% of the annual supply arrives between June and August. The rest (11%) arrives from September to November. These are average figures, including years with extremely low as well as those with extremely high summer floods. Surplus water in periods of high water flux that could not be used for irrigation of arable land was drained into the foreland of the oasis in former times. There it has supported the existence of an indigenous vegetation that served as a shelter against sand drift and as a source of livestock feed as well as of fuel and construction material. This vegetation belt is an essential component of the oasis ecosystems of the Taklamakan Desert and is indispensable for their functioning. But since about the end of the 1950s, however, population growth has resulted in a change of the traditional water and vegetation management and in a progessive deterioration of the protective vegetation.

Those processes which act together to promote the desertification under the given conditions as a consequence of population growth are compiled in Fig. 1. Direct consequences of population growth are increased water consumption for household needs, an increased area under cultivation, and increased exploitation pressure on the protective vegetation. Reduction of its area, concentration of exploitation on the smaller remaining area, and decrease in water supply act together in destroying the protective vegetation. As an immediate consequence, desertification by sand drift is promoted.

The number of inhabitants of Cele Oasis increased between 1973 and 1998 from 87 000 to about 130 000. In all probability,an earlier sudden increase in population size had already occurred during the 1950s when the "march into the deserts" was propagated in China.

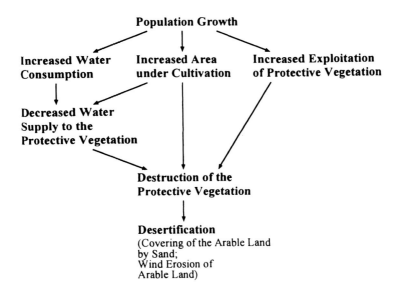

Fig. 1. Processes leading to desertification as a consequence of population growth

More details of the development, described above, and of the problems that arose from it have been described by Xia et al. (1993). Until 1980, dunes advancing from the NW since about 1958 had covered about 5 km and had approached to only 1.5 km from the town's centre by 1980. About 25% of the formerly arable land has been devastated. This caused the Chinese Academy of Sciences to set up a research station at Cele in 1983. Scientists were charged with the task of developing and implementing countermeasures in cooperation with the local authorities. This task has been successfully accomplished (H Zhang 1995a; X Zhang this vol). The further advance of dunes has been stopped, and by 1988 a part of the protective vegetation and about 400 ha of arable land had been regenerated.

However, population growth is continuing, and the need for livestock feed, fuel, and construction material is growing continuously with it. In the meantime, the resulting increase in exploitation intensity is again endangering the existence of a sufficiently large and dense protective vegetation and is reducing its shielding effect against sand drift. A vegetation management program that meets the population's growing requirements and avoids the negative use consequences is urgently needed.

Water availability is the limiting factor with regard to area as well as productivity of this vegetation (H Zhang 1995b). The decision as to the ratio in which the available water is allotted among different needs (according to Fig. 1) and which share is to be used for maintaining a certain area of protective vegetation is ultimately a political one; but this decision has to be based on economic as well as on ecological considerations, and the latter are of particular weight, as a sufficient protection against sand drift is an indispensable precondition for the existence of the oasis. However, in view of the shortage of water, its use in the management of the protective vegetation should be as efficient as possible.

Project Aims

Our project is designed to contribute to the fundamental knowledge for making proper decisions on water use in regenerating, maintaining, and managing protective vegetation. In order to achieve these objectives, the following investigations are being performed:

- Population biological investigations cover seed bank analyses at different sites and attempts to regenerate the indigenous vegetation from diaspores and to favour the settlement of certain species of particular economical and ecological value (key species, see below). Methods: Seedling appearance from soil samples is recorded under field conditions with irrigation. Appearance and mortality of seedlings after flooding of research plots and during the subsequent drying period is continuously monitored.
- Research concerning sustainable use includes measurements of biomass stock and increment under different water supplies and harvest intensities as well as investigations of the nutrient demand and turnover of stands of selected species. Additionally, nitrogen fixation in an important key species (*Alhagi sparsifolia*) is being investigated. Methods: Biomass and its increment is calculated on the basis of relationships between biomass and easily measurable plant variables (allometric regressions) on irrigated and non-irrigated plots. Nutrient demand and turnover is derived from biomass increment, litterfall and element concentration of the individual fractions. ^{15}N measurements of leaves and xylem sap of *Alhagi sparsifolia* from the research area are being performed in order to determine nitrogen fixation. Additionally, plants are being cultivated under controlled conditions and are being supplied with inorganic nitrogen of known isotopic signature.
- Water limitation of growth is being investigated under several aspects. Water use efficiency (WUE) is determined on different levels: (1) Biomass increment of key species is related to the quantity of irrigation water used; (2) the ratio of annual phytomass increment to the quantity of water transpired in the same period gives the WUE of production (WUE_P), which is determined under different water supplies; (3) the WUE of photosynthesis (WUE_{Ph}) is to be determined using measurements of the CO_2 and H_2O gas exchange during certain time periods; (4) the WUE is to be calculated from carbon-isotope ratios. Investigations into water potential, osmotic potential, pressure potential, and elasticity modulus are performed mainly in order to characterize the capability of key species to withstand drought. As an indication of the differing ability of the species to exploit

soil water, their rooting depth and intensity is to be established. Methods: Transpiration is derived from continuous recordings of sap flow during the vegetation period. In addition, steady state porometer measurements are performed in shorter time periods. Parameters of plant water status are calculated from pressure-volume curves, established with pressure chambers. Water stress metabolites are analyzed using GC and HPLC. As far as possible, root systems will be dug out.

- Interest of the oasis population in the utilization of the vegetation is considered by obtaining information on total quantitative requirements and preferences for certain species. Methods: Information is gained through interviews and questionaires.

- Vegetation mapping is being conducted on two scales. Fine-scale maps of the research plots provide detailed pictures of habitat-dependent plant distribution and its short time changes. A broad-scale map will record the given dimensions and densities of different vegetation types and is intended to allow estimations of the total usable phytomass. Moreover, an attempt will be made to identify areas where the settlement of key species could be promoted. Methods: Fine-scale (high resolution) topographic maps have been made separately for every subplot (19 in total) using a laser water level on a reference post. Within each of these subplots the exact location of every plant together with their crown projection was determined on a 10 x 10-m area. The broad scale map is based on satellite images and aerial photographs, supplemented by reference plots, the exact location of which is determined with the help of GPS.

- Necessary background information is provided by continuous recordings of climatic variables and of the soil water content in different soil depths. Supplementary measurements of soil water content are performed regularly, but not continuously, at several sites. Soil texture is analyzed in order to establish whether or not inter- and intrasite uniformity exists. Determining the nutrient content of the soil not only provides a general characterization of the vegetation's nutrient supply but indicates possible imbalances as well. Methods: Climatic variables are measured with automatic weather stations at two sites using standard methods. Volumetric soil water content is continuously recorded using TDR sensors. Soil texture and nutrient content of the soil is analyzed by using standard methods (sieving, sedimentation, fractional extraction, colorimetry, atomic absorption spectrometry).

Members from the following institutions are cooperating in the project:
Albrecht-von-Haller-Institute of Plant Sciences, Dept. Ecology and Ecosystem Research, University of Goettingen; Institute for Plant Physiology, Dept. Chemical Physiology, University of Vienna; Institute of Ecology and Geography, Chinese Academy of Sciences, Urumqi; State Key Laboratory of Arid Agroecology, Lanzhou University; Economic Institute, Xinjiang Academy of Sciences, Urumqi.

Investigated Species

The investigations focus on four bush or tree species that are particularly important for the prevention of sand drift as well as for utilization purposes (key species):

Alhagi sparsifolia Shap. (Fabaceae) is of interest in several ways. On our investigation plots the thorny bushes grow up to a height of about 1.5 m and form dense stands in which the sand is efficiently fixed. The above-ground parts die off in winter with the onset of the first frosts, but also the dense dead bushes are effective in wind-breaking and in the prevention of sand drift. In spring they sprout again from underground parts. Most probably it is a nitrogen-fixing species and contributes to nitrogen input into the nutrient-poor desert soils. It is harvested in great amounts during summer and autumn and is used as livestock feed during winter. *Tamarix ramosissima* Ldb. (Tamaricaceae) forms bushes with individual shoots reaching a height of up to 2 m above soil surface. The bushes favour the formation of dunes. By forming new shoots near the soil surface they are able to keep pace with the growing dune up to a certain level. Long shoots are used as construction material for house walls, roofs and fences. Shoots as well as old roots that are dug out from dunes are used as fuel. *Populus euphratica* Oliv. (Salicaceae) is mainly to be found along river beds and under optimal conditions can reach a height of 20 m, but stands of it can also be found outside recent river valleys, as is the case on our research plots. Sometimes, groups of trees are to be found even on top of dunes. The stems are used as construction material, although their form is often rather irregular. Moreover, stems and twigs are used as fuel and the leaves as livestock feed. *Calligonum cf. caput-medusae* Schrenk. (Polygonaceae) is not indigenous to the research area but is wide spread in Central Asia and has been planted with good success at the Cele Oasis. The bushes reach a height of about 3 m from the soil surface. They are also able to follow a continuous height increase of dunes by forming new shoots. The shoots are mainly used as fuel. *Tamarix ramosissima* and to a lesser degree also *Populus euphratica* are salt-tolerant but seem to be dependent on groundwater (cf. Devitt et al. 1997). In our research plots groundwater has been established at about 5.70 m below the soil surface in one of the *Tamarix* plots and at about 3.60 m below it in one of the *Populus* plots. Further investigations will be performed in order to establish the maximal distance from groundwater table that can be tolerated by these species. In our *Alhagi sparsifolia* plots groundwater has not been found down to a depth of 7.75 m. In the *Calligonum caput-medusae* plot drilling deeper than 4.60 m was impossible up till now because of an impenetrable soil layer in that depth.

Present State of the Project

At the beginning of the 1999 vegetation period all research sites were fenced and could be equipped with measuring devices according to plan. Those plots that had been assigned for this purpose have been irrigated in the vegetation period, and variants of different utilization intensity have been arranged. Climatic variables and soil water content could be recorded continuously during the vegetation period without major disturbances and interruptions. The physiological measurements could begin as had been planned. The initally open question as to whether the key species are appropriate for establishing allometric regressions could be answered positively: close linear correlations between the logarithm of an easily measurable plant parameter and the logarithm of dry weight have been found in every species. Thus, biomass per plant as well as per area can be calculated at all sites, and the repeated measurement in 2000 will allow calculation of the above-ground biomass increment under different conditions. Inquiries into the population's demands and preferences have been made and are now supplemented by evaluating statistics. Fine-scale maps have been set up. Satellite images and aerial photographs have been made available, and first reference plots have been established and used to construct a preliminary version of a broad-scale map. The vegetation period 2000 will be used for continuing the data recording as well as for repeating and supplementing measurements. Some investigations will also be continued in the vegetation period of 2001. If no major disturbance occurs, we expect to be able to prepare a synthesis of the results from all groups by autumn 2001, and to be able to derive recommendations for a sustainable management of the protective vegetation concurrently.

Acknowledgements. The project is funded by the European Community (ERB IC18CT980275). The authors are responsible for the content of the publication.

References

Coque R, Gentelle P, Coque–Delhuille B (1991) Desertification along the piedmont of the Kunlun chain (Hetian–Yutian sector) and the southern border of the Taklamakan desert (China): preliminary geomorphological observations (1). Rev Géomorphol Dyn 40:1–27

Devitt DA, Piorkowski JM Smith SD Cleverly JR, Sala A (1997) Plant water relations of *Tamarix ramosissima* in response to the imposition and alleviation of soil moisture stress. J. Arid Environ. 36:527–540

Walter H, Box EO (1983) The deserts of Central Asia. In: West NE (ed) Ecosystems of the World 5: Temperate deserts and semi-deserts. Elsevier, pp 193–236

Xia X, LI C, Zhou X, Huang P, Pan B (1993) Desertification and control of blown sand disasters in Xinjiang. Science Press, Beijing, 298 pp

Yang X (1991) Geomorphologische Untersuchungen in Trockenräumen NW-Chinas unter besonderer Berücksichtigung von Badanjilin und Takelamagan. Göttinger Geographische Abhandlungen 96. Goltze, Göttingen, 124 pp

Zhang H (1995a) The comprehensive controlling and demonstrating study on the transition zone at the southern margin of Taklamakan desert. Arid Zone Res 12 (4):1–9

Zhang H (1995b) Relations between soil moisture and natural vegetation around Cele oasis. Arid Zone Res 12 (4): 41–43

The Control of Drift Sand on the Southern Fringe of the Taklamakan Desert – an Example from the Cele Oasis

Ximing Zhang, Xiaoming Li and Henian Zhang

Keywords. Cele Oasis, effects of controlling drifting sand, methods of controlling drifting sand

Abstract. The Cele Oasis is located on the southern edge of the Taklamakan Desert with adverse natural conditions and serious hazard from wind-drift sand. Since starting this study work in 1983, very good results have been obtained. This cchapter introduces methods for controlling drifting sand and the results of the project.

Introduction

The Taklamakan Desert is the second largest mobile desert in the world; its area is 337 600 km^2 and mobile sand dunes occupy 82.2% of the total area. The desert constantly encroaches southward under the action of prevailing northwest and northeast winds. The desert has expanded tens to a 100 km southward and covered the famous Silk Road and many towns. Over a million people of seven counties and a city, living on the approximately 1000-km blown-sand belt of the southern margin of the desert, have suffered deeply from blown-sand disasters for a long time. During the 20th century, the desertification has speeded up further, with the increase of irrational human social and economic activities. At the beginning of the 1980s, blown-sand in Cele (Qira) city had moved three times during the recorded period, and mobile sand dunes had again moved to 1.5 km place near the city, presenting the dangerous situation of the city being confronted with desert. In the 30 years from the 1950s to the 1980s, shifting sand had moved about 5 km in the village of Cele located on the margin of the Cele county oasis, so that 1.333 ha farmland was desertified. The area of farmland desertification occupied 25% of the total cultivated area of this village, and 60 households were compelled to move to other places because their houses were buried by sand. Under the great pressure of the desert and sand encroachment, the social economy and living standard were backward and in a state of poverty. Tackling shifting sand, controlling drifting sand and averting poverty were collective wishes of various nationalities, at the same time, were the difficult problems faced with the local government.

The Control of Drift Sand on the Southern Fringe of the Taklamakan Desert 351

In October of 1982, to counter this serious situation, the People's Government of the Xinjiang Uygur Autonomous Region convened an on-the-spot meeting attended by the leaders of respective departments of the Xinjiang People's Government, the Xinjiang Branch of the Chinese Academy of Sciences and the Hetian People's Government. Together, they discussed a plan for harnessing the blown sand. The summary of the meeting determined the establishment of the Cele Desert Research Station, which is jointly led by the Hetian People's Government and the Xinjiang Institute of Biology, Pedology and Desert Research. An Experimental Study on Sand Control at Cele County was taken as the key project of Xinjiang. The Science and Technology Commission, the Water Resources department, the Forestry Department and other departments of Xinjiang gave crucial financial aid in the costs of research, drilling wall and afforestation, respectively.

The project, An Experimental Study on Sand Control in Cele County, was signed as a contract. The time limit of the study, stipulated by the contract, was from 1983 to 1988, its executive area is on the margin of the Cele Oasis; the scale and aim were to establish 666.6 ha sample land of controlled sand, and to rehabilitate 4000 ha of natural vegetation, reaching an average of 30–40% vegetation cover (Zhang 1988a).

Natural condition

The project area is situated at the juncture of the Cele village oasis and the Taklamakan Desert. The southern part of Cele County ajoins the Kunglun mountain, the northern part joins the vast Taklamakan Desert. The area of Cele county covers N 35°17'55" – 39°30'00" and E 80°03'24" – 82°10'34". The total area of the county is 32 000 km^2, mountain land occupies 40.5% of the total area, desert and Gobi occupy 56.6%, so that oasis area is only 1.6%. The climate in Cele County is inland warm desert climate, having sufficient light and heat; the yearly average hours of sunshine is 2697.5 h, the rate of sunshine is 61%. Annual total radiation is 144.4 kcal cm^{-2}. The yearly average temperature is 11.9 °C and the extreme highest temperature is 23.9 °C. The yearly average precipitation is 35.1 mm, the yearly evaporation reaches 2595.3 mm. There are nine seasonal rivers in Cele County, the yearly total runoff is 5.85 x 10^8 m^3, but the amount of water is not well distributed throughout the seasons. Water occupies 9.3% in the spring and 76.8% in the summer. There is a great difference between flood season and low water season. The main typesof soil in the project area are aeolian sandy soil, brown desert soil and irrigation-warping soil (Sulaiman and Yu 1988).

Environmental Problems in the Area

Based on textural research, Cele County Town during the historical period (including the township under Yutian County before Cele County was established) was compelled to move three times because of shifting sand encroachment. From the 1950s to the beginning of the 1980s, the desert in the northwest part of the county town moved and expanded by 80–100 m every year, the desert of Cele village moved 5 km between 1957 and 1980; mobile sand dunes were at a distance of only 1.5 km from the county town. The cause was the destruction of about 2666.6 ha *Populus euphratica* forests and many *Tamarix* spp. by blind reclaiming and excessive cutting from the end of the 1950s to the beginning of the 1960s. The increasing desertification caused the fragile ecological environment to deteriorate further. Before the project started, the area of farmland desertification occupied 25% of the total cultivated area of this village; 60 households were compelled to move to other places because of sandburial. Table 1 shows changes in social economy, population and production of Cele County, including the project area (according to data of the Statistic Department of Xinjiang Uygur Autonomous Region). From this table, we can clearly see the change in economic status around the project area; planting structure has been adjusted, the economy has developed remarkably and the average income of every farmer has greatly increased.

Table 1. Changes in economic status in the project area

Year	1973	1982	1988
Cultivated land area (1000 ha)	2.269	2.299	1.993
Crop area (10 000 ha)	1.969	1.698	1.333
Crop production (10 000 tons)	2.313	3.624	2.615
Cotton area (10 000 ha)	0.106	0.139	0.244
Cotton production (10 000 tons)	0.027	0.050	0.095
Vegetation oil area (10 000 ha)	0.023	0.018	0.084
Vegetation oil production (10 000 tons)	0.063	0.140	0.065
Population (10 000)	8.7	9.9	11.2
Netto average income of every farmer (Yuan RMB)	70 (1978)	85 (1980)	295
Industrial and agricultural production (10 000 Yuan RMB)	1703	2984	9696

The vegetation changes in the project area are the same as those in Cele County, that is, they are closely related with the activities of man utilizing, demanding, protecting and restoring vegetation. The vegetation change can be divided into several stages. In 1949 to 1957, there were big trees with diameters of tens to hundreds cm around villages and beside farmland, and green could be seen everywhere. For a long time, lands had been cultivated individually by each farmer, so that the types, height and size of planted trees were different, and 80% of artificial forests were overgrown. At that time, fuel-wood consumption was low, so fuel-wood nearby was sufficient for local use. Because transport was difficult, fuel-wood could be transported to outside for selling and its price was lower, so firing wood did not need to be taken to the Gobi and desert. The natural vegetation of *Populus euphratica*, *Tamarix* spp., etc. were basically in a state of natural maintenance. From 1958 to 1962, owing to the destruction of the forest for reclamation, the vegetation in the desert was badly destroyed. The artificial forests inside oases were also basically cut down. According to the recorded data, in this period, over 700 000 large trees were cut down in the village of Cele. At this stage in 1963 to 1967, forestry was quickly developed, better forest belts were greatly practised inside oasis farmland, and artificial forests were to some extent restored; but the years from 1968 to 1976 were a new cutting stage. After this, artificial forests in the oasis were again in a developmental stage. However, in the 1970s, the damage to *Populus euphratica* forests in the desert was at a serious stage, because many trees and shrubs inside the oasis were cut down, and the consumption of timber and fuel wood increased considerably with the development of town and population. Once again, people transferred fuel-cutting from vegetation in the oasis to *Populus euphratica* and *Tamarix* spp. in the Gobi and desert, and began to cut the desert vegetation unrestrainedly. Since the beginning of the project, we first rehabilitated and protected natural vegetation to counter this situation. After many years' effort, natural vegetation has quickly recovered. At the same time, the local government propagated the importance of protecting vegetation to the masses, and made relevant local laws and regulations, and arranged special supervision of the forests. Thus, remarkable results have been achieved.

The type of land degradation in the project area is desertification, whose main cause is excessive human economic activities, caused by a rapid population growth and a low level of productivity. Thus, to achieve the necessary agricultural production, blind reclaiming caused the natural vegetation to be destroyed, the newly cultivated land was quickly discarded, and desertification occurred owing to lack of protective systems or water resources. The animal husbandry in the agricultural area was developed unsuitably, due to the lack of balance between grass and livestock, with the result that the desert vegetation was destroyed. Population growth caused a great demand on living resources, the result was chopping trees and shrubs in a plundering way. Secondly, concentrating on agricultural production alone, devoting all attention to agriculture, underestimating forestry and neglecting the ecological effects and oasis protection caused an irrational distribution of the water resources, so that there was no water for the oasis ecology. A protective system for the oasis was, moreover not planned, executed or established under scientific direction, so that the process of desertification due to natural conditions (mainly climate factors) was also speeded up. These human and natural factors are the basis for desertification in the project area.

Approach to the Project

Situated on the southern fringe of the Taklamakan Desert, Cele County belongs to the extraarid desert zone. Being very poor in water resources for planting trees and grass in spring and autumn, the vegetation cover of in this region is only

0.24%. With fine particles of sand material, rich sand sources and frequent winds and sand activities, unreasonable human activities accelerate the process of desertification. According to the natural condition of the local area, starting with restoring natural vegetation and developing artificial vegetation through making full use of rich floodwater resources, this project solved the difficult problem of forestation in summer, and set up a comprehensive protective system possessing original characteristics. It not only probed ways to extend vegetation cover and control the expansion of drift sand, relieve the threat of the city from sand siege, but also protected and extended the oasis. It should be mentioned that both the sand-control system based on the original characteristics in the studies on a protective system for control of blown-sand disasters and artificial direct seeding over large areas using summer floodwater in forestry are creative works, which reflect and represent most modern research level and latest achievements of our country at that time in studies on the protective system and methods of forestation. The points of methods and measures are as follows:

Design and Foundation of a Comprehensive Sand-Control System

Through 6-year efforts, a comprehensive sand-control system over an area of 10 000 ha has been set up on the periphery of the oasis, forming a 4.5 km-wide green barrier, mainly taken up by grass and bushes (accounting for 77%). The system is made up of the sand-retaining river, a belt of sand-fixing grass and bushes, an artificial bush forest and a network of narrow-belt and multibelt sand-break forest (Zhang 1988). Since each of these components reduced wind force successively and obstructed sand-drift, the system effectively prevented the expansion of sand -drift, which played a remarkable role in withstanding strong wind damage and improving micro-climate conditions of the oasis farming ecosystem.

Study on Forestation with Summer Floodwater

Spring techniques are common knowledge. However, there are no reports from scientific and technological workers on achievements in artificial forestation over large areas in summer, which is considerably more difficult. This experiment succeeded in artificially planting desert shrub forest with summer floodwater. Meanwhile, detailed research on the technology of forestation, such as mode of forestation, treatment of seeds, etc., was conducted, gathering rich experience, and opening a new way for planting desert forest in regions where there are floodwater resources in summer. Especially direct seeding with summer floodwater, being a simple method, with costs and rapid results, is easy to propagate.

Artificially Promoting the Restoration of Natural Vegetation and its Rational Utilization

To restore vegetation, a grand total of 13.2 km of trunk canal, 29 km of lateral canal and branch canal, 153 sluice gates and 44 bridges had been built by the end of 1987, forming an irrigation system. By the method of irrigation with floodwater, artificial reseeding, and so on, the area of restored vegetation was 10 000 ha, the degree of vegetation cover increased to 50–60% in 1987 from 3–5% formerly. Among the restored vegetation, the community with *Alhagi sparsifolia* as the dominant species covered an area of about 3333.3 ha, having 90–95% coverage, other plant communities covered an area of 666.6 ha, having a coverage of 60–80%, the secondary young forest of *Populus euphratica* covered an area of 2670 ha, consisting of 40 million protected trees with a height of 6–10 m; the community with *Tamarix* spp. as the dominant species was distributed over an area of 3300 ha, with a coverage of 60–80%. The biomass per unit area of the artificially restored vegetation increased by more than 50 times. Under the prerequisite of guaranteeing its function of drift-sand control, the restored vegetation was used for strip-selection cutting, which played a certain role in relieving the situation of insufficient fodder and firewood.

Improvement of Sandy Land

After driftsands had been basically fixed, fodder and green manure, such as lucerne, etc., were grown on the open ground of the shelter belt network. This resulted in improved aeolian soil structure and enhanced land fertility. On this basis, horticulture and farming were developed. By 1988, an area of 68 ha had been used for growing fruits, consisting of 49 000 protected trees, and 200 ha of the land improved by planting grass had been turned into cultivated land.

Plantation Technology and Rational Utilization of the Wind-Breaking and Sand-Fixing Fuel Forest

To meet the need both for sandfixing and fuel wood, superior sandfixing shrubs that have quick growth, high value as firewood and strong regeneration were selected to be planted in large density and strip mixing. As a result, the total area of forestation reached 200 ha, firewood was provided by selection cutting, and selection cutting in strips with regulatory rotation over 5 years was proposed.

Main Results of the Project

The project mainly consists of two parts. (1) To restore and protect 1000 ha natural vegetation (including 2666.6 ha *Populus euphratica* forests, 3333.3 ha

Tamarix spp., 3333.3 ha *Alhagi sparsifolia* and 666.6 ha other plants) with only 3–5% vegetation cover which had been severely damaged by lack of water. Since the project, the vegetation cover has reached an average of 50–60%. (2) To establish a comprehensive 860.6-ha protective sample land which consists of a sand-breaking forest network, sand-fixing firewood forests, economic forests and cultivated land (including 588.6 ha sand-breaking forest network, 71.1 ha economic forests, 200 ha sand-fixing firewood forests and 37.5 ha cultivated land). The aim is to make desertified land develop from uncultivated -> planting corn -> planting wheat -> planting cotton. With the 6-year efforts, distinct social and economic benefits have been achieved:

Social benefits

The first model of large-scale shifting-sand control was established in southern Xinjiang, and plays a typical demonstrative role on the control of wind and sand calamity, the danger has been removed of the county town of Cele being engulfed by sand, and, with the people's panic gone, social stability has been ensured. Several tens of peasants have moved home. Meanwhile, villages are protected and the stable development of agricultural and stock-raising production is ensured.

Ecological Benefits

Each element of a comprehensive sand-control system can reduce 30–50% of wind velocity, recover vegetation cover to 60%, the mixed layer of sand and leaves on the surface over 6 cm, and effectively control surface wind erosion and sand movement. Relative humidity increased (5.5–17.5%), evaporation decreased (75–189 mm), the temperature of the surface, air and upper soil layer can be reduced, and the environment improved (Zhang 1988a, b).

References

Ybulayin, Sulaiman, Qili, Yu (1988) Types of soil of Cele sand land and their improvement and utilization. Arid Zone Res (Suppl edn):28–35
Zhang, H (1988a) An experimental study on sand control at Cele County, Xinjiang. Arid Zone Res 5(3):1–8
Zhang, H (1988b) Protective and economic benefits of biological devices adopted in the sand-control project of Cele County. Arid Zone Res 5 (Suppl edn):48–52
Zhang, X (1988) The construction of the sand-control system on the fringe of Cele Oasis and its effects of protection. Arid Zone Res 5(Suppl edn):11–18

The Role of Biological Soil Crusts on Desert Sand Dunes in the Northwestern Negev, Israel

Maik Veste, Thomas Littmann, Siegmar–W. Breckle and Aaron Yair

Keywords. Biological crusts, soil lichens, sand stabilization, combating desertification

Abstract. Biological soil crusts are important microphytic communities and significantly influence both structure and processes within the ecosystem. They are built up from cyanobacteria, green algae, fungi, mosses and lichens. Various crust types could be found, depending on dune slope aspect and dewfall availability. In the sand dunes of the northern Negev they cover large areas and stabilize the sand surface against wind and water erosion. Free-living and symbiontic cyanobacteria are capable of nitrogen fixation and are important nitrogen sources in the desert sand dunes. As biological crusts enhance the surface stability and soil fertilization, they are to be considered a key factor in the protection of arid and semiarid ecosystems and, thus, in combating desertification in terms of sand dune remobilization.

Introduction

Shifting sands are one of the major problems of desertification and land degradation in arid and semiarid areas. Nearly 20% of the world's arid zones are covered by aeolian sand (Pye and Tsoar 1990). Such fragile ecosystems are very sensitive to land use practices. Grazing and trampling by livestock and agricultural use destroy the vegetation cover and enhance sand mobility, thus accelerating desertification processes. In most arid regions of the world, sand dune movement is a threat to irrigated farmlands, villages, railways, highways and other infrastructures. In the deserts of the Middle East, sand dunes in Egypt cover 16% of the area (Misak and Draz 1997), 21% in the Sinai and 13% of the Negev (Danin 1996). The sand dunes of the northwestern Negev are the eastern extension of the northern Sinai sandfield (Veste 1995, Veste and Breckle 2000) and are characterized by linear forms in the southern part and complex dunes in the northern part.

The sand dune field has been inhabited by Bedouin nomads, who use the interdunes for crop growing, while goats, camels and donkeys graze the dunes

(Tsoar and Møller 1986). The intensive grazing led to a decrease in the vegetation cover on Egyptians territory (Tsoar et al. 1995), whereas on the Israeli side, the vegetation remained undisturbed from 1948 to 1967 and again since 1982. The differences could be observed in satellite images (Otterman and Waisel 1974, Danin 1996). The higher surface wind speed in combination with a high relief energy results in increased sand mobility. In 1967, the border between the Negev and the Sinai was opened, and the vegetation on the Israeli side was severely overgrazed (Tsoar et al. 1995). After the reestablishment of the borderline and land use changes in 1982, the vegetation recovered on the degraded Negev areas. Along with vegetation recovery, a thin biological soil crust established itself on the sand dunes. The structure of these biological soil crusts and their function for the ecosystem processes were investigated in the linear dunes near Nizzana and along the geoecological gradient (Fig. 1).

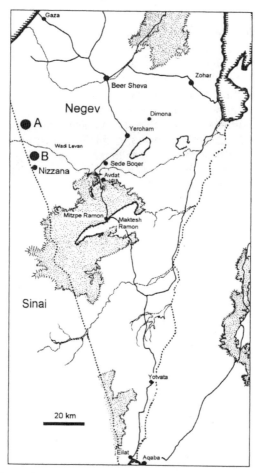

Fig. 1. Map of the Negev desert and location of the investigation sites. *A* Halzua sand field, average annual rainfall 120 mm; *B* Nizzana, average annual rainfall 90 mm.

Biological Crusts

Biological soil crusts are common cryptogamic communities in various arid and semiarid regions. They are reported, for example from Australia (Eldrige and Greene 1994), North America (West 1990; Johansen 1993), from the sand dunes of the South African Succulent Karoo (M. Veste, pers. observ.) and from various habitats of the Negev desert in Israel. The sand dunes of the northwestern Negev are covered over wide areas by crusts. The biological crusts are built up by soil material and by cyanobacteria, green algae, mosses, fungi and soil lichens. Most of the crust flora of the sand dunes near Nizzana are cyanobacteria and are represented by *Microcoleus sociatus*, *Calotrix perietina* and *Nostoc* sp.; the Chlorophytes are represented by *Chlorococcum* and *Stichococcus* (Lange et al. 1992). Most of the organisms inside microphytic crusts are still unidentified. In the Negev sand dunes various crust types could be distinguished following crust thickness, chlorophyll a content, and colour in the sand dunes near Nizzana (Fig. 2). Light brown crusts with an average thickness of 1–3 mm occur in the interdune and on south-facing slopes, whereas the crusts on north-facing slopes are darker. Here, the chlorophyll a content may serve as a measure for crustal biomass and the crust thickness. Mosses (*Bryum* spp.) are common on the north-facing slopes and especially at the dune base (Fig. 2). Soil lichens occur in the interdune area and on stable and flat north- and northwest facing slopes in the northern Haluza sand field (Fig. 1) and cover here up to 90% of the area. Cyanobacterial lichens like *Collema* spp. dominate the soil crusts. The green-algae lichens *Fulgensia fulgens*, *Squamarina cartilaginea*, *Squamarina lentigera* and *Diploschistes diacapsis* were observed locally on small soil mounds.

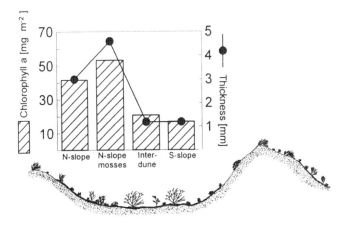

Fig. 2. Chlorophyll content and thickness of the biological crust in different geomorphological units in the sand dunes of Nizzana

Dew Fall and Crust Activity

The development of the biological crusts seems to be determined by two main environmental factors: (1) sand stability and (2) the microclimatic boundary conditions, especially the moisture regime (Veste et al. 2000). Dew is a sufficient water source for the physiological activity of the microorganisms during the entire year. Already approx. 0.1–0.12 mm of dew activate the soil crust lichens during the night and the sun-exposed soil lichens photosynthesize 1–1½ h after sunrise before drying out (Fig. 3).

Rainfall in combination with cloudy weather extends the activity time for several hours. But high amounts of rain can suppress the photosynthetic CO_2 uptake of several soil lichens species, whereas cyano-lichens are able to take CO_2 from fluid water (e.g. Lange et al. 1995). The dew distribution within the dune field is inhomogeneous and depends on the interdune morphology, albedo and slope aspects. Deep hollows show a higher frequency of stable air layers at night time and consequently low dewpoint temperature differences and dark surfaces show more extensive radiation input at night. In winter, larger amounts of dewfall remain 1–2 h longer at the surface as compared to south-facing slopes, while in summer the duration of surface wetting shows no difference between the slopes (Fig. 4). The reason for this phenomenon is to be seen in the seasonally different radiation geometry of north- and south-facing slopes (Littmann and Kalek 1998). This indicates that especially deep interdunes and north-facing slopes are more favourable habitats for biological crusts. However, the frequency of dewfall is higher in summer than in winter, but does not contribute to enhanced crustal growth, as the critical time for crustal photosynthetic activity is limited by rapid drying after sunrise (Fig. 3; Veste et al. 2000).

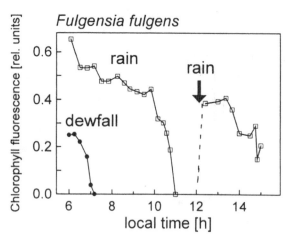

Fig. 3. Time course of photosynthetic activity in soil crust lichen *Fulgensia fulgens* after nocturnal dewfall (●) and rain (□). Activity was measured *in situ* with chlorophyll fluorescence

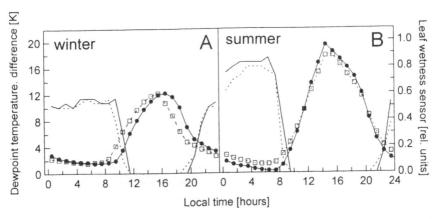

Fig. 4. Dewpoint difference at north-facing slope (solid line) and south-facing slope (dashed line) and leaf wetness sensors on north- (●) and south-facing slopes (□) in Nizzana

Nutrients

Arid and semiarid soils are low in nutrients, especially phosphorus and nitrogen (Buckley 1983, Evans and Ehleringer 1993). Under good water conditions there are indirect indications that nitrogen could limit plant growth even in arid regions (Floret and Pontanier 1982). In the sand dunes of Nizzana total nitrogen content directly under the biological crusts varied between 26 and 45 mg kg^{-1} (Fig. 5).

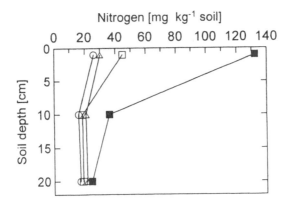

Fig. 5. Total nitrogen in the sand under different soil crust types in Nizzana (open symbols) and Haluza sand field (closed symbol). Sampling location: south-facing slope (○), interdune (△), north-facing slope (□), soil crust covered with cyanobacterial lichens (■)

Particularly under cyanobacterial lichen crusts, nitrogen content increased up to 132 mg kg^{-1} soil. The source of the nitrogen into the ecosystem can be either atmospheric deposition or biological nitrogen fixation by free-living or symbiontic cyanobacteria. The atmospheric nitrogen deposition in the sand dunes of the northwestern Negev is generally low and varies between 0.5 and 2 kg ha^{-1} a^{-1} (Littmann 1997, T. Littmann, unpubl. data). Under such conditions, biological fixation is an important nitrogen input source for the desert soils. West (1990) reported nitrogen-fixation rates of biological soils crusts from 2 to 41 kg ha^{-1} a^{-1}, while the atmospheric deposition in North American deserts is below 10% of this value (Evans and Ehleringer 1993) Even when the nitrogen fixation rates under natural conditions are still unclear and most of the measurements under laboratory conditions overestimate the fixation rates due to methodological problems (West 1990), the biological crust is a considerable nitrogen source in various desert ecosystems. Moreover, the biological crust contributes to the carbon input and soil development.

Hydrology

Crusting of soil surfaces generally leads to a reduction of infiltration and an increase in runoff. In sandy areas the infiltration rates of rainwater are normally high due to the coarse-grained texture of sand in comparison to loamy soil, where runoff is generated. However, on the sand dunes of the Negev runoff may occur after intensive rainfalls (Fig. 6; Yair 1990).

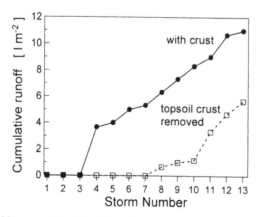

Fig. 6. Runoff yield on crusted plots (●) and plots where topsoil crust had been removed (□) in the Nizzana sand dunes.

Flow frequency and runoff rates were higher on the north-facing slopes than on the south-facing slopes in the Nizzana area. The lack of correlation between runoff volume and storm rain amount and peak rain intensity highlights several aspects. Runoff generation over microphytic crusts is often explained by the hydrophobic,

The Role of Biological Soil Crusts on Desert Sand Dunes in the Northwestern Negev 363

water-repellent properties of the crust that greatly enhance runoff generation (Bond 1964; De Ploey 1977). It is clearly shown that the crust is able to absorb high water amounts when dry and infiltration decreases with time until the crust is saturated. A proper understanding of runoff generation in the area considered requires some knowledge regarding the combined physical-biological processes that take place under wetting conditions. An analysis of the pore size distribution shows the large predominance of pore diameters below 40 μ (Verrechia et al. 1995). Upon wetting two combined processes take place (1) water absorption and swelling of silty and especially clayey particles and (2) swelling of the microbiological elements. The filamentous sheaths may absorb water five to eight times of their dry weight and increase their volume up to ten times. Swelling of the biological elements from 2.5–4 μ to 20–50 μ is sufficient to fill up most of the voids. These combined processes result in a substantial decrease in crust pore size and limit water infiltration. However, infiltration is still high, because water can still infiltrate along the larger pores that remained unclogged. Results obtained by field experiments in Nizzana indicate that water uptake time, until saturation, is in the order of tens of minutes up to several hours. This relatively long time explains why under dry surface conditions short-intensity rain showers are unable to initiate runoff. The time factor for reaching saturation is far more important than the amount of rain. The biological properties and the biomass content of the topsoil influence the infiltration rates. Topsoils covered mainly by soil lichens shows the lowest infiltration (Fig. 7). Also in other arid ecosystems the limiting infiltration by the biological crusts are well known. In South Australia a soil lichen crust on sandy soil reduces infiltration by 50% (Rogers 1977). In the sand dunes of Nizzana the runoff frequency is relatively high in wet years, but runoff yields are low to very low. This is due to the very short duration, and limited rain amounts, of the effective rain showers.

Stability

The biological crusts enhance the aggregation of soil particles, and are thus a key feature for the surface stability. The filaments of cyanobacteria, excreted exopolysaccharides (Mazor et al. 1996) and fungal hyphae stick the sand grains together and prevent saltation. Surface stability is an important factor in deserts to prevent wind and water erosion (Booth 1941; Williams et al. 1995) as well the establishment and spatial distribution of higher vegetation (Kadmon and Leschner 1995). The fragile biological crust, and therefore the surface stability, are endangered by trampling of livestock, people and, in some deserts, by four-wheel -drive vehicles.

The effects of the disturbance of the crust on the entire ecosystem depend on the scale and frequency of disturbance. Small disturbances, like footprints, can be closed by the mobile cyanobacteria within days after rainfall, whereas disturbance like the removal of large areas of the biological crusts or frequent trampling needs years to recover. This will enhance sand erosion and desertification processes, as could be seen on the Egyptian side of the Sinai–Negev sand field, where trampling by goats and camels destroyed the biological crust and high-intensity sand dune mobilization led to a further decrease in vegetation cover.

Combating Desertification

The fixation of drifting sands is a major task in the frame of combating desertification. Several experiments have been made regarding this problem worldwide. Planting methods are a traditional means to control drifting sands and are widely practiced in the Middle East, Central Asia, China and North Africa. However, sand dune fixation by vegetation is only successful when wind speed is greatly reduced and sand movement is minimized. This will allow for the establishment of new vegetation and an increase in the surface roughness. Very often binding or liquid materials, e.g. petroleum products, clay material treated with polymers, are used to stabilize the sand surface (Veisov et al. 1999). From the environmental viewpoint these methods are unacceptable for sand fixation. For a more natural surface stabilization, the biological crust can play an important role. Sand dune mobility as well dune type and morphology depend on the wind direction and wind speed, which are largely controlled by the vegetational roughness (Tsoar and Møller 1986, Littmann and Gintz 2000). Planting vegetation or building up 20–30-cm-high pressed straw checkerboards will reduce wind speed. This will allow the biological crust to establish itself and in a feedback the surface stability will increase. This was sucessfully tested in the sand dunes of northern China to protect the Baotou–Lanzhou railway. Between the checkerboards a biological crust with lichens developed. Also in an experiment under natural conditions conducted in the Nizzana area, the biological crust recovered after nearly 3 years (Fig. 8).

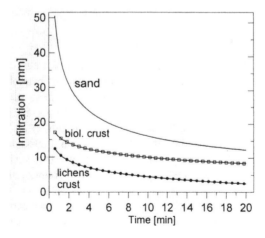

Fig. 7. Infiltration on sand, biological crust on north-facing slope and soil lichen crust in the Haluza sand field.

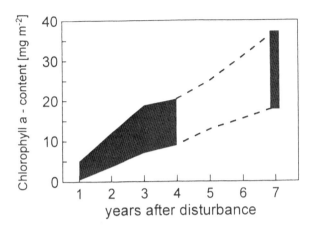

Fig. 8. Regeneration of the biological crust after topsoil crust was removed

After the disturbed topsoil is stabilized by a physical rain-crust, cyanobacteria are the first colonizers and they increase the stability of the topsoil. Only in later stages of the succession are soil lichens able to colonize the biological crusts (Eldrige and Greene 1994). Growth and succession depend on the microclimatic boundary conditions. Sufficient dewfall is important for the growth of microphytic crusts in deserts.

Conclusions

Micropyhtic crusts provide the only natural protective cover on the sand surface. An intact and living biological crust prevents wind and water erosion, enhances soil fertilization and is an important tool to combat desertification in arid, semiarid and subhumid regions.

Acknowledgments. This research is part of the BMBF project Ecosystem Processes in the Linear Dunes of the Northern Negev funded by the German Federal Ministry for Education and Research [BMBF grant 0339495A (Bielefeld), 0339692J (Halle), DISUM 23 (Jerusalem)]. Thanks are due the German–Israeli Arid Ecosystems Research Center for the technical and organizational assistance.

References

Bond RD (1964) The influence of the microflora on physical properties of sand. Effects associated with filamentous algae and fungi. Aust J Soil Res 2:111–122

Booth WE (1941) Algae as pioneers in plant sucession and their importance in erosion control. Ecology 22:38–46

Buckley R (1983) Soil nitrogen requirements of tropical sandridge plants. Biotropica 15:77–78

Danin A (1996) Plants of Desert Dunes. Springer, Berlin Heidelberg New York

De Ploey J (1977) Some experimental data on slope wash and wind action with references to Quaternary morphogenesis in Belgium. Earth Surf Process 2:101–115

Eldrige DJ (1998) Trampling of microphytic crusts on calcareous soils, and its impact on erosion under rain-impacted flow. Catena 33:221–239

Eldrige DJ, Greene RSB (1994) Microbiotic soil crusts: a review of their roles in soil and ecological processes in the rangelands of Australia. Aust J Soil Res 32:389–415

Evans RD, Ehleringer JR (1993) A break in the nitrogen cycle in arid lands? Evidence from $\delta^{15}N$ of soils. Oecologia 94:314–317

Floret C, Pontanier RS (1982) Management and modeling of primary productions and water use in a south Tunesian steppe. J Arid Environ 5:77–90

Johansen JR (1993) Cryptogamic crusts of semiarid and arid lands of North America. J Phycol 29:140–147

Kadmon R, Leschner H (1995) Ecology of linear dunes: effects of surface stability on the distribution and abundance of annual plants. Adv GeoEcolo 28:125–143

Lange OL, Kidron GJ, Büdel B, Meyer A, Kilian E, Abeliovich A (1992) Taxonomic composition and photosynthetic characteristics of the biological crusts covering sand dunes in the western Negev. Funct Ecol 6:519–527

Lange OL, Reichenberger H, Meyer A (1995) High thallus water content and photosynthetic CO_2 exchange of lichens. In: Daniels FJA, Schulz M, Peine M (eds) Festschrift für Gerhard Follmann. Geobotanical and phytotaxonomical study group, Botanical institute, University of Cologne, Germany, pp 139–153

Littmann T (1997) Atmospheric input of dust and nitrogen in into the Nizzana sand dune system, north-western Negev, Israel. J Arid Environ 36:433–457

Littmann T, Gintz (2000) Eolian transport and deposition in a partially vegetated longitudinal sand dune area. Z Geomorphol Suppl 121:77–90

Littmann T, Kalek J (1998) Microclimate as controlling factor for ecosystem processes in an arid sand dune area (northwestern Negev, Israel). Hallesches Jahrb Geowiss 20:77–92 (in German)

Mazor G, Kidron GJ, Vonshak A, Abeliovich A (1996) The role of cyanobacterial exoploysaccharids in structuring desert microbiological crusts. FEMS Microbiol Ecol 21:121–130

Misak RF, Draz MY (1997) Sand drift control of selected coastal and desert sand dunes in Egypt: case studies. J Arid Environ 35:17–28

Otterman J, Waisel Y (1974) Observation of desertification in the Israeli ERTS-1 program. XIV Convegno Int Technico-Scientifico Sullo Spazio, Rom, March 1974, pp 199–205

Pye K, Tsoar H (1990) Aeolian sand and sand dunes. Unwin & Hyman, London

Rogers RW (1977) Lichens in hot arid and semi-arid lands. In: Stewart MRD (ed) Lichen ecology. Academic Press, London, pp 211–252

Tsoar H, Møller JT (1986) The role of vegetation in the formation of linear sand dunes. In: Nickling WG (ed) Aeolian geomorphology. Allen & Unwin, Boston, pp 75–95

Tsoar H, Goldschmith V, Schoenhaus S, Clarke K, Karneli A (1995) Reversed desertification on sand dunes along the Sinai/Negev border. In: Tchakerian VP (ed) Desert aeolian processes. Chapman & Hall, London

Veisov SK, Cherednichenko VP, Svintsov IP (1999) The fixation of drifting sands. In: Babev AG (ed) Desert problems and desertification in Central Asia. Springer Berlin Heidelberg New York, pp 143–153

Verrechia E, Yair A, Kidron G, Verrechia K (1995) Physical properties of the psammophile cryptogamic crust and their consequenses to the water regime of sandy soils, north -western Negev, Israel. J Arid Environ 29:427–437

Veste M (1995) Structures of geomorphological and ecological units and ecosystem processes the linear dune ecosystem near Nizzana/Negev. Bielefelder Ökol Beitr 8:85 –96

Veste M, Breckle S–W (2000) Negev – pflanzenökologische und ökosystemare Analysen, Geogr Rundsch 9:24–29

Veste M, Littmann T, Friedrich H, Breckle S–W (2000) Activity of soil crust lichens and its microclimatic boundary conditions in a desert sand dune ecosystem of the northern Negev (Israel) (submitted)

West NE (1990) Structure and function of microphytic soil crusts in wildland ecosystems of arid to semi-arid regions. Adv Res 20:179–223

Williams JD, Dobrowolski JP, West NE, Gillette DA (1995) Microphytic crust influence on wind erosion. Am Soc Agric Engin 38(1):131–137

Yair A (1990) Runoff generation in a sandy area – processes in the Nizzana sands, eastern Negev, Israel. Earth Surface Landforms 15:597–609

Restoration of Disturbed Areas in the Mediterranean – a Case Study in a Limestone Quarry

C. Werner, A.S. Clemente, P.M. Correia, P. Lino, C. Máguas, A.I. Correia and O. Correia

Keywords. Limestone quarry, Mediterranean vegetation, mycorrhizer, reclamation, revegetation, soil degradation

Abstract. Limestone quarrying activities have extremely strong environmental impact, since they imply vegetation clearing and loss of soil. A reclamation project was conducted in a limestone quarry of the Serra da Arrábida (southwest Portugal), a natural park with a dense evergreen sclerophyllous shrub community. The successive revegetation of quarry terraces results in distinct plant communities of different age and cover. In this work we examined five different terraces, which were revegetated at 3-year intervals, to evaluate the establishment and growth of introduced species as well as colonization and succession of natural vegetation.

New revegetation techniques were also evaluated using three native species (*Olea europaea* var. *silvestris*, *Pistacia lentiscus* and *Ceratonia siliqua*). Different treatments were applied in a randomized block design to improve the revegetation process: (1) fertilization, to overcome growth limitation due to nutrient deficiencies; (2) mycorrhization, to improve nutrient uptake by plants, as well as their competitive capacity for other resources; (3) addition of a long-term water -holding polymer to reduce water stress. Growth and vigour of the plants was monitored and ecophysiological studies were conducted, comprising water relations and fluorescence measurements. The results of the first year revealed species-specific differences during the adaptation processes: O. europaea was the most robust species. C. siliqua was the most sensitive to the transplantation stress, but after the initial adaptation the highest growth rates were found in this species, where fertilization significantly increased growth rate. Survival rate during the first summer was high in all species, being lowest in C. siliqua and highest in O. europaea (95 and 98%, respectively).

Introduction

Desertification and degradation of Mediterranean soils is an important environmental problem demanding urgent and appropriate solutions. The Mediterranean Basin contains extensive karst landscapes, some of which have experienced the longest histories of intensive human occupation of any karst terrains on earth (Gams et al. 1993). A particular kind of soil degradation is due to opencast mining, which has become a major environmental problem and affects more land surface every year.

Limestone mining causes soil denudation and often a great modification of the original land relief (Sort and Alcaniz 1996). The exploitation in platforms increases the drainage and, therefore, the physical and chemical erosion of the substrate, hindering the natural germination and settlement of young plants, which delays recolonization. Water and nutrient stresses, characteristic of the mediterranean ecosystems, are additional problems for pioneer plants under such conditions. Natural processes of plant colonization in abandoned limestone quarries are very slow and act over the time scales of a primary succession. The length of time taken for successful establishment of plant species may be reduced if artificial revegetation is considered.

Artificial revegetation of quarry terraces can be used not only for a rapid establishment of attractive shrublands, which disguise the visual impact, but also to circumvent colonization problems, enhancing seed retention, germination and seedling survival of herbaceous and indigenous shrub species. Revegetation can further accelerate the process of soil formation and buffer the impact of erosion agents such as rain and runoff. Addition of treatments to overcome water and nutrient deficiencies may lower environmental stresses, improve the stabilization of new plant communities and accelerate the revegetation process. One possibility is the inoculation with mycorrhizal fungi, which play an important role in the nutrient uptake by plants, especially of nitrogen and phosphorus, which are often limiting in the mediterranean ecosystem (Herrera et al. 1993). The use of mediterranean evergreen sclerophylls for revegetation has the advantage that they are perfectly adapted to the climatic and edaphic conditions.

The main purpose of this project is to obtain biological and ecological information about mediterranean species able to colonize these degraded areas through the evaluation of plant cover and diversity in revegetated quarries and to explore new revegetation techniques.

Material and Methods

The sites were located in the Serra da Arrábida Natural Park in southwest Portugal, a small chain of limestone outcrops with maximum elevation of 500 m (see Catarino et al. 1982). Mean annual rainfall is about 650 mm with a pronounced dry period from June to September (Correia and Catarino 1994). This study was performed in one of the largest limestone quarries of this region

(SECIL, Outão). The limestone is extracted in 20-m-high terraces with nearly vertical slopes. Revegetation was conducted by SECIL, with 2-year-old container-grown plants, introduced successively after exploration with 3-year intervals, from 1983 to 1995 (Fig. 1). The indigenous mediterranean species used for revegetation were *Ceratonia siliqua, Olea europaea, Juniperus phoenicea, Phillyrea angustifolia, Arbutus unedo, Quercus coccifera, Q. faginea, Myrtus communis* and *Spartium junceum. Pinus halepensis*, though not natively abandoned at this site, was introduced as a fast-growing species. Plants were raised by nursery techniques from stored seeds indigenous to the side. Introduced plants were irrigated during the first year.

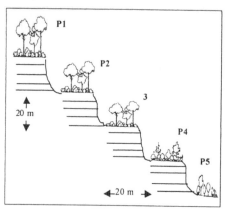

Fig. 1. Schematic representation of the recovery of vegetation on the quarries terraces in Outão (Serra da Arrábida). P1–P5 are the platforms at an elevation of 335, 315, 300, 280, and 260 m, respectively. Revegetation was initiated *from the top to the bottom* in 1983, 1986, 1989, 1992 and 1995 in *P1–P5*, respectively

The establishment of these container-grown plants and the natural recruitment of herbaceous and shrub species were studied in five revegetated terraces during spring and summer of 1998. The vegetation was sampled within 40 3 x 3-m plots, positioned along a belt transect parallel to the quarry wall. The percentage of bare soil, number of individuals, canopy shrub cover and height of adult shrubs per plot were recorded. The presence of all herbaceous species was recorded. Species diversity was measured according to Whittaker et al. (1965), where the rate (d) at which species are added as the area (A) increases is related to the species number (S) by a semilog expression (Shmida 1984).

A new revegetation program was conducted in 1998, on a recently explored terrace, with 2-year-old nursery plants (of ca. 50–120 cm height) of three native evergreen sclerophylls *O. europaea* var. *silvestris* Brot., *C. siliqua* L., and *P. lentiscus* L. The following treatments were applied: (1) addition of 100 g hydrogel per plant (long-term water-holding polymer, STOCKOSORB 400K, Stockhausen, Degussa-Huls), (2) fertilization with 120 g of a slow-release NPK-fertilizer, (3) mycorrhization with 85 g Mycor-Tree (a mixture of end- and ectomycorrhizal

Restoration of Disturbed Areas in the Mediterranean 371

fungus inoculates and Terra-Sorb) per plant (Plant Health Care, Inc., Pittsburgh, PA, USA).

A marl layer of about 1 m in depth was added on the bare rock. Nursery plants were planted in February 1998 in holes of about 50 cm depth, filled with organic sandy soil. Organic soil analysis and roots of nursery plants showed that they were partially mycorrhized. A randomized plot design was used, with three replicates (blocks) placed along the longitudinal direction of the platform. Each plot contained five plants of one species and one treatment (18 different plots per block, total number of plants 270).

Ecophysiological measurements were conducted on 3 consecutive sunny days (16–18 September 1998) during midday (11–15 solar time). Photoinhibition was determined from chlorophyll a fluorescence measurement of maximum quantum yield of PS II (F_v/F_m), which is a good indicator of environmental stresses on photosystem II. Measurements were conducted simultaneously with portable, pulse-modulated fluorometers (2 Mini-PAM and a PAM-2000, Walz, Effeltrich, Germany). Two horizontally orientated sunlit leaves per plant were dark adapted for 15 min prior to F_v/F_m determination (see Werner et al. 1999) from three plants per plot ($n = 18$). Water potentials (Θ) were determined with two Scholander-type pressure bombs (Manofrigido, Portugal) on small terminal shoots from two to three plants per plot ($n = 6$ to 9). Vigour was determined as adapted from Sousa Santos and Moura Martins (1993) by a defoliation index (0–5) with 0: a healthy green plant, 1: 0–10%, 2: 10–25%, 3: 25–50%, 4: 50–100% defoliation, 5: dead plant. Growth was estimated as the increment of the main shoot (cm), which was marked on each plant at the beginning of the experiments. Data were tested for significance with an ANOVA and Duncan post-hoc tests using the Statistica package (StatSoft, Tulsa, USA).

Results

Analysis of the different revegetated terraces indicated that the use of mediterranean sclerophyllous plants resulted in low mortality and rapid plant growth and establishment. *P. halepensis*, although introduced at a density similar to that of the sclerophyllous species (of around 10%) in the first two terraces (P1 and P2) (Fig.2) presented a higher growth rates. A high cover of about 50% was attained in *P. halepensis*, while the sclerophyllous species showed a very low cover of about 10%. The diversity index measured as the slope of the semilog species-area relation showed a higher diversity in terraces 1 and 5 (with a slope of 0.16 and 0.23, respectively) (Fig.3). Terrace 2 showed a slope of 0.12 while the other terraces showed values lower than 0.1.

In the new revegetation experiment a deterioration in plant vigour and some leaf shedding was observed during the first 3 months, which was probably due to transplantation stress (Fig. 4). Plant status improved and all species recovered to good vigour after 1 year. During the first year after planting, *O. europaea* was the most robust and *C. siliqua* the most sensitive species. Survival rate after first summer drought was high in all species with 95, 97 and 98% for *C. siliqua*, *P. lentiscus* and *O. europaea*, respectively.

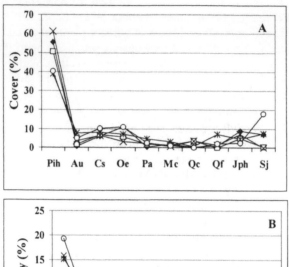

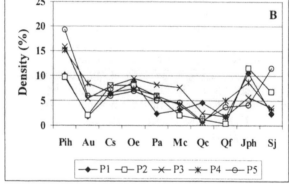

Fig. 2. Relative cover (**A**) and density (**B**) of the main mediterranean species introduced. Pih *Pinus halepensis*; Au *Arbutus unedo*; Cs *Ceratonia siliqua*; Oe *Olea europaea*; Pa *Phillyrea angustifolia*; Mc *Myrtus communis*; Qc *Quercus coccifera*; Qf *Quercus faginea*; Jph *Juniperus phoenicea*; Sj *Spartium junceum*; *P1, P2, P3, P4, P5* platforms 1, 2, 3, 4, 5 respectively

At the end of the dry season, in September, a relatively high maximum quantum yield of PS II (F_v/F_m) was found (Fig. 5), indicating successful plant establishment. Only *P. lentiscus* revealed significantly lower values than *O. europaea* and *C. siliqua* (mean F_v/F_m of 0.65±0.11, 0.7±0.1 and 0.73±0.06, respectively, p<0.001). However, species responded differently to the applied treatments and a high variability was found. While mycorrhization significantly improved the F_v/F_m in *O. europaea*, the reverse pattern was found in *P. lentiscus*. Addition of fertilizer significantly increased F_v/F_m in *C. siliqua*. Significant effects of treatments on midday water potential (Ψ) were found only in *O. europaea*, where addition of gel, mycorrhizer or a combination of both increased Ψ. This species exhibited the lowest Ψ (of –3.6 ± 1.2 MPa, $p<0.001$). A positive effect on the growth rate was found for the fertilizer treatment in *C. siliqua* (Fig. 6), revealing the highest increment of main shoot (as well as increase in stem diameter, not shown).

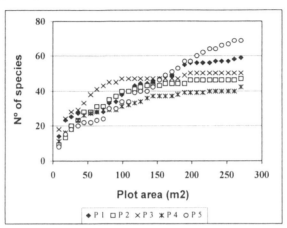

Fig. 3. Species-area (S-A) relation according with Whittaker's plant diversity sampling method. The plant species numbers in different plots when plotted on semilog scale gives the slope, calculated by a least-squares regression, i.e. the rate at which species are added as the area increases. *P1, P2, P3, P4, P5* platforms 1, 2, 3, 4, 5 respectively. *P1* S=20.11+0.163logA (R2=0.95); *P2* S=21.82+0.115logA (R2=0.77); *P3* S=31.3+0.089logA (R2=0.67); *P4* S=19.99+0.092logA (R2=0.85); *P5* S=10.08+0.232logA (R2=0.99)

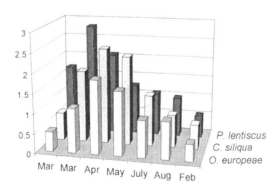

Fig. 4. Seasonal variation in vigour (defoliation index). Treatments were pooled for each species ($n = 90$)

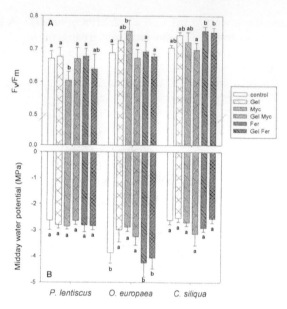

Fig. 5. A Chlorophyll fluorescence measurements of maximum quantum yield (F_v/F_m) at midday ($n = 13$–15, $\pm$ SE). **B** midday water potential (Ψ) ($n = 7$–9, $\pm$ SE) in September. *Different letters* indicate significant differences at $p<0.05$ between the treatments for each species

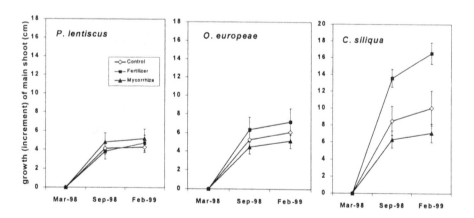

Fig. 6. Growth of main shoot (cm) for each species. No significant differences were found for the gel treatments and data were pooled ($n = 30 \pm$ SE)

Discussion

Reclamation of quarry-degraded areas is a difficult task, since the limestone exploitation results in a complete removal of vegetation and loss of soil. An addition of a soil layer is the first step to provide an environment which allows recolonization and establishment of new plant communities. However, erosion and runoff of bare soils counteract restoration aims.

The results of this study suggest that artificial revegetation favours a quicker establishment of species and lowers the visual impact on the landscape. This also contributes to minimize the soil erosion and runoff (Muzzi et al. 1997) and to promote favourable microclimatic conditions for the colonization by other species. The species used for revegetation seemed to be able to increase the total vegetation cover and the diversity. However, higher cover values of the pine species determine a high competition for space and light at later successional stages. If rapid regreening of the area is the first goal, then fast-growing species (e.g. *P. halepensis*) could be used at the first stage of revegetation.

The revegetated platforms were floristically diverse and highly variable in composition, with higher diversity in the oldest and in the youngest terrace due to the proximity of natural vegetation and the introduction of lime and soil from other regions.

We consider it a primary aim to reestablish the natural plant community and, therefore, only native mediterranean species were used in the new revegetation program. Native species have the advantage of being highly adapted to the local edaphic and climatic conditions, which are characterized by strong seasonality, and they can withstand years of extreme climatic events.

The experiments indicated that the added treatments may improve the revegetation process. Fertilizer significantly enhanced plant growth during the first year, though with different intensity in each species. In this nutrient-poor system, a rapid response to fertilizer was expected, and might help to overcome growth limitations during plant establishment, and may function as an initial catalyst for plant growth.

While the effects of the mycorrhizal treatment are expected to become evident only after a prolonged inoculation time (e.g. after 1 year of growth), they might improve plant nutrient uptake on a long-term scale. Hererra et al. (1993) showed that inoculation of seedlings with selected rhizobia and mycorrhizal fungi improved outplanting performance, plant survival and biomass development. The scarcity of propagules in eroded soil can become a handicap to plant establishment, because the formation of dynamic rhizosphere is critical, particularly in low-nutrient ecosystems (Haselwandter and Bowen 1996).

The high variability of the results may be at least partially explained by the fact that plants may initially invest more in below-ground growth, and only after successful establishment an enhanced growth may be visible above ground. This is especially important in an environment where growth limitations may occur mainly by water and nutrient supply, as was the case in the quarry. During the first years of revegetation there may be little above ground competition.

In a future project these studies will be extended in quarries of several Mediterranean countries, to address the ecological aspects of plant establishment, competition for resources, soil processes and other factors which determine recolonization and plant establishment in degradated areas.

Acknowledgements. This work was funded by PRAXIS XXI/PCNS/C/BIA/ 180/96, Portugal. The authors wish to thank SECIL for its permission and active cooperation in the use of this site and all logistic support. The help of the students of Plant Ecology during the vegetation sampling is gratefully acknowledged.

References

Catarino FM, Correia OCA, Correia AIVD (1982) Structure and dynamics of Serra da Arrábida mediterranean vegetation. Ecol Mediterr 8: 203–222

Correia OA, Catarino FM (1994) Seasonal changes in soil-to-leaf resistance in *Cistus* sp. and *Pistacia lentiscus*. Acta Oecol 15: 289–300

Gams I, Nicod J, Julian M, Anthony E, Sauro U (1993) Environmental changes and human impacts on the Mediterranean karsts of France, Italy and the dinaric region. In: Williams PW (ed) Karst terrains-environmental changes and human impact. Catena, Cremlingen, pp 59–98

Haselwandter K, Bowen GD (1996) Mycorrhizal relations in trees for agroforestry and land rehabilitation. For Ecol Manage 81: 1–17

Herrera MA, Salamanca CP, Barea JM (1993) Inoculation of woody legumes with selected arbuscular mycorrhizal fungi and rhizobia to recover desertified mediterranean ecosystems. Appl Environ Microbiol 59: 129–133

Muzzi E, Roffi F, Sirotti M, Bagnaresi U (1997) Revegetation techniques on clay soil slopes in northern Italy. Land Degrad Dev 8: 127–137

Shmida A (1984) Whittaker's plant diversity sampling method. Isr J Bot Vol 33: 41–46

Sort X, Alcaniz JM (1996) Contribution of sewage sludge to erosion control in the rehabilitation of limestone quarries. Land Degrad Dev 7: 69–76

Sousa Santos MND, Moura Martins AM (1993) Cork oak decline, notes regarding damage and incidence of *Hypoxylon mediterraneum*. In: Luisi N, Lerario P, Vannini A (eds) Recent advances in studies on oak decline. Proc Int Congr. Selva di Fasano, Brindisi, Italy, 13–18 Sept. 1992, pp 115–121

Werner C, Correia O, Beyschlag W (1999) Two different strategies of Mediterranean macchia plants to avoid photoinhibitory damage by excessive radiation levels during summer drought. Acta Oecol 20: 15–23

Whittaker RH, Niering WA, Crisp MO (1965) Structure, pattern, and diversity of a mallee community in New South Wales. Vegetatio 39: 65–76

Indigenous Agroforestry for Sustainable Development of the Area Around Lake Nasser, Egypt

Irina Springuel

Keywords. *Balanites aegyptiaca, Faidherbia albida*, roots architecture, Wadi Allaqi

Abstract. Lake Nasser was created as a result of the construction of the Aswan High Dam, which was completed in 1969. The water of the lake has covered the whole Nubian Nile Valley and penetrated deep into the desert through tributary wadis. This lake is about 500 km long, of which 291.8 km lies in Egypt. Since its formation, the quantity of water in the lake has varied dramatically, determining the dynamics of the lake and its ecotone zone and hence the development programmes of the area and their rationale. The position of the lake, bounded by Arabian rocky desert on the east side and Libyan sandy desert on the west, creates a large habitat diversity that provides an opportunity for integrated development of the area around the lake.

The downstream part of Wadi Allaqi, which is the largest of the wadis in the southern part of the eastern desert of Egypt and drains to the lake, was selected as the area for in situ research. In our research we have tested the possibilities for sustainable development of the area by enriching the diversity and productivity of the natural vegetation in the ecotonal zone. The concept of agroforestry could be applied to this type of land management. Economically important indigenous desert plants, *Balanites aegyptiaca* and *Faidherbia albida*, were selected for cultivation in ecologically favourable habitats.

An experimental farm with 600 trees of *Balanites aegyptiaca* and 200 trees of *Faidherbia albida* was set up in the main channel of Wadi Allaqi. The results obtained on the growth rate show that with sufficient water supply the desert plants grow fast and some individuals of *Balanites* and *Faidherbia* reached a height above 2 m in less than 2 years. Surface and subsurface irrigation schemes were used. The homogenous growth of trees was observed at subsurface irrigation. The architecture of the roots supports the idea of subsurface irrigation, which facilitates the vertical growth of the roots.

This is a long-term experiment which is still in progress. It could be considered complete when plants are self-sustaining and survive without irrigation and protection from grazing.

Introduction

The High Dam Reservoir is one of the largest man-made reservoirs, created as a result of the construction of the Aswan High Dam, which became operational in 1964 (Abu–Zeid and Abdel–Dayem 1992) and was completed in 1969. The water of this reservoir has covered the whole Nubian Nile valley in Egypt and Sudan and penetrated deep into the desert through tributary wadis. It is about 500 km long, of which the 291.8 km in Egypt is known as Lake Nasser (Fig.1). The shoreline of this lake, bounding 5300 to 7800 kilometres of desert lands (the perimeter length depending on reservoir storage, White 1988), is subject to riparian vegetation (Springuel 1994), in contrast to the previous narrow ribbon of largely cultivated vegetation along the river (White 1988).

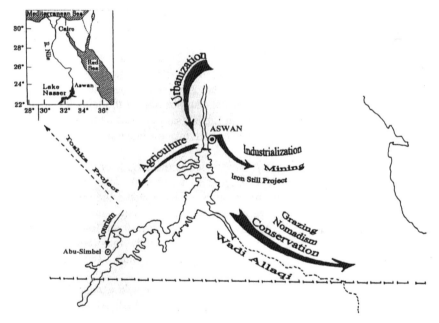

Fig. 1. Location map and development around Lake Nasser. After Bornkamm (1997)

The position of the lake bounded by the Arabian rocky desert on the east and Libyan sandy desert on the west, creates a large habitat diversity that provides an opportunity for integrated development of the area around the lake (Fig. 1). Since the lake's formation, most attention for development areas around the lake has been paid to its western part, with a low slope gradient and numerous broad khors, than to the eastern rocky side. Numerous agricultural projects were set up on the western side of the lake supported by different national and international organizations. A recent example is the national Tushka project. The canal, named after Sheikh Zayed, the President of the United Arab Emirates, will extend about 550 km, taking water through the extremely arid Nubian Western Desert to the Kharga Oasis, and will pass to the New Valley, along the chain of oases in the Western Desert. This is a long-term 20-years project. In its first phase, the canal will extend 60 km and carry water to irrigate 540 000 acres of desert land (Southern Egypt Development Project 1998, The Economist 1999).

The Situation is different on the eastern side of the lake. Only a few small farms cultivating mainly vegetables were set up in different places on its shores. Development is oriented to the exploitation of mineral resources. At present there are about 200 quarries and mines extracting different minerals and ornamental stones. The most recent example is the large Aswan Iron–Steel project, which based on rich iron ore, lies 50 km east of the lake.

In spite of the importance of Lake Nasser as the main source of water for all Egypt, little attention has been paid to the conservation of quantity and quality of the water in the lake. Only one conservation area (Wadi Allaqi) was set up on its eastern side, that includes part of the desert, its marginal zone with the lake and a small part of the lake itself. The creation of a conservation area makes it possible to emphasize sustainable resource use, maintenance of ecological processes and conservation of biotic diversity, including the indigenous plants, which are of potential economic value (Belal and Springuel 1996).

The present study is part of the research work being conducted in Wadi Allaqi Biosphere Reserve by Allaqi project, under the umbrella concept of sustainable development within a conservation context (Belal 1993).

Description of the Area and its Socio/Economic Status

Wadi Allaqi is a major dry river which drains from the Red Sea hills to Lake Nasser in southern Egypt. It crosses the hyperarid area with sparse and unpredictable rainfall, varying from less than 5 mm in its downstream part to 50 mm on the Red Sea hills; however, a few years may pass with no rainfall. As a result of the formation of the reservoir behind the High Dam, water entered the mouth of Wadi Allaqi, which thus became part of the lake. Because of the shallow gradients of the wadi bed, the high fluctuation water level in the lake results in significant lateral movement of the water. The annual variation caused the shoreline to move up to 10 km, while long-term fluctuations (32 m between minimum and maximum level was recorded update) have resulted in changes of the order of tens of kilometres. About 80 km of the wadi was inundated and remained under water for a few years at the first peak (178 m asl) of the high water level in Lake Nasser in 1977–1980. When the lake level began to fall in the following decade, reaching 150.6 m asl in 1988, the water receded about 40 km and remained low for a few years. A second sudden rise in lake level began in

1996 and reached its maximum of 182.3 m asl in 1998, adding 15 km more to the first inundation.

From total of 100 km of wadi that were affected by the lake, only 40 km is permanently inundated. The rest of the land, that is subjected to periodical inundation, as well as adjusted desert, where arid conditions are altered to more mesophytic habitats by the increase in humidity and water seepage from the lake, represents the ecotonal system (Springuel 1994). The pioneer stage in ecotone formation is characterized by fast colonization of exposed land covered with silt deposit by *Tamarix nilotica* (Ehrenb) Bunge, which forms mono stands with almost 100% cover and a height of about 7–8 m at 10 years of age. Dense carpets of annuals dominated by *Glinus lotoides* L. cover the frequently inundated land. In spite of a few plant species recorded (Springuel 1996), the vegetation is characterized by a huge biomass that attracts numerous predators. Freshwater, abundant pastures, available fuel-wood and shelter for animals attract the nomads from the eastern desert, who began to semi-settle on the shores of the lake and even make some attempt at small-scale cultivation, still, however, keeping their traditional pastoral livelihood without causing significant environmental disruption (Fahim 1979; White 1988; Briggs et al. 1993; Springuel 1997; Belal et al. 1998).

The ongoing research programme in Wadi Allaqi is implemented in cooperation with national and overseas scientists and reveals the considerable economic potential of natural resources.

However, the use and management of these resources to meet the criteria of sustainable development is a very complicated process, due to the high instability and vulnerability of the ecosystem. Even at the beginning of the lake formation, White (1988) mentioned that the large and unpredictable fluctuation of water level in the lake has discouraged agricultural development on its shores, and this is still true at present. Alternating periods of flooding and exposure have a major effect on the vegetation, soil composition (Springuel and Ali 1990; Pulford et al. 1992; Briggs et al. 1993; Springuel 1996) and amount of water stored in the wadi deposits (Springuel et al. 1992).

Recommendations for sustainable uses of natural resources of the marginal zone and areas close to the lake have been given in a number of publications (Pulford et al. 1992; Belal 1993; Briggs et al. 1993). All authors agreed that integrated development schemes could be indigenous agroforestry, animal husbandry and ecotourism, but in a harmonious, environmentally sound and socially acceptable manner, and for the benefit of the local population. Rangeland management should include seeding in ecologically suitable habitats where soil moisture is sufficient for plant growth, cultivating deeply rooted highly palatable indigenous trees and shrubs which could be self-sustaining, constructing fences to prevent overgrazing and developing wildlife habitats.

This chapter deals with only one of the aspects of the management, the establishment of indigenous economically important desert trees. The development of plantations as an alternative to the exploitation of wild stocks has been proposed for fuel-wood, building material and medicinal plants. We expect that after a short nursing period, with irrigation provided, the trees could be self -sustaining and regenerate naturally in ecologically suitable habitats.

A simple subsurface irrigation system was designed based on ecological studies of desert plants in their natural habitats. We anticipated that the percolation of water downward through the sandy soil will facilitate the elongation of the roots. As soon as roots reach the underground water the irrigation could be stopped.

Selection of Plants for Indigenous Agroforestry

To fulfil the objective of establishment of indigenous agroforestry, the selection of plants was restricted to phanerophytes growing in the Wadi Allaqi area. Furthermore, the selection of plants was based on ecological and sociological criteria. Under ecological criteria we understand the ability of plants to tolerate the extreme conditions (drought and flood) characterizing the ecotonal system. Plants successfully growing in the Nile Valley and in ecologically favourable desert habitats will fulfil the criteria for ecologically suitable plants. The sociological aspect related to the value of plants in relation to human use was based on indigenous knowledge of the nomads living in the Eastern Desert. The studies conducted by the Allaqi Project indicated that *Balanites aegyptiaca* (L.) Delile has the highest economic value of the plants growing in Wadi Allaqi and is also the most tolerant of both drought and flooding (Hall and Walker 1991; Brunelle 1998). *Balanites* is recognized by IPALAC (The International Program for Arid Land Crops) as an important crop that makes a great contribution to the economy of the region where it is grows. Despite its nutritional and medicinal value and other uses, *Balanites* is practically unutilized in Egypt, probably because its natural growth is relatively small and scattered, making it difficult to establish a *Balanites* industry.

Mixed plantation with *Acacia* could help to improve the soil fertility. *Acacia* trees are particularly important because they are drought-reserve fodder in the Eastern Desert, fed to stock only at times when other food is very scarce (Springuel and Mekki 1994). The cultivation of *Balanites aegyptiaca* in the Lake Nasser area could be important for the sustainable development of this area and its importance will be increased in relation to the Sheikh Zayed Canal that will connect Lake Nasser and Baris (in the southern part of Kharga oasis), where a natural population of *Balanites* still exists.

Balanites Farm

The Allaqi experimental/demonstration farm was established in the wadi bed, at 181 m asl that was not inundated by the lake water prior to the establishment of the farm. The soil in the farm is alluvial deposit consisting of alternating layers of silty loam and loamy sand with a layer of silt on the surface. According to the soil survey of the area, soils are non-saline with little carbonate and a pH value from 7 to 9. Chemical analysis shows that soils contain appropriate amounts of calcium and magnesium and are poor in respect to soluble potassium (Cultivation of Medicinal Plants 1997).

An area of approximately four feddans (acres) was surrounded by an iron fence to protect the seedlings from grazing by wild and domestic animals. When the trees grow to a sufficient height, the fence could be removed and reused for a new farm. About 600 seedlings of *Balanites aegyptiaca* and 200 seedlings of *Faidherbia albida* (Delile) A.Chev. were planted in eight experimental plots in spring 1997, when the size of most seedlings was not more than 20 cm in height. The growth of the plants was monitored regularly until the whole farm was inundated in October 1998.

Table 1. Irrigation requirements for different crops in Egypt

Crop	Irrigation requirements for 1 month (m^3 feddan^{-1})	Method of irrigation
Wheat	777	Surface
	444	Sprinkler
Horse beans	520	Surface
	340	Drip
Corn	750	Surface
	400	Drip
Sunflower	1000	Surface
	833	Drip
Tomato	500	Surface
	400	Drip
Grapes	208	Surface
	158	Drip
Citrus	333	Surface
	166	Drip
Balanites [a]	19	Subsurface

[a] 200 trees feddan^{-1}

The irrigation system was based on a drip irrigation scheme; however, to avoid the filter installations, the drips were removed and water was added on the soil surface directly from the tubes. A simple subsurface irrigation system was designed and applied on two experimental plots. PVC tubes (4" diameter), open at the top and the bottom, were installed vertically in the ground to a depth of 1 m, while half a metre of the tube remained above the surface. Three- to four-month-old seedlings of *Balanites* and *Faidherbia* were planted close to the tubes. At the beginning of growth, surface irrigation was applied to allow the seedlings to

Indigenous Agroforestry for Sustainable Development of the Area around Lake Nasser 383

establish themselves in the soil and to enlarge the root system. When new foliage appeared, taking about 2 to 3 months, water was added directly to the tube, bringing the water close to the roots of the trees. Water was added to tubes twice in week to fill it up to the top. With a tube diameter of 10 cm and length of 150 cm ($\pi r^2 l$), each tree received about 12 liters of water, i.e. monthly about 96 liters. Altogether 200 trees planted in 1 acre will require 19 m^3 of water monthly, which is a very small amount compared with the main crop requirements in Egypt (Table 1).

Results of observations on the growth of the trees are shown in Table 2. The survival rate of 15-months-old seedlings was high, above 95% in most plots, except for *Faidherbia* in plot 8. In spite of the common belief that the growth of desert trees is slow, our results show that in 1.5 years the trees could reach a height of more than 1 m, with an average annual growth above 60 cm. However, some individuals could reach a height above 2 m and maximum heights of 242 and 350 cm were recorded for *Balanites aegyptiaca* and *Faidherbia albida*, respectively. The high standard deviation indicates considerable heterogeneity in the individuals within each plot.

Table 2. Growth parameters (height and diameter) of 15-month-old *Balanites aegyptiaca* and *Faidherbia albida* trees and growth rate (GR) of both species in the period from July 1997 to August 1998

Irrigation	Subsurface		Surface	Surface	Surface
Plot no.	2		4	6	8
Survival (%)	96	100	95	97	83
Name	B	F	B	B	F
Height (cm) + SD	130 + 33	82 + 31	124 + 48	98.5 + 50	131 + 75
Diameter (cm) + SD	1.2 + 0.5	0.9 + 0.49	1.3 + 1.02	1.1 + 0.9	2.11 + 2.03
GR (cm)	76	60	70	58	98
Plot no.	1	3	5	7	
Survival (%)	97		97	95	97
Name	B		B	B	B
Height (cm) + SD	141 + 40.11		120 + 41	110 + 49.91	144 + 50
Diameter (cm) + SD	1.38 + 0.6		1.2 + 0.9	1.1 + 0.95	1.95 + 1.7
GR (cm)	77		62	62	98

B Balanites aegyptiaca; *F Faidherbia albida*; *SD* standard deviation; *GR* growth rate

The results of our study show the large variation in the growth parameters of trees growing in the different plots. The growth of *Faidherbia* and *Balanites* was faster in the two plots (plots 7 and 8) situated in a depression in the northern part of the farm; both these plots were additionally characterized by a high variability in the trees and lowest survival of *Faidherbias* compared with other plots. Fast growth of the trees was observed in plots in the southern part of the farm (plots 1 and 2) where subsurface irrigation was applied. In order to compare the effect of subsurface and surface irrigation on the growth of *Balanites* trees, while avoiding the effect of the soil, the neighbouring trees growing on the homogenous substrata in the two adjacent plots (1 and 3) with subsurface and surface irrigation were examined. The mean height of 42 trees under subsurface irrigation was 151 cm, while trees were smaller with surface irrigation (mean height of 42 trees was only136 cm). Most probably, the subsurface irrigation supports better growth of *Balanites*; however, further experiments are needed to support this statement.

The architecture of the roots of *Balanites aegyptiaca* under surface and subsurface irrigation varies greatly (Fig. 2a, b). Under subsurface irrigation the main root subdivided into two main branches, which penetrate deeply close to the PCV tube or even twining on the tube. Many lateral roots are developed as soon as the main root has reached the end of the tube at a depth of 1 m. The volume of the soil occupied by the roots is about 1.5 m of vertically and less than 0.5 m of horizontally.

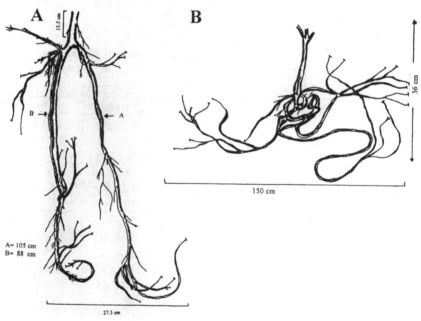

Fig. 2.a.,b. Roots of 15-months-old *Balanites aegyptiaca*. **A** Subsurface irrigation; **B** surface irrigation

Under surface irrigation the two main root branches are short and give rise to many lateral branches which spread horizontally at a depth of 20–50 cm. Often, roots are found close to the place where water was applied. The volume of the soil occupied by roots is mainly horizontal (1.5–2 m) with a vertical depth up to 50 cm.

The vertical dimension in root distribution could be much more advanced for agroforestry compared with the horizontal distribution of the roots. First of all, the deeply rooted trees can absorb the subsurface water and hence need little or even no irrigation. Second, the roots with vertical distribution occupy a smaller area compared with horizontally spread roots; hence more trees could be planted on the same area. This will greatly affect the costs of the farm in places where protection from grazing is needed. The present study shows that the architecture of the roots could be altered by applying a different irrigation technique.

Conclusions and Recommendations

The results presented in this chapter are the first and very initial stage in the long -term experiment conducted in Wadi Allaqi on the shores of Lake Nasser. However, even in this initial stage, the obtained results throw light on possibilities of cultivation of indigenous plants as well as give guidance for continued work toward the sustainable development of the area. Particularly the support given to the project by the local nomadic and seminomadic populations indicates that the indigenous agroforestry could reasonable supplement indigenous practices. Trees selected for cultivation (*Balanites aegyptiaca* and *Faidherbia albida*) are important indigenous trees with high economic value, protected by Bedouins in their natural habitats (Springuel and Mekki 1994).

The present experiment shows that indigenous trees could be easily cultivated and grow fast with sufficient water supply. Irrigation should be applied; there is particularly good evidence for subsurface irrigation; however more studies are needed to support the idea of subsurface irrigation. The importance of root architecture in the agroforestry industry is emphasized in the present study. The architecture of the roots supports the idea of subsurface irrigation, which facilitates the vertical growth of the roots.

In accordance with the Egyptian policy for conservation of freshwater, the cultivation of desert trees with low water requirements such as *Balanites aegyptiaca* and *Faidherbia albida* could be very important in remote areas on the shores of Lake Nasser. More indigenous trees need to be tested for cultivation on the shores of the lake, giving preference to trees with proved ecological amplitude. The present study showed that young *Faidherbia* trees could survive a long inundation period while completely covered by water. In contrast, only mature trees of *Balanites aegyptiaca* could survive inundation, while trees up to 4 years old died. Because of the instability of the environmental conditions, when flood and drought periods alternate, a mixture of trees with different water requirements could be recommended for cultivation in the same plot. The ecology, ecophysiology and ecosociology of the selected plants should be studied in detail and the field study should be supported by laboratory experiments.

Acknowledgements. The present work was possible by joint effort of many people and financial support from many organizations. The author wishes to thank all the staff of the Unit of Environmental Studies and Development (UESD), South Valley University, and particularly Professor Belal, Director of Unit, for encouraging the work in all stages. Efforts of the enthusiastic team of researchers from UESD in establishing the experimental farm and taking the measurements of the trees are greatly acknowledged. Special thanks are to Mohamed Soghir Badri, who was involved in all stages of this work, and his death in a tragic car accident was a great loss to the team.

References

Abu–Zeid MA, Abdel–Dayem S (1992) Egypt's Programmes and Policy Options for Low Nile flows. In: Abu–Zeid MA, Biswas AK (eds) Climatic Fluctuations and Water management. Butterworth–Heinemann, London

Belal AE (1993) Sustainable Development in Wadi Allaqi. Desertification Control Bull 23: 39–43

Belal AE, Springuel I (1996) Economic Value of Plant Diversity in Arid Environments. Nat Resour 32, 1: 33–39

Belal AE, Leith B, Solway J, Springuel I (eds) (1998) Environmental valuation and management of plants in Wadi Allaqi, Egypt. Final Report IDRC. File: 95-1005-01/02 127-01. South Valley University, Aswan, Egypt

Bornkamm R (1997) International field study for biological diversity, conservation and sustainable use and development in Wadi Allaqi Biosphere Reserve. Report. Distribution limited. UESD, Aswan, Egypt

Briggs JA, Dickinson G, Murphy KJ, Pulford ID, Belal AE, Moalla S, Springuel I, Ghabbour S, Mekki AM (1993) Sustainable development and resource management in marginal environments: natural resources and their use in the Wadi Allaqi region of Egypt. Appl Geogr 13: 259–284

Brunelle R (1998) Comparative analysis of the ecology and exploitation of *Balanites aegyptiaca* in Egypt. PhD Thesis, University du Quebec A, Montreal

Cultivation of Medicinal Plants (1997) Mid–Year Report. Submitted to Ministry of International Co-operation, Egypt. Assiut and South Valley Universities

Fahim H (1979) Regional Sociological Impacts of the Aswan High Dam. In: Driver RE, Wunderlich WO (eds) Environmental Effects of Hydraulic Engineering Works. Tennessee Valley Authority, Knoxville, pp 479–87

Hall JB, Walker DH (1991) *Balanites aegyptiaca*. A Monograph. School of Agricultural and Forest Sciences Publications Number: 3, University of Wales, Bangor, UK

Pulford ID, Murphy KJ, Dickinson G, Briggs JA, Springuel I (1992) Ecological resources for conservation and development in Wadi Allaqi, Egypt. Bot J Linn Soc 108: 131–141

Southern Egypt Development Project (1998) Arab Republic of Egypt, Ministry of Public Works and Water Resources National Water Research Centre, Cairo, Egypt

Springuel I (1994) Riparian Vegetation in the Hyper-Arid Area in Upper Egypt, Lake Nasser Area, and Its Sustainable Development.. In: Proc Int Workshop on the Ecology and Management of Aquatic-Terrestrial Ecotones, University of Washington, Seattle, pp 107–119

Springuel I (1996) Environment and Producers in Wadi Allaqi Ecosystem. Arid Ecosyst 3: 43–57

Springuel I (1997) Management of Biosphere Reserves Towards Sustainable Development. In: Belal AE, Springuel I (eds) Proc Worshop on the Arab-MAB Network of Biosphere Reserve, UNESCO Cairo Office, Egypt

Springuel I, Ali MM (1990) Impact of Lake Nasser on desert vegetation. In: Bishay A., Dregne H (eds) Desert Development. Part I: Desert agriculture, ecology and biology, vol 5. Proc 2nd Int Desert Development Conf, Cairo, January 1987. Harwood Academic Publishers, Chur, Switzerland, pp 557–568

Springuel I, Mekki AM (1994) Economic value of desert plants: *Acacia* trees in Wadi Allaqi Biosphere Reserve. Environ Conserv 21, 1: 41–48

Springuel I, Radwan UA, Biswas PK, Hileman D, Huluka G, Alemayenu M (1992) Water conditions of the soils in Wadi Allaqi, Lake Nasser Region. In: Kishk MA (ed) Proc 1st Nat Conf on the Future of Land Reclamation and Development in Egypt. Minia, Egypt, pp 307–326

The Economist (1999) A survey of Egypt, New and Old. Economist March: 1–18

White GF (1988) The Environmental effects of High Dam at Aswan. Environment, 30, 7: 5–11 and 34–40

Ziziphus – a Multipurpose Fruit Tree for Arid Regions

S.K. Arndt, S.C. Clifford and M. Popp

Keywords. Biogeography, physiology, drought stress, fodder, traditional medicines

Abstract. The progressive desertification in many semiarid regions of the world increases the need for plants that can cope with arid environments and meet peoples' requirements for food, fodder and fuel. Species of fruit trees in the genus *Ziziphus* represent examples of such multipurpose plants with great potential for selection and use in drought-prone regions.

Ziziphus trees and shrubs inhabit arid environments on every continent due to their versatility in being able to adapt to drought stress. They play an important role in the conservation of soil, with their strong root system which stabilizes the soil and protects it from erosion. The leaves provide fodder for livestock, the hard wood is used for turning, making agricultural implements, fuel and high quality charcoal. In many regions, *Ziziphus* is grown as a hedge, with its spines creating effective live-fencing, and with its highly nutritious fruits providing a valuable source of energy, vitamins and also income when sold on local markets. In addition, extracts from fruits, seeds, leaves, roots and bark of *Ziziphus* trees are used in many traditional medicines to alleviate the effects of insomnia, skin diseases, inflammatory conditions and fever. For these reasons, *Ziziphus* trees have an important role to play in the integrated economy of the arid lands.

Introduction

Nearly one third of the Earth's surface or ca. 49 million km^2, excluding polar regions, has been classified as arid land according to soil type, vegetation and climate (Kozlowski et al. 1991). The dynamics of arid land ecosystems is controlled by the limiting availability of renewable resources such as water, soil, nutrients, flora and fauna, which result in their low primary productivity. For centuries mankind has tried to confront the problems of arid lands and to revegetate the deserts. Historically, the main focus has been on the development of new irrigation techniques and to select fruit varieties that could cope with poor water quality, with great effort invested in changing the desert to suit the crops, rather than the other way round (Cherfas 1989). However, a lot of promising drought-tolerant plants are growing naturally in drought-prone areas around the world and these may be readily exploited to address the problems in those regions. These natural resources provide a good basis for the selection and development of new perennial plants for arid areas that can meet the escalating demands of ever-growing human and livestock populations with regard to food, fuel, fertilizer, fibre, shelter and medicare. A good example of naturally occurring multipurpose plant species with potential for arid regions are the shrubs and trees of the genus *Ziziphus*.

Description and Distribution

The genus *Ziziphus* (ber, jujube) belongs to the buckthorn familiy (Rhamnaceae). It is a genus of about 100 species of deciduous or evergreen trees and shrubs distributed in the tropical and subtropical regions of the world (Johnston 1963). Some species, like *Z. mauritiana* and *Z. jujuba*, occur on nearly every continent, whereas other species, like *Z. nummularia, Z. spina-christi and Z. mucronata*, are restricted in their distribution to distinct areas (Table 1). *Ziziphus* species can grow either as trees and shrubs (*Z. mauritiana, Z. rotundifolia, Z. jujuba, Z. mucronata*) or exclusively as small shrubs or bushes (*Z. nummularia, Z. lotus, Z. spina-christi, Z. obtusifolia*).

The fleshy drupes of several species are rich in sugars and vitamins, and this fact has made *Ziziphus* species important fruit trees for many centuries. In both China and India, *Ziziphus* trees have a long tradition of selection and cultivation, with the result that the species occurring in these countries (*Z. mauritiana, Z. jujuba*) are better known and more widely researched than those in other regions. *Z. mauritiana* is an example of an extremely drought-hardy species, and is a dominant component of the natural vegetation of the Indian desert (Cherry 1985). These trees are well adapted to seasonal drought and hot conditions. In India during the summer months of May and June, *Z. mauritiana* enters into dormancy by shedding its leaves. These trees perform well even on marginal and inferior lands where most other fruit tree species either fail to grow or give poor performance (Jawanda and Bal 1978). In India, the scions of varieties which have been selected to improve the yield and fruit quality are routinely grafted on to the vigorous rootstocks of wild species to provide a reasonable cash crop on land which is unsuitable for other forms of cultivation (Cherry 1985). Improved Indian cultivars like Gola and Seb have been imported to Israel and Africa, where they have been grafted onto native rootstocks of *Z. spina-christi* and *Z. abyssinica*, respectively (Cherfas 1989). The same technique was successfully used in Zimbabwe to produce high-quality Indian selections on the native *Z. nummularia* rootstock species (Kadzere and Jackson 1997).

Jujubes (*Z. jujuba*) were eaten by the ancients of the chalcolitic age (1500–1000 b.c.), and the fruits have been in cultivation for the past 400 years in both India and China (Anonymous 1976). At about the beginning of the Christian era, the Chinese jujube was imported into Europe and is now widely distributed throughout Persia, Armenia, Syria and the Mediterranean regions in Spain and

France. Reports of early writers and explorers emphasized the heat and drought tolerance of jujubes, and probably because of this, jujubes were thought most likely to succeed in the dry regions of the southwestern US (Locke 1947).

Table 1. Some important *Ziziphus* species and their occurrence

Continent	Species	Region
Africa	*Z. abyssinica* Hochst.	Tropical Africa
	Z. lotus Lamk.	Northern Africa
	Z. mauritiana Lamk.	Tropical Africa, Sahel Zone, Zimbabwe
	Z. mucronata Willd.	Southern Africa
	Z. spina-christi Willd.	Middle east
Asia	*Z. jujuba* Mill.	China, India, Korea, Malaysia
	Z. mauritiana Lamk.	China, India, Pakistan, Malaysia
	Z. nummularia W.i.A.	India
	Z. oenoplia Mill.	Tropical Asia
	Z. rotundifolia Lam.	India
	Z. rugosa Lam.	India
	Z. sativa Gaertn.	Pakistan
	Z. spina-christi Willd.	Middle east
	Z. xylopyra Willd.	India
Australia	*Z. mauritiana* Lamk.	
Europe	*Z. jujuba* Mill.	Mediterranean
	Z. lotus Lamk.	Mediterranean
	Z. mauritiana Lamk.	Mediterranean
	Z. sativa Gaertn.	Mediterranean
North America	*Z. amole* M.C.Johnst.	Mexico
	Z. celata J.i.H.	USA
	Z. jujuba Mill.	USA
	Z. mexicana Rose	Mexico
	Z. obtusifolia Gray	Mexico, USA
South America	*Z. cinnamomeum* Tr.&Pl.	Venezuela
	Z. mistol Griseb.	Argentinia, Paraguay
	Z. joazeiro Mart.	Brazil, Paraguay
	Z. oblongifolia S.Moore	Brazil

Physiological Characteristics

Currently, most cash crop fruit production in semiarid regions relies on species such as peach, which require relatively intensive management and high irrigation inputs for successful establishment and fruit development. Compared to other more commonly cultivated fruit tree species like peach, *Ziziphus* species have

Ziziphus – a Multipurpose Fruit Tree for Arid Regions 391

several physiological and morphological characteristics that may contribute to their ability to adapt to arid environments (Table 2).

Table 2. Physiological parameters of well watered *Z. mauritiana* and *Prunus persica* (peach) trees measured in the field[a] or in the glasshouse[b], respectively.

Parameter		*Z. mauritiana*		*P. persica*	
[a] A	(μmol m^{-2} s^{-1})	20.4	c	9.4	c
[a] g_s	(mol m^{-2} s^{-1})	0.5	c	0.1	c
[b] NRA_{leaf}	(μmol g^{-1} FM)	3.8	f	0.8	d
[a] Nitrogen	(% leaf DM)	3.7	f	3.5	f
[a] Starch	(% root DM)	29.7	e	15.0	e
[a] Root:shoot ratio (DM basis)		1.9	e	0.5	e
(1 year old trees)					

A, net carbon assimilation; g, stomatal conductance; NRA, nitrate reductase activity; c, Clifford et al. (1997); d, Bussi et al. (1997); e, Jones et al. (1998); f, unpublished

Ziziphus plants typically develop a deep and extensive root system that ensures its ability to exploit deep water sources, thereby maintaining a sufficient water and nutrient supply for prolonged periods when the upper soil layers are drying out. An indication of the importance of the root is the high root-to-shoot ratio of *Z. mauritiana* and deep rooting which has been reported as a characteristic of both *Z. nummularia* (Anonymous 1976) and *Z. mauritiana* (Depommier 1988). Under ideal environmental conditions, *Z. mauritiana* exhibits very high rates of net photosynthesis and stomatal conductance. Any surplus of assimilated carbohydrates that is not invested in growth is stored as starch in the roots, leading to very high reserves of carbohydrate in the below-ground structures. Compared to other species for which data are available, the nitrate reductase activity (NRA) in leaves of *Z. mauritiana* is very high, with nitrate reductase activity of 1 μmol NO$_2^-$ g FM^{-1} h^{-1} measured in leaves of drought-stressed plants (pre-dawn Ψ_{leaf} 2.0 MPa, unpubl. data from our lab). Nitrate reductase activity is usually very sensitive to drought stress, and high NRA levels are a prerequisite for rapid growth during favourable conditions to meet the plant's high demand for nitrogen for production of amino acids, proteins and nucleic acids. Consequently, total nitrogen content of the leaves is very high, which is a mesic character of *Ziziphus* leaves, that lack xeromorphic adaptations such as heavy cuticularization, or deep folds in their surfaces with sunken stomata. The combination of high levels of NRA and net photosynthesis results in a high relative growth rate, essential if these plants are to compete effectively during brief periods of active growth. The large carbohydrate reserves in the roots contribute to the strong regeneration potential of *Ziziphus* plants. *Z. mauritiana* is reported as having a great power to recover from injury of any kind, including fire, and thrives on burnt grassy tracts (Anonymous 1976; Grice 1996). After such events, most plants of *Z. mauritiana* resprouted vigorously within 3 months and by the 4th month after fire, burnt and unburnt plants were similar with respect to the distribution of individuals and their physiological characteristics (Grice 1997). Because of its ability to resprout from both crown and roots, along with its resistance to herbicides, *Z. obtusifolia* has demonstrated its ability to increase its cover on Texas rangeland after the release of competition from other woody vegetation by bush control treatments (Speer and Wright 1981). There are reports that wild jujube plants (*Z. lotus*) also have the ability to resprout vigorously even after being cut to groundlevel in the previous fall in Morocco (Regehr and El Brahli 1995).

Ziziphus plants are cross-pollinated and are highly outbreeding, and as a result of this, the natural population, which largely regenerates through seeds, exhibits a vast range of genetic heterogeneity. The potential of this variability has been severely underutilized. In contrast, the commercial cultivars which are clonally propagated via budding on a suitable rootstock, have retained their genetic fidelity. In evolutionary terms, this genetic variability may well benefit *Ziziphus* in harsh environments by allowing rapid adaptation to changing environmental conditions at a population level.

Mechanisms of Drought Resistance

A wide range of physiological studies conducted on *Ziziphus* have demonstrated that *Ziziphus* species are tolerant of high temperatures and drought conditions encountered in arid regions (Clifford et al. 1997, 1998; Arndt et al. 2000). This research revealed that *Ziziphus* species are very flexible in response to drought and that they exihibit a range of reaction mechanisms (Fig. 1).

The first mechanism is one of drought avoidance through production of an extensive root system maintaining contact with deep water sources. In times of moderate water limitation, stomatal control of water loss maintains plant water status as the soil dries. Results from glasshouse experiments confirmed the high sensitivity of stomatal closure in *Ziziphus* during drought, with significant reductions in conductance occurring before any change in leaf water potential could be detected (Clifford et al. 1998). When water stress became more severe, osmotic adjustment occurred. Active accumulation of solutes in the cell sap contributed to turgor maintenance, this being a prerequisite for continued growth during drought (Hsiao et al. 1976). As a consequence of osmotic adjustment in *Ziziphus*, many metabolic functions can continue even under severe drought stress. When water is severely limiting, *Ziziphus* trees are able to selectively shed their mesophytic leaves. Leaf loss in water-stressed plants can be regarded as a beneficial adaptation that reduces water loss and in the short-term prolongs survival. *Ziziphus* trees are able to survive long periods without leaves, with high levels of mucilagenous substances in the twigs and stems which may act as a water capacitor (unpubl. data from our lab). This, together with the high carbohydrate reserves in the root, supports the maintenance of metabolism during drought periods and may enable quick, vigorous resprouting, as soon as water is available again.

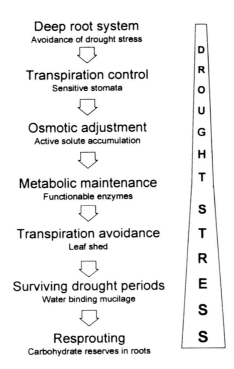

Fig. 1. Mechanism of drought resistance in *Ziziphus rotundifolia*

Consequently, Ziziphus trees exhibit a flexible combination of mechanisms of drought avoidance and drought tolerance. They can respond in various ways to drought stress and this makes them well adapted to changing environmental conditions typical in semiarid and arid regions, where the patterns and amounts of rainfall are extremely erratic.

Use as Multipurpose Trees

A great majority of the rural population in arid regions meet their daily household requirements through biomass or biomass-based products such as food, fuel (firewood, cowdung, crop wastes), fodder, fertilizer (organic manure, forest litter, leaf mulch), building materials (poles, thatch) and medical herbs. *Ziziphus* species meet many of these needs and they can be used for a variety of purposes in arid regions.

Shelter Vegetation

Ziziphus species can contribute to control the rate of desertification. Soil erosion in desert areas is largely due to the removal of structureless topsoil by the wind and rain. This can largely be checked by planting wind breaks, creating shelter belts and stabilizing sandy tracts and dunes with adapted grasses and shrubs like *Ziziphus* (Khoshoo and Subrahmanyam 1985). *Z. nummularia* shrubs have been shown to effectively check wind erosion, help in deposition of soil, and bring about a change in the microhabitat, causing favourable conditions for the appearance of successional species such as perennial grasses (Anonymous 1976). In the Sahelian climate, *Z. mauritiana* plays an important part in the conservation of soil because of its abundance and vigorous root systems (Depommier 1988).

Fruits

The fruits of all *Ziziphus* species are edible and can be prepared for consumption in many ways. The drupes are eaten either fresh, pickled, dried or made into confectionery, and the juice can be made into a refreshing drink (Khoshoo and Subrahmanyam 1985). In Zimbabwe, *Ziziphus* fruits are used to produce jam and kachaso, a crudely distilled spirit of consirable potency (Coates Palgrave 1990).

Though the fruit does not find much favour with the upper classes (poor man's fruit, Khoshoo and Subrahmanyam 1985), it has a high nutritional value and a great commercial potential. Consequently, in many regions of the world, *Ziziphus* fruits are sold on local markets, generating cash income for people of rural areas and improving family nutrition. In Oman, the most widespread indigenous fruit tree that grows in wadis throughout the country is *Z. spina-christi*, and the small brown fruits are sold in the suk (Lawton 1985). In Zimbabwe, the people from the rural areas pick the fruits of the wild-growing *Z. mauritiana* for sale in the urban markets.

Fruits of *Z. mauritiana* have higher contents of protein and vitamins A and C than apples (Anonymous 1976). The fruits contain between 70 and 165 mg ascorbic acid per 100 g of pulp, which is two to four times higher than the vitamin C content of citrus fruits. The mineral content of calcium, phosphorus and iron in *Z. mauritiana* fruits is also reported as being higher than in apples and even oranges (Jawanda and Bal 1978).

For farmers, *Z. mauritiana* is an interesting crop because it is so fast-growing and bears fruits within 2–3 years (Jawanda and Bal 1978). The yield varies according to precocity and bearing potential of different types, but it increases progressively with the age of the trees. In India, Umran is the highest yielding cultivar of *Z. mauritiana* in Punjab, giving an average yield of 80–200 kg ha^{-1}. (Anonymous 1976). In Australia, some large specimens of *Z. mauritiana* were reported as producing more than 5000 seeds per plant in 1 year, but with great variation between individual shrubs and between seasons (Grice 1996). In China, *Z. jujuba* plays an important role as fruit tree. It produces very large, tasty fruits, the so-called Chinese dates, which have an importance similar to that of dried plums in Europe (Warburg 1916). In the late 1950s, *Z. jujuba* was ranked as the most important fruit tree crop in China (Pieniazek 1959).

Fodder, Fuel and Fencing

Besides the fruits, nearly every organ of *Ziziphus* plants can be utilized. The leaves and twigs of most *Ziziphus* species can be used as high nutritional fodder for livestock. Due to their high protein content per dry weight, they are an important nitrogen source for the animals. *Z. nummularia* is prized for its leaves, which provide fodder for livestock in the summer months, particularly in the fodder-deficient areas of Punjab, Haryana and Rajasthan in India, where the average total yield of forage was ca. 1000 kg ha^{-1} (Anonymous 1976). In the Sahelian climate, the young leaves of *Z. mauritiana* are commonly eaten as vegetables, with the older foliage used as fodder (Depommier 1988). Sena et al. (1998) pointed out that leaves of *Z. mauritiana* are an important famine food in Niger because they are an excellent source of the essential fatty acid linoleic acid and several metals including iron, calcium, magnesium and zinc.

The wood of the *Ziziphus* tree species is dense and compact. It is used for various purposes in the everyday life of the people including the production of tools, poles, toys, and for turning. *Z. mauritiana* is an excellent fuel–wood tree and makes a good charcoal, with a heat content of 4900 kcal kg^{-1} (Khoshoo and Subrahmanyam 1985). In the Sahel Zone, it is considered good both as firewood and charcoal (Depommier 1988). *Z. nummularia* has also been reported to produce high-quality hardwood with high calorific value, making it an ideal source of fuel and charcoal (Anonymous 1976).

The spiny branches are widely used as living fences. The twigs of *Z. nummularia* are also used as fencing material and for making cool-airscreens for use on the windows (Jones et al. 1998). In an agroforestry project, *Z. mauritiana* was successfully introduced as a living fence in Burkina Faso.

Medicinal Properties

Most of the drugs from higher plants which have become important in modern medicine had a folklore origin and are traditional in systems of medicine such as the Ayurvedic Pharmacopoeia (Khoshoo and Subrahmanyam 1985). *Ziziphus* species are used for many medicinal purposes in folk medicines all over the world (Table 3). In India and China, *Ziziphus* species in particular have been used to treat different diseases and ailments. *Ziziphus* extracts have also featured in homeopathic folk medicines in other regions such as the Middle East, Southern Africa and South America. Almost every part of a *Ziziphus* plant has been used for medicinal purposes, as listed for *Z. jujuba* in Table 4. The wide variety of medicinal properties of *Ziziphus* plants is suprising, with uses against skin diseases, diarrhea, fever and insomnia. A general characteristic of *Ziziphus* extracts used for medicines is their antiinflammatory and antibacterial properties.

Table 3. Medicinal properties of different *Ziziphus* species and their use in traditional medicines

Species	Traditional Medicine	Medicinal use	Reference
Z. spina-christi	Egypt	Treatment of different diseases	Glombitza et al. 1994
Z. abyssinica	Zimbabwe	Treatment of tonsilitis, pneumonia, gonorrhoea, infectious diseases	Gundidza and Sibanda 1991, Coates Palgrave 1990
Z. joazeiro	Northeastern Brazil	Lessening of inflamation, relief of pain, reducing secondary infections, remedy against fever	Fabiyi et al. 1993, Nunes et al. 1987
Z. jujuba	Turkey	Hypoglycaemic agent at diabetics	Erenmemisolglu et al. 1995
Z. jujuba	China	Treatment of bronchitis, insomnia, diarrhea, ulcers, wounds and fever, sedative activity	Kustrak and Males 1987, Tanaka and Sanada 1991
Z. mucronata	Southern Africa	Treatment of diarrhea, dysentery, lumbago and skin infections, remedy agains pain	Auvin et al. 1996, Coates Palgrave 1990
Z. nummularia	India	Treatment of skin diseases, colds and coughs	Dwivedi et al. 1987
Z. rugosa	India	Treatment of diarrhea, menorrhhagia and infection of teeth	Acharya et al. 1988
Z. sativa		Treatment of ulcers, wounds, eye diseases and bronchitis	Shah et al. 1985
Z. spina-christi	Saudi Arabia	Treatment of wounds, skin diseases, inflammatory conditions, fever; diuretic agent	Tanira et al. 1988
Z. spina-christi	Bedouines	Birth control	Shappira et al. 1990
Z. spinosa	China	Treatment of insomnia, neurasthenia	Zeng et al. 1987

Data from our lab demonstrated the biological antibiotic and fungicidal activity of extracts of Z. jujuba, Z. mauritiana and Z. nummularia leaves, stems and roots (unpubl. data). The work suggested that some of the Ziziphus extracts had a higher antimicrobial activity than Penicillin G and Nystatin, and they were also effective against multiresistant strains of Aspergillus and Candida species, but further work is required to identify the active constituents of the extracts. It has also been reported that some Ziziphus species may contain the potential anticancer agent betulinic acid. In seeds of Z. spinosos (Zeng et al. 1987), stem bark of Z. joazeiro (Barbosa–Filho et al. 1985), Z. nummularia (Maurya et al. 1989) and Z. vulgaris

Ziziphus – a Multipurpose Fruit Tree for Arid Regions

(Li and Zhang 1986), betulin or betulinic acid has been detected. These data indicate that Ziziphus might have commercial potential as a source of much -needed new antibiotic and antitumour agents as well as other medically effective substances.

Table 4. Medicinal use of the different plant organs of *Z. jujuba*

Plant organ	Medicinal uses or properties
Leaves	Astringent, febrifuge, promote growth of hair, used in form of plaster in the treatment of strangury
Fruits	Anodyne, anticancer, pectoral, refrigerant, sedative, stomachic, styptic, tonic, considered to purify blood and aid digestion, used internally in the treatment of chronic fatigua, loss of appetite, diarroea, anaemia, irritability and hysteria
Seed	Hypnotic, narcotic, sedative, stomachic and tonic, used internally in the treatment of palpitations, insomnia, nervous exhaustion, night sweats and excessive perspiration
Root	Used in the treatment of dyspepsia, treatment of fevers, powdered roots are applyed to old wounds and ulcers

according to: Grieve (1971), Duke and Ayensu (1985), Bown (1995), Chopra et al. (1986)

Conclusion

Due to their characteristics as a drought-adapted fruit tree with multipurpose uses, *Ziziphus* species are promising plants for arid regions, and especially in developing countries. Local varieties can be used as shelter vegetation, fuel, fodder and for fencing, with improved varieties imported and used as stock for valuable fruit crops. The successful introduction of *Ziziphus* trees to Israel and Zimbabwe has provided a good model for the development of this species as a successful perennial crop for other drought-prone regions of the world.

Acknowledgements. The authors would like to thank the European Commission/BBSRC for funding the research (contract TS3*-CT93-0222)

References

Acharya SB, Tripathi SK, Tripathi YC, Pandey VB (1988) Some pharmacological studies on *Zizyphus rugosa* saponins. Indian J Pharmacol 20:200–202
Anonymous (1976) The wealth of India. A dictionary of Indian raw materials and industrial products, vol XI:X-Z. Council of Scientific and Industrial Research, New Dehli, pp 111–124

Arndt SK, Wanek W, Clifford SC, Popp M (2000) Contrasting adaptations to drought stress in field-grown *Ziziphus mauritiana* and *Prunus persica* trees: water relations, osmotic adjustment and carbon isotope composition. Aust J Plant Physiol 27: (in press)

Auvin C, Lezenven F, Blond A, Augeven–Bour I, Pousset JL, Bodo B, Camara J (1996) Mucronine J, a 14-membered cyclopeptide alkaloid from *Ziziphus mucronata*. J Nat Prod 59:676–678

Barbosa–Filho JM, Trigueiro JA, Cheriyan UO, Bhattachargya J (1985) Constituents of the stem bark of *Ziziphus joazeiro*. J Nat Prod 48:152–153

Bown D (1995) Encyclopaedia of herbs and their uses. Dorling Kindersley, London

Bussi C, Gojon A, Passama L (1997) In situ nitrate reductase activity in leaves of adult peach trees. J Hortic Sci 72:347–353

Cherfas J (1989) Nuts to the desert. New Sci 19:44–47

Cherry M (1985) The needs of the people. In: Wickens GE, Goodin JR, Field DV (eds) Plants for arid lands. Unwin Hyman, London

Chopra RN, Nayar SL, Chopra IC (1986) Glossary of Indian medicinal plants. Council of Scientific and Industrial Reasearch, New Dehli

Clifford SC, Kadzere I, Jones HG, Jackson JE (1997) Field comparisons of photosynthesis and leaf conductance in *Ziziphus mauritiana* and other fruit tree species in Zimbabwe. Trees 11:449–454

Clifford SC, Arndt SK, Corlett JE, Joshi S, Sankhla N, Popp M, Jones HG (1998) The role of solute accumulation, osmotic adjustment and changes in cell wall elasticity in drought tolerance in *Ziziphus mauritiana* (Lamk.). J Exp Bot 49:967–977

Coates Palgrave K (1990) Trees of Southern Africa. Struik Publ, Cape Town, pp 549–552

Depommier D (1988) *Ziziphus mauritiana* Lam. Bois For Trop 218:57–62

Duke JA, Ayensu ES (1985) Medicinal plants of China. Reference Publications Inc., Algonac, Michigan

Dwivedi SPD, Pandey VB, Shah AH, Eckhard G (1987) Cyclopeptide alkaloids from *Ziziphus nummularia*. J Nat Prod 50:235–237

Erenmemisoglu A, Kelestimur F, Koker AH, Ustun H, Tekol Y, Ustdal M (1995) Hypoglycaemic effect of *Ziziphus jujuba* leaves. J Pharm Pharmacol 47:72–74

Fabiyi JP, Kela SL, Tal KM, IstifanusWA (1993) Traditional therapy of dracunculiasis in the state of Bauchi, Nigeria. Dakar-Med 38:193–195

Glombitza KW, Mahran GH, Mirhom YW, Michel KG, Motawi TK (1994) Hypoglycemic and anti-hyperglycemic effects of *Ziziphus spina-christi* in rats. Planta Med 60:244–247

Grice AC (1996) Seed production, dispersal and germination in *Cryptostegia grandifolia* and *Ziziphus mauritiana*, two invasive shrubs in tropical woodlands of northern Australia. Aust J Ecol 21:324–331

Grice AC (1997) Post-fire regrowth and survival of the invasive tropical shrubs *Cryptostegia grandifolia* and *Ziziphus mauritiana*. Aust J Ecol 22:49–55

Grieve A (1971) A modern herbal. Dover Publications, New York

Gundidza M, Sibanda M (1991) Antimicrobial activities of *Ziziphus abyssinica* and *Berchemia discolor*. Cent Afr J Med 37:80–83

Hsiao TC, Acevedo E, Fereres E, Henderson DW (1976) Water stress, growth and osmotic adjustment. Philos Trans Roy Soc Lond, Ser B 237:479–500

Jawanda JS, Bal JS (1978) The ber, highly paying and rich in food value. Indian Hortic Oct–Dec 19–21

Johnston MC (1963) The species of *Ziziphus* indigenous to United States and Mexico. Am Jour Bot 50:1020–1027

Jones HG, Jackson J, Popp M, Sankhla N, Clifford SC, Arndt SK, Corlett JE, Joshi S, Kadzere I (1998) Final report: selection of drought-tolerant fruit trees for summer rainfall regions of Southern Africa and India. EU project TS3-CT93-0222

Kadzere I, Jackson JE (1997) Indigenous fruit trees and fruits in Zimbabwe: some preliminary results of a survey in 1993–94. In: Jackson JE, Turner AD, Matanda ML (eds) Smallholder horticulture in Zimbabwe. University of Zimbabwe publications, Harare, pp 29–34

Khoshoo TN, Subrahmanyam GV (1985) Ecodevelopment of arid lands in India with non -agricultural economic plants – a holistic approach. In: Wickens GE, Goodin JR, Field DV (eds) Plants for arid lands. Unwin Hyman, London

Kozlowski TT, Kramer P, Pallardy SG (1991) The physiological ecology of woody plants. Academic Press, Sand Diego

Kustrak D, Males Z (1987) Fitokemijski pregled cicimaka i srodnih vrsta roda *Zizyphus* Juss. Farm Glas 17:145–153

Lawton RM (1985) Some indigenous economic plants of the Sultanate of Oman. In: Wickens GE, Goodin JR, Field DV (eds) Plants for arid lands. Unwin Hyman, London

Li SY, Zhang R (1986) Quantitative determination of betulinic acid in *Zizyphus vulgaris* by TLC-colometry. Zhongyao Tongbao 11:683–685

Locke LF (1947) The chinese jujube: a promising tree for the southwest. Okla Agric Exp Stn Bull B 319:78–81

Maurya SK, Devi S, Pandey VB, Khosa RL (1989) Content of betulin and betulinic acid, antitumour agents of *Zizyphus* species. Fitoterapia 60:468–469

Nunes PH, Marinho LC, Nunes ML, Soares EO (1987) Antipyretic activity of an aqueous extract of *Zizyphus joazeiro* Mart. (Rhamnaceae). Braz J Med Biol Res 20:599–601

Pieniazek SA (1959) The temperate fruits of China. Fruit Var J 14:29–33

Regehr DL, El Brahli A (1995) Wild jujube (*Zizyphus lotus*) control in Morocco. Weed Technol 9:326–330

Sena LP, Vanderjagt DJ, Rivera C, Tsin ATC, Muhamadu I, Mahamadou O, Millson M, Pastuszyn A, Glew RH (1998) Analysis of nutritional components of eight famina foods of the Republic of Niger. Plant Foods Hum Nutr 52 (1):17–30

Shah AH, Pandey VB, Eckhard G, Tschesche R (1985) Sativanine-E, a new 13-membered cyclopeptide alkaloide containing a short side chain, from *Zizyphus sativa*. J Nat Prod 48:555–558

Shappira Z, Terkel J, Egozi J, Nyska A, Friedman J (1990) Reduction of rodent fertility by plant cosumption: with particular reference to *Ziziphus spina-christi*. J Chem Ecol 16:2019–2026

Speer ER, Wright HA (1981) Germination requirements of lotebush (*Zizyphus obtusifolia* var. *obtusifolia*). J Range Manage 34:365–368

Tanaka Y, Sanada S (1991) Studies on the constituents of *Zizyphus jujuba* Mill. Shoyakugaku Zasshi 45:148–152

Tanira MO, Ageel AM, Tariq M, Mohsin A, Shah AH (1988) Evaluation of some pharmacological, microbiological and physical properties of *Zizyphus spina-christi*. Int J Crude Drug Res 26:56–60

Warburg O (1916) Die Pflanzenwelt, 2. Band, Dikotyledonen. Bibliographisches Institut, Leipzig, pp 366–369

Zeng L, Zhang RY, Wang X (1987) Studies on constituents of *Zizyphus spinosus* Hu. Acta Pharm Sin 22:114–120

Root Morphology of Wheat Genotypes Grown in Residual Moisture

Günther Manske, Nigatu Tadesse, Maarten van Ginkel, Mathew Reynolds and Paul L.G. Vlek

Keywords. Wheat, residual moisture, drought, root morphology, genotypic differences

Abstract. Breeding for drought tolerance and improved water-use efficiency is important in order to use the water for irrigation more efficiently. Twelve wheat genotypes were grown under conditions of late drought in Northwest Mexico. Under these conditions, the extent of root exploration for the available soil moisture reserves is often a major determinant of drought tolerance. Distribution of root–length density (RLD) was assessed in the soil profile (0–100 cm soil depth).

Most of the RLD was accumulated in the upper soil layers, however, with genotypic differences. Higher RLD across all soil depths was not responsible for improved water-use efficiency. Averaged across soil depth, the largest root fresh weight was observed in a drought-susceptible check. Grain yield was negatively correlated with RLD in the upper soil layer, but was not correlated with RLD in the deeper soil. Drought-susceptible genotypes had most of their roots restricted to the upper soil, while drought-resistant genotypes had high RLD deeper in the soil profile. Genotypes were identified to be used in a breeding program as donor parents to increase RLD for utilizing subsoil moisture and enhancing grain yield in late-drought environments.

Introduction

Water for irrigated wheat cultivation is becoming a limited resource in many areas of the tropics because of increasing urban water consumption and increasing cropping intensity. Therefore, drought stress is an important limiting factor for wheat production in semiarid regions, where not sufficient or no water for irrigation is available. Rajaram and van Ginkel (1996) have identified three types of drought for wheat based on stages of plant development at which drought is most severe. In the case of late drought, wheat is produced on subsoil-stored moisture, accumulated during the rainy season before or at the beginning of the cropping cycle. Under these conditions, the extent of root exploration for the available soil moisture reserves is often a major determinant of drought tolerance

(Bai et al. 1997). The labour-intensive root measurements in germplasm that are becoming possible by new techniques are increasingly in demand by breeders. The objective of this study was to evaluate wheat genotypes for root morphological traits that contribute to differences in drought tolerance.

Methods

This study was conducted during the winter of the 1996/1997 season at the CIANO/CIMMYT field station near Ciudad Obregon in Northwest Mexico (27°N 109' W, 40 m asl). Obregon has a semiarid climate, and wheat is usually grown under intense irrigation during the winter seasons. Twelve genotypes representing drought-resistant and -susceptible wheat germplasm (*Triticum aestivum* L.) were grown under conditions of residual moisture simulating subcontinent-type drought (Rajaram and van Ginkel 1996). The wheat plants received only a single irrigation immediately after planting. Soil moisture content was monitored during the crop cycle. Soil samples were taken from four different layers (0–30, 30–60, 60–90, 90–120 cm). The experiment was laid out as an alpha-lattice design with three replications. At hard dough growth stage, roots were sampled by using a root auger.

Soil cores of 750 cm^3 (20 cm deep, diameter 8 cm) were taken down to 100 cm depth. Six subsamples per plot assured significant differences between genotypes. The subsamples of each plot were bulked – but separately for each soil layer – and the roots were separated from the soil by washing in a root-washing machine (hydromatic eludration system, Smucker et al. 1982). The root-length density was measured by the gridline intersect method (Tennant 1975) through an image -analysis system. Root-length density (RLD) was calculated by the root length divided by the volume of the cores. Two weeks after anthesis, scores were taken for intensity of the green colour of the canopy (leaves, spikes and stems) (score 1 – yellow to 5 – completely green) by visual observation from each plot. Data were analyzed using GLM procedure of SAS (1985).

Results

During the crop cycle there was a decline in soil moisture, especially in the upper soil layers (0–30 and 30–60 cm). The soil water table rose in the deeper layers due to irrigation in adjacent fields (Fig. 1) The experimental station in Obregon is known for high yields in wheat, and CIMMYT´s wheat breeders select at this location for high yield potential. However, the grain yields in this trial were very low (experimental average 2.15 t ha^{-1}) because of the drought stress (Table 1). The wheat genotypes Sujata and Pastor had the highest grain yields, Check 3 the lowest. Also Seri, Nesser, the synthetic line 2 and Pavon had relatively low grain yields. Plant height tended to be positively related with grain yield (r = 0.45, not significant, Table 1).

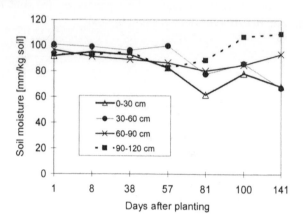

Fig. 1. Development of residual moisture in different soil depths of wheat grown under conditions of residual moisture (only one irrigation after planting), in Obregon, Northwest Mexico

Scores for stay grain were not clearly related to grain yield. Some of the genotypes with low yield potential like Check 3 and Synthetic 2 stayed green for a long time (Table 1). However, the low-yielding genotypes Sonalika and Pavon yellowed rapidly, whereas the high-yielding Sujata and Synthetic 1 stayed green longer. Genotypic differences for root fresh weight were significant, but not for RLD (Table 2). The latter showed significant differences among soil depths and genotype x depth interactions. Most of the RLD was measured in the upper soil layers, regardless of genotypic differences (Fig. 2). Root length was twice as high as on the average of the total examined profile (0–100 cm depth), and almost four times higher than in the deep layer (80–100 cm) (Table 1). Improved total root volume and weight in the total profile (0–100 cm) were not responsible for improved water-use efficiency. Averaged across soil depth, the largest root fresh weight was observed from the drought-susceptible Check 3. Grain yield was negatively correlated ($r = -0.70^{**}$) with RLD in the upper soil layer (0–20 cm), but not with RLD in the deeper soil. Drought-susceptible genotypes tended to have most of their roots restricted on the upper subsoil, while drought-resistant genotypes had high RLD on deeper soil profile (Fig. 2). The high-yielding Pastor and Sujata had a deeper-penetrating root system compared, for instance, to the lower-yielding Pavon. Pastor and Sujata had equal RLD distributions of 50% within the top 60- and bottom 40-cm soil layers, whereas about 70% RLD of Pavon was in the top 60-cm soil layer (Fig. 2).

Scores for staying green were positively correlated ($r =0.69^{**}$) with RLD in the

Root Morphology of Wheat Genotypes Grown in Residual Moisture

Table 1. Grain yield, scores for staying green, plant height, root fresh weight (RFW) and root length density (RLD) at different soil layers of 12 wheat genotypes grown at residual moisture

Genotypes	Grain yield t ha^{-1}	Score[a] staying green	Plant height cm	RFW 0–100 cm g sample^{-1}	RLD in diff. soil layers		
					0–100 cm cm cm^{-3}	0–20 cm cm cm^{-3}	80–100 cm cm cm^{-3}
Sonalika	2.29	1	80	3.98	0.95	1.92	0.40
Baviacora	2.51	3	70	4.12	0.98	1.85	0.43
Pavon	2.15	1	67	4.74	1.11	2.45	0.45
Nesser	2.07	2	58	4.24	1.00	1.82	0.45
Tevee	2.29	3	65	3.50	0.88	1.92	0.49
Seri	1.76	2	62	4.69	1.08	2.09	0.56
Check 3	0.71	5	65	5.36	1.14	2.71	0.61
Synthetic 1	2.44	4	75	3.76	0.92	1.76	0.65
Synthetic 2	2.15	5	77	4.82	1.04	1.89	0.68
Dharwar	2.20	3	85	4.42	1.09	2.25	0.70
Sujata	2.65	5	90	4.15	0.88	1.42	0.71
Pastor	2.52	3	70	4.39	1.01	1.62	0.72
Mean	2.15	3	72	4.35	1.01	1.97	0.57
LSD 5%	0.38	0.93	6.56	0.70	0.18	0.72	0.24

[a] Score staying green after anthesis, 1<5 for increasing green

deep soil layer (80–100 cm depth). However, RLD in the deep soil layer (80–100 cm) was not correlated with grain yield, because Check 3 and Synthetic 2 had high RLD in the deep soil layer without high yields because of low genotypic yield potential. In contrast, Baviacora could obtain high grain without having a deep -penetrating root system (Table 1) because of a high yield potential and other unknown morphological and physiological traits. Possibly Baviacora had a higher water-utilization efficiency (yield/water absorbed). It seems to be that a high genotypic yield potential is a prerequisite for improved yields under drought. Without this, increased water-uptake efficiency due to an improved deep -penetrating root system cannot be converted into higher yields under drought. Therefore, future breeding for improved drought tolerance has to combine both high yield potential and improved root systems. Pastor and Sujata had the highest RLD in the deepest soil layer (80–100 cm) and high yields. Both genotypes could be used in a breeding program as donor parents to increase RLD for utilizing subsoil moisture and enhancing grain yield in late-drought environments or under reduced-irrigation management.

Acknowledgements. This study was financially supported by BMZ, Germany, and the National Science Foundation, US.

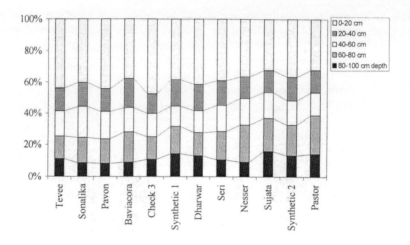

Fig. 2. Distribution of root length density at 20-cm intervals from 0–100-cm soil depth of 12 wheat genotypes grown at residual moisture, in Obregon, Northwest Mexico

Table 2. Analysis of variation: mean squares and probability by F-test for root fresh weight and root length density at five soil depths at 20-cm intervals from 0 to 100-cm depth of 12 wheat genotypes grown at residual moisture

Source of variation	DF[a]	Root fresh weight		Root length density	
Genotype (G)	11	0.16	**	0.12	NS
Soil Depths (D)	4	3.42	**	11.06	**
G x D	44	0.06	NS	0.12	*
Rep	2	0.15	*	0.22	*

*, ** significant at $P = 0.05$ and 0.01, respectively; NS = no significant differences; [a] degree of freedom

References

Bai QA, Sinclair TR, Ray JD (1997) Variation in transpirational water use efficiency and root length development among wheat cultivars. Soil Crop Sci Soc Florida Proc 56:14–19

Rajaram S, van Ginkel M (1996) Wheat Spec Rep No 5. CIMMYT, DF Mexico

SAS Institute (1985) SAS user guide: statistics. Version 5 SAS Inst, Cary, NC

Smucker AJM, McBurney SL, Srivastava AK (1982) Quantitative separation of roots from compacted soil profiles by the hydro-pneumatic eludration system. Agron J 74:500–503

Tennant D (1975) A test of modified line intersect method of estimating root length. J Ecol 63:995–1001

Field Studies in Solar Photocatalysis for Detoxification of Organic Chemicals in Water and Effluents

Lea Muszkat, Leonid Feigelson and Lena Bir

Keywords. Solar photocatalysis, TiO_2, organic pollutants, water treatment

Abstract. TiO_2 – photocatalytic degradation of organic contaminants in water included studies of water samples taken from a heavily polluted well under exposure to natural sunlight; waters from liquid commercial pesticide formulations, simulating treatment of rinse waters of agricultural sprayers; in these studies most compounds were appreciably degraded within 3.5 h but more complete detoxification required longer exposure. Photocatalytic oxidation of dye chemicals from textile industry resulted in complete degradation of selected azo and thiazine dyes. In the case of dyes the photocatalytic oxidation is accompanied by a reaction of dye sensitization. Enhanced degradation of colourless refractory pollutants has been observed in the combined process. The main advantage of the combined approach is the utilization of visible light for the degradation of refractory organic pollutants, both coloured and colourless, which cannot be photosensitized in the visible region.

Different reaction patterns of photooxidation has been witnessed for the herbicides metribuzin (4-amino-6-tert-butyl-4,5-dihydro-3-methylthio-1,2,4 -triazine-5-one) and bromacil (5-bromo-3-sec-butyl-6-methyluracil). In metribuzin oxygen has a pronounced effect on the rate of photooxidation, while the influence of hydrogen peroxide is quite moderate. The photolytic process in this case would apparently start via a reaction of the excited herbicide molecule with hydrogen peroxide or with oxygen. In the case of bromacil, oxygen does not have a pronounced effect on the rate of photooxidation, which however is considerably enhanced by hydrogen peroxide. The reaction is initiated by hydroxyl radicals generated by hydrogen peroxide photolysis. These conclusions are supported by the different effects of isopropanol inhibition.

Introduction

The removal of organic pollutants from water and effluents is of special significance for water quality preservation. Many of the organosynthetic chemicals are persistent, tend to accumulate in living organisms and are capable of penetration into groundwater (Cohen et al. 1986; Muszkat et al. 1993). Some of them exhibit long-term toxicity. The conventional methods for the remediation of organic pollutants are unsatisfactory: biodegradation is ineffective at the sub-ppm levels. Other treatments, like adsorption on activated carbon, air stripping and pyrolysis, are non-destructive processes, where only phase transfer of the contaminant occurs. These limitations led to increased interest in the application of advanced oxidation processes (AOPs) to the treatment of wastewater and polluted drinking water. Solar-titanium dioxide photocatalyzed degradation (SPO) of organics promises to play an important role in environmental remediation of pollution by xenobiotics and by other contaminants, due to its unique advantages (Matthiews 1993; Ollis 1985; Bahnemann 1999). The SPO differs from other water treatment AOP methods in that it is capable of both oxidation and reduction. While oxidation is used to destroy organic toxicants, reduction is found to be effective for removing heavy metals from water by conversion of metal ions to their metallic form, which plate out onto the photocatalyst. Other advantages of the TiO_2-photocatalytic oxidation include the capability of complete mineralization of almost all organic compounds, owing to its very high positive oxidation potential, thus avoiding the possibility of toxic intermediates. A unique advantage of photocatalytic oxidation over other AOPs is the option of using natural sunlight as energy source. TiO_2 is capable of utilizing light below 390 nm, the portion of solar spectrum which amounts to about 5% of the solar energy reaching the Earth's surface. The method provides a "green" technology. It is self-sustaining and non-specific oxidation powered by solar energy, a renewable energy source. TiO_2 photocatalyzed degradation studies of several dyes show that the process is of wide application: Reeves et al. (1992); Vinodgopal et al. (1996); Zhang et al. (1998). Today, one of the still unresolved problems which prevent large-scale application of the TiO_2-photocatalyzed degradation is the low quantum yield of the process. In the present chapter we report on a new possibility of enhancing the rate (and the apparent quantum yield) of the process by involving in this process photooxidant species external to the TiO_2 photocycle.

A general remark concerns the effect of pollutant concentrations. Due to the apparent first-order kinetics, the same amount of energy is required to go from 100 ppm to 1 ppm, as to to go from 100 to 1 ppb . Therefore, at higher concentrations, the amount of pollutants destroyed per unit energy is much greater, which is of significance in economic considerations. The scientific literature reports on the complete degradation of many organic substances at both low levels of the $\mu g\ l^{-1}$ range, as well as at high levels of hundreds of $mg\ l^{-1}$. While the destruction of pollutants at the $\mu g\ l^{-1}$ level is of great significance in the purification of drinking water and also in obtaining ultra-pure water, the destruction of pollutants at sub 10 $mg\ l^{-1}$ level is by itself of special importance, since it allows the treatment of hazardous industrial wastewater. This application of solar energy seems to be quite promising. In this case, photocatalytic treatment involves addition of oxidants. Scientific literature reports the photocatalytic mineralization of almost all classes of hazardous chemicals under laboratory conditions. Mechanistic studies have not been extensive and relatively few experiments have been performed on a pilot scale. As already mentioned, the disappearance of pollutants

Field Studies in Solar Photocatalysis for Detoxification of Organic Chemicals 407

apparently follows first-order kinetics. Therefore, at higher concentrations, the amount of pollutants destroyed per unit energy is much greater.

Results and Discussion

In the present chapter some of our field studies are summarized. They demonstrate specific applications of the solar photocatalytic technique:

Groundwater Samples

Groundwater samples from a heavily polluted well were subjected to photocatalytic treatment in the presence of TiO_2 and H_2O_2 (Muszkat et al. 1992). Results from this treatment are presented in Tables 1 and 2. The pollutants, in the $\mu g\ l^{-1}$ range, included a wide range of aliphatic and aromatic halogenated compounds, herbicides from the series of triazines, acetamides, bromouracils, hydroxybenzonitriles and ureas, and also several groups of industrial organics. Efficient decontamination was achieved by solar photocatalytic treatment. Most of the contaminants were appreciably degraded (ca. 70% of decomposition) within 3.5 h by natural sunlight and complete decontamination required a longer exposure period. The compounds were found to undergo complete solar photodegradation to less than 0.1 ppb

Halocarbon Solvents. The concentration of the widely abundant solvents 1,2 -dichloroethane, trichloroethylene and tetrachloroethylene is decreased by several orders of magnitude. The results for the very volatile chloroform were erratic (not shown), probably due to evaporation losses. At least 15 h of sunlight were required in order to reduce the extremely high initial concentrations of some of these solvents (about 2000 ppb) by more than 1 order of magnitude.

Halobenzenes. The initial concentrations of several halobenzenes in the well water were in the range 0.5–20 ppb. As shown in Table 1, exposure shorter than 3.5 h was sufficient for complete detoxification of most compounds of this group.

Aromatic Hydrocarbons. The two aromatic xylenes were completely degraded within less than 3.5 h to below 0.1 ppb.

Triazine Herbicides. The triazines detected in the well water were atrazine, trietazine, propazine, terbutryn and an isomer of atrazine (m/z = 215). These molecules were decomposed to less than 0.1 ppb. It can be seen (Table 1) that 15 h exposure was sufficient to reduce the level of atrazine, of its isomer and of propazine to less than 0.1 ppb, but trietazine and terbutryn required a longer time.

Acetamides, Bromacil and Urea Herbicides. Three widely used herbicides, alachlor, propachlor and bromacil, were found at high levels in the contaminated well. Exposure of 3.5 h resulted in complete disappearance of the urea derivatives methobromuron and 1-methoxy-1-methyl-3-phenylurea, but in the case of alachlor and bromacil the initial concentration was relatively high and complete degradation required a longer exposure time.

Table 1. Solar photocatalytic degradation of industrial organics in polluted well water; treatment with TiO_2 and H_2O_2

Compound	Concentration (ppb)			
	Untreated	Exposure time (h)		
		3.5	15	75
Organochlorine solvents				
1,2 Dichlorethane	1950	750	70	10
Trichlorethane	1900	160	4	1
Tetrachlorethane	2400	160	4	nd[a]
Chloro and Brombenzenes				
Chlorobenzene	21	nd	nd	nd
Dichlorobenzenes	16	0.6	nd	nd
Bromoxylenes	1	nd	nd	nd
Trichlorbenzenes	20.5	1.5	nd	nd
Alkyl benzenes				
Xylenes	9	nd	nd	nd

[a] Not detected

Table 2. Solar photocatalytic degradation of herbicides in polluted well water; treatment with TiO_2 and H_2O_2

Compound	Concentration (ppb)			
	Untreated	Exposure time (h)		
		3.5	15	75
Atrazine	43	8.5	4	nd[a]
Trietazine	23	9.5	5	0.3
Alachlor	640	3.5	nd	nd
Bromacil	115	40	6	nd
Metobromurone	22	nd	nd	nd
Chloroxynil	6	4.5	nd	nd
Bromoxynil	3	nd	nd	nd

[a] Not detected

Chloroxynil, Bromoxynil and Some Benzene Derivatives. The herbicides chloroxynil and bromoxynil (dichloro- and dibromo- chloroxynil, bromoxynil and some benzene derivatives hydroxybenzonitrile, respectively) and also dichloro -benzaldehyde and di-tert-butyl hydroxy toluene (BHT), which were detected at levels of 0.5–6 ppb, were all completely degraded (to < 0.1 ppb) within 15 h.

Phthalate Esters and Fatty Acid Derivatives. The long-chain carboxylic acids, tetradecanoic, hexadecanoic and octadecanoic acid, which appeared in rather minute concentrations (< 12 ppb) in the contaminated well, were essentially unaffected by photocatalysis. Octadeceneamide was degraded within 3.5 h to

Field Studies in Solar Photocatalysis for Detoxification of Organic Chemicals 409

below 0.1 ppb, possibly by photolysis to octadecenoic acid. Diethyl and diisooctyl esters of phthalic acid were not appreciably degraded.

Hydroxymethyl Coumarins and Dibutoxy Methane. These compounds were completely removed during illumination. The two isomers of hydroxymethyl coumarin and butylmethyl pyridinedione (butyl methyl uracil) were degraded within 3.5 h but dibutoxymethane (formaldehyde dibutyl acetal) disappeared only after 15 h.

TiO$_2$- Solar Photocatalysis in the Absence of H$_2$O$_2$

Field studies of the polluted well water point to the possibility of inhibition of the photodegradation by some unknown factors. This inhibitory effect is prevented by hydrogen peroxide (Muszkat et al. 1995), which leads to effective solar photodegradation, presumably due to oxidation of inhibitory constituents in the polluted water. In the absence of hydrogen peroxide, significant degradation occurred only during the first hour, and then the process underwent a strong retardation and the level of pollutants stabilized, owing to system deactivation. The data on photodegradation of dilute malathion and atrazine solutions (Muszkat et al. 1995) reveal that effective photodegradation occurs also in the absence of H$_2$O$_2$. Comparison with the direct excitation, non-catalytic photodecomposition emphasizes the decisive contribution of TiO$_2$.

Treatment of Rinse Water from Agricultural Spray Containers

The potential of photocatalytic oxidation carried out in the absence of hydrogen peroxide has been examined for purification of rinse water from agricultural spray containers, mainly from aircraft sprayers. These waters are significantly polluted with pesticide formulations, at levels of 10–100 ppm. Thus treatment is required prior to the disposal of the rinse waters. To simulate such conditions, the photocatalytic treatment of pesticides in their formulation mixtures was examined (Muszkat et al. 1995).

We see that 90% decomposition was achieved within less than 5 h of solar exposure, while sub-ppm level has been reached within 18 h. Satisfactory results have also been obtained in closed systems, covered with polyethylene film. The covering of sun-irradiated basins with a polyethylene film provides a simple solution to problems inherent from semivolatile contaminants such as pesticide formulations.

Solar Photocatalytic Decomposition of Textile Dyes

A complete degradation of selected azo- and thiazine dyes has been achieved by TiO$_2$-solar photocatalytic oxidation (SPO). Here, the SPO is accompanied by a reaction of dye sensitization. Enhanced degradation of colourless refractory pollutants has been observed in the combined process (Muszkat et al. 1999). The main advantage of the combined approach is the utilization of visible light for the degradation of refractory organic pollutants, both coloured and colourless, which cannot be photosensitized in the visible region. Although little is known on the subject, the present results on a photo-enhanced photodegradation demonstrate that the method is promising in the removal of undesirable toxic chemicals from textile industry wastewater.

Reaction Mechanism

Our recent studies revealed the existence of two different patterns in the self -substrate-photosensitized oxidation of pesticides, in the absence of the TiO_2 photocatalyst (Muszkat et al. 1998). The specific role of oxygen in some of the non-catalytic reactions has been noticed, where a reaction of the excited pesticide (metribuzin case) occurred with an oxygen molecule. In the other reaction type a reaction of the excited hydrogen peroxide (bromacil case) with pesticide occurred. The recognition that several direct photooxidation mechanisms could be involved in the treatment of polluted water is of significant practical importance. It may allow the exact photooxidation procedure to be tailored to a specific case of pollutant.

Conclusions

In spite of the great number of publications dealing with AOPs for water treatment, there is an evident lack of information on the scaling-up potential of the laboratory experiments. Such information is essential for the evaluation of the technical and economical feasibility of a given process. In order to achieve a reliable evaluation, more experimental work is necessary to clarify the main factors which affect the process and accelerate the degradation. Laboratory and field studies point to the promising potential of solar photocatalytic oxidation. However, several topics have to be extensively investigated, to turn these possibilities into reality. One such point is the development of more active and less costly photocatalysts, either in suspension or in supported form. Increase in the efficiency of sunlight utilization is undoubtedly required. In addition, further progress is required in large scale field studies, to allow the design and evaluation of photoreactor systems and of catalyst suspension designs, and, certainly, the solution of the photocatalyst fouling problem, which is crucial for the implementation of this technology. Thus, while the outlook for the commercial application of TiO_2 photocatalytic degradation seems good, much further work is needed to develop this much-needed environmental approach.

References

Bahnemann D (1999) Photocatalytic detoxification of polluted waters. In: Hutzinger O (ed) The Handbook of Environmental Chemistry, Vol II, Part L, pp 285–351

Cohen ZZ, Eiden C, Lorber MN (1986) In: Garner WY (ed) Evaluation of pesticides in ground water, ACS Symp Ser 315, American Chemical Society, Washington DC, pp 170–196

Matthiews RW (1993) Photocatalysis in water purification. In: Ollis DF, Al–Ekabi H (eds) photocatalytic purification and treatment of water and air, Elsevier, Amsterdam, pp 121–136

Muszkat L, Halmann M, Raucher D, Bir L (1992) Solar photodegradation of xenobiotic contaminants in polluted well-water. J Photochem Photobiol A Chem 65:409–417

Muszkat L, Bir L, Raucher D, Magaritz M, Ronen D (1993) Unsaturated zone and groundwater contamination by organic pollutants in a sewage-effluent irrigated site. Ground Water 31:556–565

Muszkat L, Bir L, Feigelson L (1995) Solar photocatalytic mineralization of pesticides in polluted waters. J Photochem Photobiol A Chem 87:85–88

Muszkat L, Bir L, Feigelson L, Muszkat KA (1998) Reaction patterns in photooxidative degradation of two herbicides. Chemosphere 36:1485–1492

Ollis D (1985) Contaminant degradation in water of halogenated hydrocarbons. Environ Sci Technol 19: 480

Reeves P, Ohlhausen R, Sloan D, Pamplin K, Scoggins T, Clark C, Hutchinson B, Green D (1992) Photocatalytic destruction of organic dyes. Solar Energy 48:413–420

Vinodgopal K, Wynkoop D, Kamat PV (1996) A photosensitization approach for the degradation of a textile azo dye. Environ Sci Technol 30:1660–1666

Zhang F, Zhao J, Shen T, Hidaka H, Pelizzetti E, Serpone N (1998) TiO_2-assisted photodegradation of dye pollutants. Appl Catalysis B Environ 15:147–156

Part VI:

National Programs

Activity of the Consulting Centre to Combat Desertification in Turkmenistan

Ch. Muradov

Keywords. Combat desertification, solar and wind energy, ecology, sustainable development

Abstract. The Desert Research Institute implements an active and important part in the international efforts to research and combat desertification. At present, the Consulting Centre had prepared a number of project proposals concerning desertification and a convention of combating desertification.

The Convention to Combat Desertification stresses the necessity to encourage the responsibility of those people who are affected by desertification, to make them involved in all the different stages of implementation of the Convention, to emphasize the role of the local authorities' function, which should be oriented toward reducing the risk of losing livelihood and attracting all relevant social groups and structures to direct participation. The National Action Programme to Combat Desertification (NAPCD) in Turkmenistan was prepared in 1996, and after confirmation by the State Commission on NAPCD development, it was submitted to the Government of Turkmenistan for consideration. The state programme 10 years of stability initiated by the President of Turkmenistan, Saparmurat Turkmenbashi, promoted sustainable utilization of natural resources and NAPCD was prepared within the framework of this programme.

The NAPCD pursues the following objects: assessment of the physical and economic potential of Turkmenistan by transition to market relations, identification of criteria of land degradation, development of new modern technologies by rehabilitation on degradation lands and sustainable utilization of natural resources. To solve the problems set by the Convention, we need fresh views and relationships, innovative approaches and a strategy for the realization of the new programmes, particularly in the issues of bottom-up-approach implementation. Certian measures for the preparation of the first National Forum should be carried out. The National Forum is called upon to encourage the development of partnerships and to serve with corresponding agreements as a mechanism of consultation, coordination, controll and analysis during the whole process of preparation and realization of National Action Plan to Combat Desertification in Turkmenistan.

For several years, the Institute of Deserts, jointly with the German Agency on Technical Cooperation (GTZ) and the German firm GEOPLAN, has been carrying out consultations and seminars on desertification problems. A number of seminars with participation of the Institute's staff were held to extend the consultative service of the Institute in Turkmenistan and in foreign countries. New trends of research using GIS–technology have been discovered, and special attention has been paid to the socio-economic aspects of desertification in Turkmenistan using PRA (Participatory Rapid/Rural/Relaxed Appraisal) methods.

A number of seminars and competitions among local populations have been held in the Erbent farm association (EFA). An art competition among the pupils was devoted to nature protection. Interviews with local inhabitants with the help of PRA methods have been conducted in EFA. Today the project Participatory management of natural resources in Central Karakum is conducted within the precincts of the institute.

In the framework of the project, the National Institute of Deserts, Flora and Fauna (NIDFF) helps to EFA to fix shifting sands around the village of Bakhardok, to supply the Yarma settlement with electricity (for house lighting) using wind and sun energy, providing the inhabitants of Erbent with drinking water of good quality. The implementation of the project will reduce the expansion of desertification processes in the Karakum. Gas and oil extraction complexes protected from sand drifts will be able to continue normal exploitation of gas and oil fields. Stabilization of moving sand dunes will improve the ecological situation in the region. Thus, it will be a considerable contribution in realization of the National Action Programme to Combat Desertification.

The Desert Research Institute has developed a complex technique of moving sand stabilization. The technique includes construction of mechanical barriers, growing sand-binding plants and chemical sand stabilization. The present project is based on the use of this technique. Only 64% of desert rangelands in Turkmenistan are supplied with water. Rangelands located close to cultural zones are exploited severely and remote rangelands are not utilized. Practical application of solar generators will stimulate the development of remote rangelands. Animal pressure around a cultural zone will be decreased and desertification will be reduced. So the project will contribute greatly to implement the National Action Programme to Combat Desertification.

The territory of Turkmenistan has vast resources of solar energy. On the latitude of Ashkhabad, by clear skies 10 000 MgJ of sun radiation falls on $1 m^2$ of the surface located perpendicular to the sun's rays. By medium cloud cover this amount of energy is reduced to 6 500 MgJ. The amount of solar energy within the limits of Turkmenistan varies only slightly. This feature testifies to the possibility of using solar generators in Turkmenistan.

The problems of pasture management using GIS–technology, the migration of people from desert areas, the improvement of leasing conditions in livestock breeding, strengthening the role of local bodies (archyn, gengeshy) in environmental protection of desert areas, providing remote settlements with gas in EFA to decrease the shortage of saksaul for fuel are the main objects of study of the Consulting Centre.

Over the past few years, with the help of the Consulting Centre, a number of projects have been prepared with participation of scientists from other countries to

Activity of the Consulting Centre to Combat Desertification in Turkmenistan 417

obtain financial support from international funds and other donor organizations, namely:

• Caspian ecological programme EC TACIS. The regional centre to combat desertification (for five Caspian countries).
• Integrative methods of cattle-breeding development and pasture conservation in Central Asia (USA, Kazakhstan, Uzbekistan).
• Study of runoff water collection in the deserts of Central Asia to increase agricultural production (Holland, Israel, Uzbekistan).
• Effect of biogenic crusts on biomass estimation in arid regions by satellite imagery (Israel).

A number of priority projects to combat desertification in the Aral Sea Basin have been developed and submitted for consideration to the International Fund of the Aral Sea jointly with the western donors, including the Global Ecological Fund, the World Bank, etc. To extend the consultative service of the Institute, an English version of a book on the main trends of the Institute's activity and recommendations on rational use of desert territories for industrial consumers has appeared: *Desert Problems and Desertication in Central Asia*, Springer, 1999. A TV film of the Institute's activity has been released, and reports on progress have been frequently published in the newspapers, as well as in the Internet.

The regional seminar: *Realization of the UN Convention to Combat Desertification and the UN Convention on Biodiversity in Central Asia: a new approach* was organized by the NIDFF 2–5 March, 1999, in Ashkhabad, with the assistance of the German Agency GTZ. In reports and discussions, the representatives of Central Asia and Russia recommended developing a Subregional Action Programme to Combat Desertification and mitigate the effect of drought in Central Asia, and determined the NIDFF as focal point and scientific coordinating centre for this project. Cooperation with experts of the Islamic Republic of Iran has been extended include specialists with a Ph.D. in the field of desertification, as well as to make joint expeditions. Today the expedition *Ecological conditions and biodiversity of border zone between Turkmenistan and the Islamic Republic of Iran* is in a developmental stage, and joint scientific -experimental projects on agricultural production habe been developed together with experts of the NIDFF, the Turkmen University of Agriculture and the MJF Fund of Iran.

The themes of scientific research works of the National Institute of Deserts, Flora and Fauna scheduled to run during 1999–2000 take into consideration the demands of the ministries and other state organizations of Turkmenistan, and have been submitted for consideration by them and might be inserted with amendments. One such example is an elaboration of the National Action Programme to Combat Desertification in Turkmenistan with the financial help of UNEP and ESCAP, realization of which is conducted with the technical support of the German Agency on Technical Cooperation (GTZ) and the Secretariat of CCD.

In future, the Consulting Centre to Combat Desertification will continue its work, expanding its consulting services in the Central Asian region.

Desertification in China and Its Control

Sun Baoping and Fang Tianzong

Keywords. China, desertification, control, measure

Abstract. The situation of desertification in China is described in this chapter. China is one of the seriously affected developing countries with vast areas of desertification, which have very large populations and frequent disasters, and with limited arable lands. The lands affected by desertification are mainly distributed at a range of longitude of 74–119°E and latitude of 19–49°N. The total area of the arid, semiarid and dry subhumid areas is about 3.32 million km^2, of which 2.62 million km^2, covering 79.0% of the total area, and occupying 27.3% of the total land territory of China, has been desertified, is still developing at present, at an annual rate of 2 460 km^2. The authors introduce the general background on China's desertification, such as the current situation, distribution, types, and causes of desertification, also including its damages. Major measures to combat desertification, including ecological protective project systems and special control measures (especially in sandy desertified regions) are given in this chapter. In addition, a practical study case (in Daxing County, Beijing), is introduced. The development in combating desertification in China is also reported.

Introduction

Desertification of drylands is a global problem in both developed and developing countries; it damages almost 30% of the total land area of the world and is expanding at an annual rate of 50 000–70 000 km^2. Desertification brings about direct threats to nearly 1 billion people, 20% of the total global population. China is one of the developing countries affected by serious desertification problems. It is estimated that there are 2.62 million km^2 of desertification affected lands covering 27.3% of the total land territory, or 79.0% of the area of the arid, semiarid and dry subhumid area of China (the total dry land is about 3.32 million km^2). The definition of desertification is "land degradation in arid, semiarid and dry subhumid areas resulting from various factors, including climatic variation and human factors"(UNEP 1995). International communities call for worldwide social participation and wisdom to slow down the spread of desertification and control its disastrous results. On the United Nations Conference on Environment and Development (UNCED) at Rio in 1992,

Desertification in China and Its Control

combating desertification was determined as a priority area of international activities.

The Chinese people living in the desertified regions have carried on a tenacious struggle against the disasters of desertification for a long time, in order to promote international cooperation and exchange in combating desertification. This chapter mainly deals with desertification in China and its control, including the basic background and some research achievements.

Present Situation of Desertification in China

Distribution and Types

The desertified lands are mainly distributed at a range of longitude of 74–119°E and latitude 19–49°N. The total area of the arid, semiarid and dry subhumid areas is approximately 3.32 million km^2, of which 2.62 million km^2, covering 79.0% of the total area of these areas and occupying 27.3% of the total land territory of China, has been desertified. It is estimated that the severity of desertification in China is steadily accelerating and developing at an annual rate of 2 460 km^2, is due to the wide distribution of the desertified lands and the difficulties in rehabilitating the affected lands and restoring the degraded environment, caused by the irrational human economic activities on sandland, rangeland and farmland.

According to the UNEP's assessment and China's actual situation, there are two types of desertification. One is sandy desertification, occurring on sandy land surface, due to irrational land use under dry and windy climatic conditions. Another is formed on the earthy deposit or weathering crust as a result of irrational land use and water erosion. The Sandy desertification mainly has occurred in the northwest of China, its forms including (1) removal of fixed dunes (sandy lands); (2) sandy rangelands degraded by wind erosion and (3) arable lands becoming coarsened.

Causes of Desertification in China

Natural Factors

- Drought. China lies in the eastern Euro–Asia continental shelf, influenced by atmospheric circulation and also topography. Northwestern China is controlled by such dry weather. K (aridity index) is used to classify the climate, $K = 0.16 \times sum(t)/R$, where, sum(t) is the total sum of daily average temperature of ≥ 10 °C and R the sum of rainfall during the same period. K = 1.5–1.99 indicates a semiarid region, K = 2–3.99 an arid region and $K \geq 4$ an extremely dry region. Most desertification-affected regions are dry (Wang Lixian 1991).

420   S. Baoping, F. Tianzong

- Gale. Wind plays an important role in the formation of the surface morphology of desertified land. Average annual wind velocity ranges from 3.3–3.5 m s^{-1}, the maximum wind velocity occurs in spring in the arid and semiarid zones of northern China. The days when wind velocity exceeds the critical velocity of drifting sand range from 200 to 300 days within a year, of which, on 20–80 days the wind speed is over force 8 (on the Beaufort scale), and gales mainly occur in winter and spring with little rainfall and sparse vegetation cover. Owing to this, wind erosion and desertification are liable to occur.
- Source of sand. Desertification-affected regions usually lie in a basin or the steppe along a river, where a large amount of uncemented material accumulated in the basin becomes the main sand source of desertification. Based on interpretation of remote sensing and ground calibration of site investigation during different periods (Wang Lixian 1996), the area of expanding desertified lands, mainly so-called sandy desertification, developed at an annual rate of 1560 km^2, from the 1950s to the 1980s; at present, at an annual rate of 2460 km^2, including all types of desertification.
- Topographic factors. The topography in China is manifold. Uplands account for 33%, plateaus 26%, basins 19%, hilly regions 10%, and plains about 12%, mountainous and hilly regions, including upland, plateau and hilly regions, about two-thirds of the total territory of China. Undulated topography is one of the causes of soil loss. The relief is high in the west and low in the east. The surface may be divided into three levels, the first represented by the Qingzang Plateau, higher than 4000 m; the second level, the Yungui Plateau, the Loess Plateau, the Inner Mongolia Plateau, Sicuan Basin, Tarim Basin etc., between 1000 m and 2000 m high in elevation; the third level, with elevation below 1000 m, includes the Northern China Plain along the middle and lower reaches of the Yangtze River, and the southeast coastal hilly area. Thus, desertification will occur in lands with different topography.
- Storm. Monsoon influences most of China's territory; some places can concentrate so much rainfall in a certain period as to cause water erosion. The isopluvial line of 400 mm annual precipitation crossing from the northeast to the southwest of the country divides the territory into two parts. The eastern part is a humid area accounting for 90% of the cultivated land in China, and inhabited by 90% of the country's population; heavy water and soil erosion occur and cause much loss.

Artificial Factors

When wind erosion is dominant, a bad land use system, overgrazing and irrational water use can bring about desertification. For the regions where water erosion is dominant, unduly felling trees, reclamation on steep slope land and inappropriate cultivation practices are the main influential factors of desertification. In addition, bad irrigation and drainage on farmland can cause salinization or secondary salinization, which contribute to desertification.

The Damages by Desertification

Degradation of soil, such as the coarsening of soil texture, and increase in soil unit weight reduce the water-holding and fertility-holding capacity. Meanwhile, the loss of soil nutrients results in a reduction in soil fertility.

The degradation of vegetation cover and ecosystem. The restoration ability of the plant cover becomes more vulnerable and self-regulation function of ecosystem is reduced.

Arable land is covered by shifting sand, or topsoil is blown by wind force. Land productivity is reduced, pasture, towns, villages, traffic arteries and water conservancy works are jeopardized by desertification, thus drought and flood disasters occur more frequently.

All these factors jeopardize people's subsistence. It is estimated that the direct economic loss brought about by desertification amounted to more than 3 billion yuan (RMB) per year during the past decade in China, and the indirect economic loss may be as much as two or three times the direct one. Most impoverished counties lie in desertification-affected regions.

Major Measures to Combat Desertification

Management System

The Chinese Government is deeply concerned about desertification and its control. As early as the 1950s, the Government organized a large number of scientific and technical personnel to study the status of desertification, mobilize the masses and material resources to contribute to a campaign combating desertification. Many organizations are responsible for carrying out national desertification-combating activities. Because China is still a developing country, the fund from the Government budget could only afford US$ 10–20 ha^{-1} to establish forests for soil and water conservation and sand fixation. However, governments at different levels are making efforts to take part in combating desertification in various forms.

The Chinese Government has successfully formulated and promulgated laws, policies and regulations regarding the prevention and control of desertification, such as the Forest Law, the Law of Soil and Water Conservation, the Water Resource Law, the Law of Environmental Protection etc. In 1991, the China Desertification Rehabilitation and Desert Reclamation Action Programme (1991 –2000) started to operate and is one of the major engineering projects to establish the ecosystem in China (CCICCD 1997).

With the aim of promoting the development of desertification control, the Government also worked out some policy measures concerning combating desertification and favourable terms of taxation regarding the management and utilization of desert resources.

Technical Personnel Training and International Cooperation

Combating desertification needs various trained and skilled people. In addtion to some universities, colleges and institutes, there are many organizations to help in dealing with the work in desertified regions, many courses and symposia are held to meet the demands of desertification control. Soil and water conservation and sand-fixation plantations have aroused a great amount of attention in the international community. China keeps close relations with more than 70 countries and more than 10 international organizations in the world. China is also a treaty member of the Rio UNCED. With the development of China, cooperation and exchanges with the international society on desertification control will be further promoted and strengthened.

Major Ecological Protective Projects

The Chinese Government is setting up many huge protective forest systems to conserve soil and water and prevent damages from desertification. According to China's status, the major ecological protective forest systems are: Three North protective forest system, Protective forest system in the Yangtze River middle and upper reaches, Coastal windbreak system and Farmland shelter ribbon system on the plains. Because most desertification-affected regions are in northern China, the Three North protective forest system will be introduced in detail here. Three North refers to northwestern, northeastern and northern China. It extends 4480 km in length from east to west, and is 400–1700 km wide from north to south covering 551 counties in 13 provinces and accounting for 42.38% of the total territory of China. In these vast areas, the farmland suffers from wind and sand damage which amounts to 6.7 million ha, making up 40% of the total farmlands in China. Pasture affected is 3.7 million ha, and the total land lost due to desertification is 0.1 million ha per year. The loss of eroded soil from the loess plateau amounts to 1600 million t. In order to deal with such a serious situation, the Chinese Government decided to launch, on a large scale, a green project of protective forest system in 1978, which was known as the Green Great Wall. In the first phase of the project (1978–1985), 6.06 million ha of lands was planted, of which 63.8% is windbreak, the forest coverage of the Three North region is 6.2%. By the end of the second phase (1986–1995), the coverage was 7.9%. According to a survey, about 12% of desertified lands is completely under controll, and 10% of desertified lands is becoming under control (Zhang Jianlong 1996). It is estimated that by the end of whole project in 2050, 5.23 million ha of plantation will have been completed.

Desertification in China and Its Control 423

A Practical Case: Daxing County

Although desertification has developed in northern China as a whole, there are some good examples of combating desertification in China, where it has been controlled; Daxing is taken as a study case here.

Site Description

Daxing County lies 39°56' N, 116°20' E to the south Beijing city, in a dry subhumid zone. The sandy land shores the left bank of the Yongding River, which is one of the most serious blown-sand areas in Beijing; mean annual precipitation of 600 mm; stormy in July and August, severely dry in spring, windy in winter and early spring. The type of soil can be classified as damp soil and silt grain soil. The water table depth has gone from less than 1 m in the 1950s to more than 10 m at present. The total area of sandy land is 0.1 million ha, of which desertified land accounts for 60%.

Achievements

In the 1950s, Daxing County began to build a farmland shelter ribbon. In the 1970s, a huge amount of dunes were levelled into farmlands in the desertified region, more than 3000 dunes disappeared where the Three North protective project was carried out in 1981. A further large-scale mass campaign for combating desertification was launched following the requirement of "giving priority to protection and giving consideration to beautifying the environment, consideration being given to both control and use of desertified land, a combination of protective forest and fruit tree orchard" (Wang Lixian 1996). By the end of 1995, the area of woodland amounted to 22 000 ha, the forest coverage amounted to 21%. The protective system is formed basically; building shelter ribbon for 98% of the farmland, establishing protective forest ribbons along rivers, mainline channels and roads; making a traffic line with a green corridor all the year round; planting woodland of 4600 ha, and fixing sand land of 700 ha, building a forest park of 133 ha, orchards on sandy land of 3,933 ha.

Major Measures

Strengthening Administrative Function

Daxing County makes policies, plans and regulations at different administrative levels to facilitate both scientific research and practical planting for combating desertification in a reasonable way.

Comprehensive Development

Daxing is a satellite county of Beijing, so that the establishment of a protective forest system should not only serve to improve the environment of Beijing, but also aim at the development of production in sandy land practice. Its practice is as follows:

- Combination of forest and fruit trees to develop a shelter belt, both to beautify the environment and promote economic benefits.
- Devoting major efforts to planting fast growth and high yield forests to gain benefits from timber harvesting.
- Multiple management of agriculture, forestry, fruit and husbandry.

How to Control Shifting Dunes

- Combination of planting of grass, shrub and tree on shifting dunes, which forms a comprehensive protective system around sand dunes.
- Chemical methods. Macromolecular compounds are extracted for turning these materials into a viscous fluid or a solid. These macromolecular compounds can glue individual sand grains to form a cemented thin layer of sand on the sand surface while spraying. In Daxing, the cementing agents used in some experimental sites are: asphalt latex, oil-shale solution, marlite latex etc. Up to now, owing to the economic limitations, large-scale use of chemicals has been not popularized, only in those experimental sites on desertified lands.
- Others. The techniques of low-pressure irrigation, sprinkling and dip irrigation have been applied in desertification control; these water-saving techniques have made Daxing turn a new leaf in desertification control.

After several years' hard work, Daxing County can show an obvious change in both crops and ecological environment. The protective forest system has protected crops from moving dunes, and also provides protection for the south of Beijing. It is observed that the wind velocity has decreased by 40% in winter and spring, the number of blown sand days decreased by 34%; the atmospheric temperature in the forest network increased 0.4–0.9 °C in spring and fall, and decreased 0.9–1.4 °C in the summer season; relative humidity has been increased by 5–14%.

Trend to Combating Desertification

Since China signed the UN Convention to Combat Desertification, a National Action Programme to Combat Desertification in China has been formulated in time, which determines China's strategic objectives and key focus to combat desertification (CCICCD 1995). The major ideas include: efforts to put land desertification, which is growing at an annual rate of 2460 km^2 under full control,

Desertification in China and Its Control   425

and to establish step by step a comparatively complex ecosystem and an optimum industrial system in the affected areas. The specific plans include:

- By the year 2000, about 3.18 million ha of desertified lands caused mainly by wind erosion, 3.5 million ha of the Loess Plateau caused by water erosion, 10 million ha of degraded rangeland and 2 million ha of salinized rangeland in China will be rehabilitated. In addition, 6.86 million ha of land will be afforested and 165 nature reserves of various kinds with a total area of 59.5 million ha will be set up. Thus, the regional ecological environment will be improved and the issue of food and shelter for the poverty-stricken population in the affected area will be resolved by joint efforts.
- By 2010, nearly 7.45 million ha of wind-eroded lands, 0.7 million ha of water -eroded lands on the Loess Plateau and 22 million ha of degraded rangelands will be brought under control and revegetation. In addition, 6.17 million ha of the affected land will be afforested and the total area of nature reserves will be increased up to 68.68 million ha. Thus, the local ecoenvironment will be remarkably improved, and as consequence, the people's living environment and quality will be purified.
- By 2050, great efforts will be made to bring all the desertified lands under principle rehabilitation so that the ecoenvironment can be protected and economic growth be enlarged in a sound way.

Conclusion

Desertification is a serious problem in China. In spite of the great achievements in combating desertification, China is still confronted with many difficulties. The Government should take the responsibility to educate , mobilize and organize local masses and work out well-conceived plans and relevant laws and regulations. In order to realize the objectives of the National Action Programme to Combat Desertification, the following measures have been taken: to increase public awareness and education to encourage the whole nation to participate in the National Action Programme; to intensify administration and law enforcement in combating desertification and ensure the effective protection of the ecoenvironment in the affected areas; to make further efforts in the implementation of the projects to combat desertification; to encourage the dissemination of scientific findings by popularizing know-how, technology and traditional skill in tackling desertification; to make sound and wise use of natural resources in the affected area through the optimum arrangement of a desert development industry and trade service; to raise funds from various sources to increase support of combating desertification and to widen and enhance international cooperation and exchange.

References

CCICCD (1995) China National Action Programme to Combat Desertification. China Forestry Publishing House, Beijing

CCICCD (1997) Country Paper of Desertification in China. China Forestry Publishing House, Beijing

UNEP (1995) United Nations Convention to Combat Desertification in Those Countries Experiencing Serious Drought and/or Desertification, Particularly in Africa. UNEP's Information Unit, CCD/95/1, Geneva

Wang Lixian (1991) Soil and water conservation engineering. China Forestry Publishing House, Beijing

Wang Lixian (1996) The experience of combating desertification in China. In: Wang LX (ed) Combating desertification in China. China Forestry Publishinging House, Beijing

Zhang Jianlong (1996) Achievements and problems in the construction of "Three North Protective Forest System". For Econ 2:11–20 (in Chinese with English abstract)

Environmental Problems of the Southern Region of Kazakhstan

Kuralay Nukhanovna Karibayeva

Keywords. Deflation, degradation, erosion, investments, NEAPSD

Abstract. Environmental problems of the region and methods to halt the existing disbalance between the environmental situation and economic policy for the sustainable development of the region on the basis of environmental rehabilitation and rational use of natural resources in accordance with the provisions of the Agenda 21 and the Concept of the Environmental Safety of the Republic of Kazakhstan are considered.

Introduction

The increasing interest in environmental improvement, the creation of more favorable conditions for people's life and labor must become one of the most important tasks in the Republic of Kazakhstan. The deterioration of the natural environment, together with economic destabilization, lead to poor living conditions and unsustainable human development. Faults inherent to the institutional basis of social development and nonobservation of the principles of sustainable development lay the foundation for these processes. An acute environmental problem in the Southern Region is the pollution of air, water, and soil, depletion of water resources, pastures, arable lands, and the reduction of biological and landscape diversity.

There are four oblasts in the Southern Region: the Almaty, South Kazakhstan, Kyzylorda, and Jambyl oblasts with a total area of about 711.600 km^2 (Fig. 1).

The region is mainly represented by oblasts specializing in a few agroindustrial businesses.

The southeastern and southern borderlines of the region are at the same time the state borders of Kazakhstan with China, Uzbekistan and Kyrghizstan, which helps to raise its international status.

A very notable characteristic is the abundant presence of rare and endemic species of flora and fauna which predominate in the south region of Kazakhstan. Three main large rivers – the Chu, Ili, and Syrdarya – are transboundary rivers and originate in the neighboring states. In this region are the largest lakes of Kazakhstan – the Balkhash, and Alakol, and the interior Aral Sea. The preponderance of sands and alkali soils (saline lands) on the plain indicate the instability of the landscapes. In a mountainous environment, natural calamities can often be observed. The regions of potential environmental crisis are the Aral and Balkhash areas. The region is rich in coal, whose total reserves are 975 million tons, with the forecast reserves of more than 4 billion tons. The number of large deposits of granite and marble which are unique in their composition, and great reserves of salt which does not require iodine processing are full of promise for development. With its oil reserves, the region of South Kazakhstan occupies the second place in the republic (the deposit of Kumkol and others).

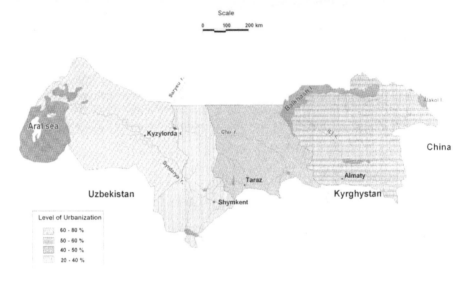

Fig. 1. Map of level of urbanization in the territory of South Kazakhstan

Natural Geographic and Environmental Characteristics

From the point of view of natural climatic characteristics, the region is located in the most environmentally extreme part of Kazakhstan: the area of deserts aggravated in the southeast by the high-altitude zones of the Tian–Shan mountains extended 2400 km along the state border (four latitude subzones and seven high-altitude zones). Because of this fact, there is great variety of hydro-thermal indices (precipitation, temperature, (Figs. 2,3), soils, (Fig. 4), and zonal ecosystems.

Environmental Problems of the Southern Region of Kazakhstan 429

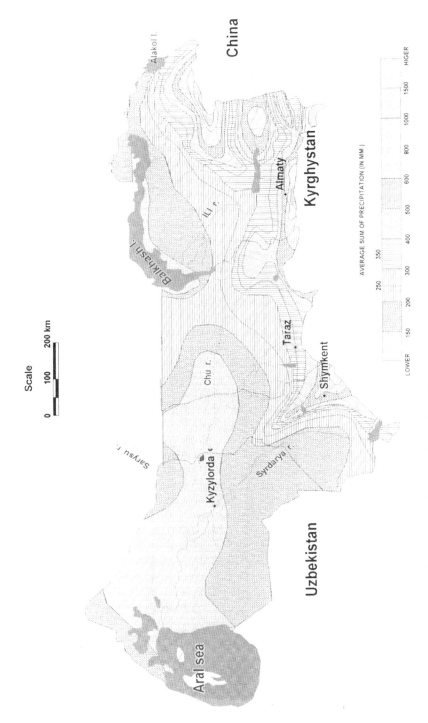

Fig. 2. Map of average total precipitation in the territory of South Kazakhstan

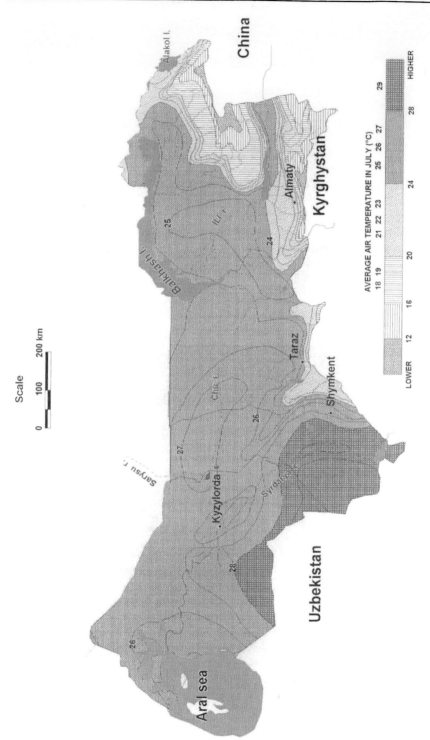

Fig. 3. Map of average air temperature in July in the territory of South Kazakhstan

Environmental Problems of the Southern Region of Kazakhstan

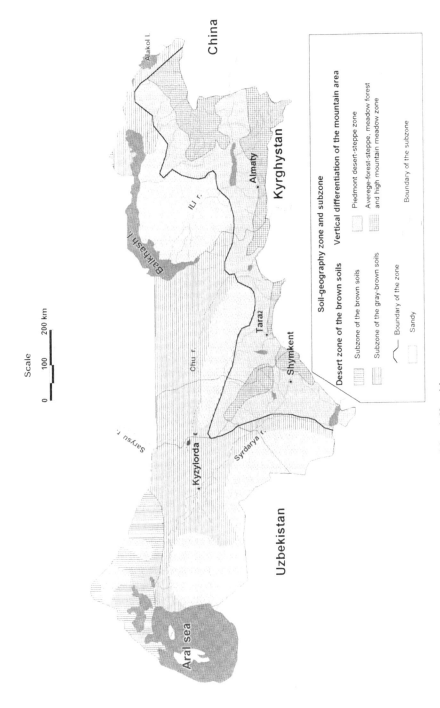

Fig. 4. Map of the soil-geographical and subzones of South Kazakhstan

Social and Economic Features

At the beginning of 1998, the population was 5 307 000 people, which is 33.8% of the total population in Kazakhstan. The added share of value of South Kazakhstan is 22% of the total republic. The largest cities and oblast centers are Almaty, Taldykorgan, Taraz, Shymkent, Kyzylorda; the average cities (with a population from 50 000 to 100 000) are Turkestan, Kentau, Baikonur, and Tekeli; there are 16 small cities. About 56% of the population in the region resides in urban areas. The greatest degree of urbanization is in the Almaty oblast, the least in the Kyzylorda oblast. In general, vast areas in the South Kazakhstan region are sparsely populated (9.2 people km^{-2}).

The main source of growth of population is its natural growth. However, in the region, the tendency is to declining birth rate and growing of the death rate, which results in the decline of the natural population growth down to 15.3 per 1000. Population dynamics is more and more influenced by the socioeconomic and environmental factors, which have mainly predetermined the trends in the migration processes.

There is still a high level of infant mortality (30.4 per 1000 new born infants).

Gross domestic product per capita in oblasts of the region is as follows: in the Almaty oblast USD1425, in the Kyzylorda oblast USD1208, in the Jambyl oblast USD816, and in the South Kazakhstan oblast USD814. The human potential index is correspondingly 0.61; 0.59, 0.58, and 0.57.

The difficult economic situation in the region has also affected the labor market, where the situation remains difficult. More than 170 000 people registered at the employment offices during the past year. This problem is most acute in the Kyzylorda and the South Kazakhstan oblasts, where only 2–7% of the applicants could find work. The level of "hidden" unemployment is also growing.

Basic Sectors of Economy

Despite the developed industry, the main factor in the economy of the South Region is agricultural production. The economic situation in the region is difficult. In the structure of agricultural gross production, livestock breeding has traditionally played a leading role. Thus, in 1990, the share of livestock breeding was 61%, and of crops 39% of the total gross production in the sector. By January 1, 1998, these indicators had changed and made up 38 and 62%, respectively together with this, a reduction in production volumes was observed in livestock breeding as well as in crops in the past 5–7 years.

The continuing reduction in number of livestock in comparison to the relative period of the previous year and decrease in livestock productivity led to shrinking of the meat market by 10%. At the same time, milk production grew by 30% and egg production by 10%. The industrial sector in the region includes the Phosphorus Joint Stock Company (City of Shymkent), the Shymkent Lead Plant and oil refinery, the lead and zinc production plant (Tekeli), and the phosphorus

Environmental Problems of the Southern Region of Kazakhstan 433

production plants (city of Taraz). Despite the wide spectrum of industrial enterprises, this region is by its orientation mainly agricultural.

The basis for livestock breeding in the region is its natural forage reserve: pastures 42.9 million ha, and hayfields 1.0 million ha. The number of livestock is 8 828 000, including cattle 1240, sheep 7081, horses 376, camels 46, pigs 85 000. Pervasive desertification can be widely observed with reduction in productivity of fodder lands, and inundation. The number of livestock has dropped by three or four times. The transition period from kolkhozes to farming units is characterized by the economic crisis.

In the region there are 1.7 million ha of irrigated lands, which is 70% of the total irrigated areas in the Republic. The region is the main producer of such valuable crops as beets, rice, grain corn, cotton, grapes, and vegetables. In 1998, about 30% of the total gross agricultural product of the Republic was produced in this region. In recent years, the areas of irrigated lands have been contracting steadily because of the unsatisfactory state of the irrigation systems, and the shortage of irrigation water, due to economic reasons. In such a situation, land withdrawn from irrigation loses its flushing regime and under an arid climate becomes quickly salinated and loses its productivity.

State of Ecosystems, Public Health and the Priority Environmental Problems

Because of the mainly agricultural specialization of the region, a steady and sufficient water supply is of great importance. However, the problems of pollution and shortage of water resources are extremely acute nowadays.

The runoff volume of the large transboundary rivers, the Ili and Syrdarya, has dropped because of water intake in the territory of neighboring states (China, Uzbekistan, Kyrghizstan). Moreover, as the result of intense economic activity in these countries, much of the drainage and sewage water is released into these rivers and spoils the quality of the river water used for irrigation and drinking water supply. Because of lack of agreements with the neighboring countries, Kazakhstan is deprived of great amounts of water. Depletion of water resources in the south and southeast parts of Kazakhstan led to severe desertification of deltas, and exhaustion and extinction of tugais and hayfields. Erosion and deflation of soils, their compaction and secondary salinization, the expansion of zones covered with sands and eventually the decline in soil productivity and yielding ability of crops are related to the deterioration of the pastures.

The Syrdarya River is the largest watercourse in the Aral Sea basin, and supplies the population in the Kyzylorda and the South Kazakhstan oblasts with water. At present, it is one of the seven most polluted rivers. The dry floor of the Aral Sea became a larger focus of salty dust storms. Annually, more than 7 million tons of sand, and 50–70 thousand tons of salts are blown out by the wind. Not only is the runoff of the Syrdarya becoming less, but the water quality is also deteriorating as the result of the disposal, on one hand, of highly mineralized drainage waters from the rice and cotton fields (containing not only nitric and phosphorus compounds but also pesticides), on the other hand because of badly purified waste waters from the inhabited centers. The situation with drinking water supply to the population is unsatisfactory. In the Kyzylorda oblast every rural citizen receives on average 15 l of water per day (125 l is the average figure for the Republic).

The investigations showed high mortality in the Aral area, with malignant tumors (four times higher in comparison with control figures), esophageal cancer (three times), mental disorders (8.9 times). Pollution, as well as the high mineralization of the drinking water, are considered to be the main reasons for such a level of mortality. Additional funds required for the construction of water pipelines are more than 900 million dollars. The high level of infectious and parasitic diseases can be explained on one hand by the low quality of water supply and sewage services, and on the other, by the unsatisfactory sanitary and technical condition of the existing water supply systems.

Reduction of the runoff of the Ili River led to a drop in the level in Balkhash by more than 2 m, which negatively affects the natural environment in the Balkhash area. Saksaul shrubs were reduced, the muskrat hunting trade disappeared, fishing volumes and productivity of reed and tugai shrubs dropped abruptly.

The deterioration of public health in the region, the drop in land productivity, and the irrecoverable losses in biological and landscape diversity which are directly dependent on the environmental situation require urgent measures in order to change the nature conservation policy, and determine new approaches to the resolution of environmental problems (Baitulin et al. 1999).

Because of the current crisis situation in the region there is need to resolve the following problems in the shortest possible time: regulation of the transboundary water utilization from the Syrdarya and Ili Rivers; reduction of water consumption and volume of pollution; prevention of desertification processes; increasing the productive capacity of the irrigated lands, productivity of pastures, conservation of biodiversity; development of renewable sources of energy; institutional problems.

In order to resolve the most acute environmental problems in the region within the framework of the long-term environmental strategy Environmental Situation and Natural Resources and the NEAP/SD plan, a number of high-priority environmental projects were developed (Table 1). Project proposals in the Southern Region (Table 1) were given support at the local level by the oblast akimats, were discussed and approved at the donor conference (in May 1998), and included into the long-term strategy of the Republic of Kazakhstan Environmental Situation and Natural Resources. The donors, EPIC, and the governments of Germany and Japan, started preparatory works for financing and implementating the projects on the Improvement of Water Resource Management of the Balkhash–Alakol River Basin (pilot project), Reduction in drinking water consumption and losses in the municipal sector (a pilot project for Almaty as a case study), and Improvement of the System for Collection, Utilization, and Storage of Solid Waste in the City of Almaty.

Environmental Problems of the Southern Region of Kazakhstan 435

Table 1. Projects of the South Region of the Republic of Kazakhstan

Project title	C	Expected results
Development and Implementation of Interstate Measures Targeted at Balance Preservation of Transboundary Water Courses Ecosystems	2.0	Conclusion agreements with China and other countries on issues of the use of transboundary water courses. Ratification of conventions, improving water resources management, reduction of deficiency and pollution. Improved water supply in Kyzyl Orda, Almaty and South Kazakhstan oblasts
Rehabilitation of water conservation zone of the Syrdarya river.	1.2	Improved water quality in Syrdarya River. Improvement of the environment in the Syrdarya River basin.
Improvement of water resources management in the Balkhash–Alakol River basin (pilot project)	7.0	Balkhash lake and Ili River Delta protection, developing mechanisms of river basins management. Establishing the Water Users' Association in the Balkhash–Alakol River basin
Reduction in drinking water consumption and losses in municipal sector (a pilot project for Almaty as a case study)	20.0	Reduction of water consumption and wastewater. Reduced threat of breakage in the Sorbulak sewage catching basin
Construction and Rehabilitation of Sewage Treatment Facilities in Kyzylorda and Shymkent (pilot project)	15.5	Reduction of wastewater. Improving the sanitary situation in the Aral region
Making inventory of environmentally affected non-fertile lands and their transformation	1.4	Correcting the cadastre of agricultural lands. No losses of natural resources. Establishing the Rehabilitation Centre for Human Health Restoration by means of less degradation of arable lands and pastures
Improvement of rational pasture use system; creation of cultivated pasture to prevent desertification process in the Kzylorda, South Kazakstan and Almaty oblasts.	7.4	Zoning of pastures, introducing the technology for sustainable grazing, training farmers, conservation of cattle number, preventing desertification. Increasing yielding capacity of pastures, conservation of biological diversity
Development and Implementation of Measures on Improving Arable Land Fertility (pilot projects in the South Kazakstan, Kzylorda oblasts	17.4	Restoration of the fertility of secondary salinizated and dehumificated soils. Prevention of desertification. Increase of carbon dioxide absorbtion by soil

Project title	C	Expected results
Mitigation of the negative impact of road transport in Almaty and Shymkent on the environment and human health (pilot projects)	25.0	Reducing the motor transport emissions. Reducing the lead concentrations in atmospheric air. Improving the quality of urban environment. Reducing the greenhouse gas emissions
Improvement of the system of collection, utilization and storage system of municipal solid waste (MSW) in the cities of Shymkent and Almaty (pilot projects)	5.6	Reduction of the environmental pollution, improving the sanitary and epidemiological situation
Improving collection, utilization and storage system for organic waste, including livestock complex waste	1.3	
Extension of forest areas for restoration and conservation of biodiversity and biocenosis	14.1	Introducing technology and rational forest use. Development of the burnt areas. Increase of carbon dioxide absorbtion by forest ecosystems at the expense of the forest planting
Development of the system of especially protected natural territories and ecotourism based on the above	8.3	Scheme of the development of the especially protected natural territories. Establishing three reserves and four national parks. Conservation of the unique objects, development of the ecotourism. Training specialists
Utilization of associated gases in the Prorva and Kumkol oil deposits (pilot projects)	60.0	Greenhouse gas emission reduction. Resource-saving, improving the energy supply
Total	186.1	

C, Project total cost (mln US $)

However, to apply the complex approach to resolve the environmental problems in the region, to increase the efficiency of nature conservation activities and of the financial resources employed, and to attract a wider circle of donors, a decision has been taken to develop an encompassing, integrating umbrella project for the Southern Region. The umbrella project is being developed within the framework of the National Environmental Action Plan for Sustainable Development (NEAP/SD) and, being one of the main mechanisms to achieve the project objectives, it conceives harmonization of basic provisions of the UN Convention on fighting desertification, conservation of biodiversity and climate change, and securing favorable conditions for vital functions of more than 5 million inhabitants in the Southern Region of Kazakhstan (Baitulin, in this volume). The projects implemented in this region, aimed at conservating water resources, envisage a set of measures for the rehabilitation of water objects. The main part of the measures is related to changing the management bodies that control river runoff and its utilization. As a result of signing interstate agreements with China, Kyrghizstan, and Uzbekistan, the transboundary water utilization will be regulated along the transboundary water runoff of the Ili and Syrdarya Rivers, irrigation of cultivated areas will be improved, transboundary pollution of the largest water objects in Kazakhstan will be lessened, and the conditions will be set for the fulfillment of commitments under the international convention on transboundary watercourses. The implementation of these projects will determine the fate of the Aral region and the Balkhash area, and the drinking and irrigation water supply in the southern oblasts (Fig. 5).

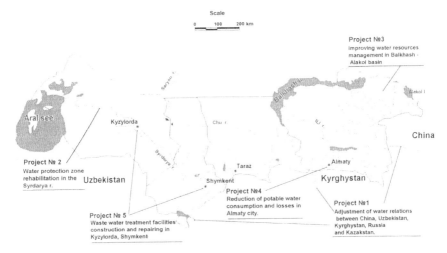

Fig. 5. Conservation of water resources of south Kazakhstan

A project on construction and reconstruction of city cleaning facilities in the cities of Shymkent and Kyzylorda is also aimed at improving the sanitary and epidemiological situation in the Aral area. The construction of facilities for purification of sewage waters will avoid further pollution of the Rivers of Badam, Arys, and Syrdarya.

The projects Improvement of Water Resource Management of the Balkhash –Alakol River Basin conceive a number of actions which would improve the situation with water consumption for agricultural purposes, and terminate the desiccation of the lakes of Balkhash, Alakol, Sasykkol, and the deterioration of the delta of the River Ili. The pilot project based on the example of the Almaly district in the city of Almaty is aimed at the reduction of water consumption for municipal services in Almaty, the prevention of overflow and rupture of the wastewater -accumulating basin of Sorbulak, to improve sanitary conditions in the oblast.

The implementation of the measures aimed at the rehabilitation of salinized and deteriorated lands in the region will lead to the prevention of the processes of desertification, with a simultaneous improvement of soil fertility. Stock taking and transformation of poorly productive lands, improvement of the systems for pasturing, formation of cultivated pastures, construction of irrigation constructions and utilization of underground waters for irrigation will help to improve the situation in the agricultural sector, raise productivity of pastures and arable lands, and subsequently lead to an increased number of livestock.

Deterioration of pastures is one of the seven prioritized problems of NEAP/SD included in the strategy for development of the agroindustrial complex of the country till the year 2030 (Fig. 6).

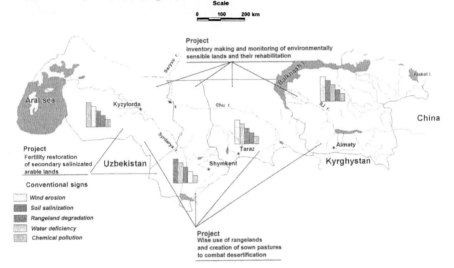

Fig. 6. Range and arable land conservation in South Kazakhstan

The deterioration of the pasture lands can be explained by different factors, the most important one being the economic activities. Extensive development of natural resources led to the situation when more than 60 million ha of pastures in the Republic are under different degrees of deterioration, including 48 million ha of deteriorated land due to nonregulated pasturing. This is the main reason for soil deflation (24.8 million ha), average and strong degeneration of pastures (24.0 million ha), the decline of the traditional sectors of pasturing and livestock breeding, and unemployment and poverty in the population involved in agricultural production.

In view of the above, the development and implementation of the project to prevent pasture deterioration and its amelioration are urgently needed. Today in the Republic, practical measures for the required legal basis for economic transformation related to land have been undertaken for the implementation of a qualitatively new stage of land reform.

First, transfer from state ownership to nonstate ownership has been implemented. All kolkhozes and sovkhozes were reorganized and on their basis the new agricultural units — production cooperatives, joint stock companies, individual farms — have been formed. Today in the region, there are 78 state entities and 14 260 individual farming entities. Second, radical changes were made in land relations; the land market is in the process of development. Third, optimization of management structure has been performed, the market infrastructure is being developed, the oblast agroindustrial exchange offices are in operation.

Fourth, market changes have been made in pricing, credit, tax policies, and the system of product sales. The State procurement prices were replaced by agreed prices. The process of attracting foreign and domestic investments in the agricultural sector has started. These changes formed the basis for the establishment of market relations in the agroindustrial complex.

Decrease in the amount of state procurement of the basic agricultural products, shortage of funds with agricultural producers, shortage of modern machinery and technologies, lack of skills for modern management with newly emerged individual farmers, have had a negative impact on the situation in the agricultural sector.

For improvement of the environmental situation in the Southern Region with the aim of sustainable development, the following actions are required:

- Establishment of a network of environmental/ecological stations; organization of monitoring over the processes of desertification; obtaining of complex geoinformation; processing, analysis, and assessment of the environmental situation for taking adequate measures/actions.
- Inclusion of desertification monitoring into a singlestate system of environmental monitoring.
- Development of criteria related to desertification of the Southern Region for utilization in the process of environmental zoning in the Republic of Kazakhstan.
- Improvement of territorial organization to prevent processes of land deterioration and to secure environmentally and economically expedient apportionment of lands based on landscape-environmental and normative approach.
- Formation of a legal framework for standardization and introduction of norms in the sphere of land utilization. Development of economic mechanisms for a temperate regime of nature utilization.
- Rational utilization of natural pastures.
- Superficial and radical improvement of deteriorated pastures and hay meadows.
- Rehabilitation of fertility of arable lands.
- Restoration of crops on arable land withdrawn from agricultural circulation.
- Prevention of erosion of soil caused by wind and water.
- Introduction of water conservation technologies for irrigation in farming different crops and inundation of pastures.
- Utilization of non-traditional sources of energy (wind, solar radiation, etc.) for the development of local stations for energy conservation in the regions affected by desertification.

- Implementation of forest restoration activities and forest growing in the areas of the state forest fund and other territories applicable for forest growing.
- Fixing migrating sands with the aim of protecting pastures, inhabited points, and the objects of national economy.
- Technical and biological recultivation of infringed lands with the aim of involving them in the agricultural or recreational circulation and sanitary-hygienic purposes.
- Melioration of secondarily salinized soils.
- Liquidation of technogenous pollution of soils.
- Planting of trees and shrubs in urban and rural areas using biologically purified wastewater.
- Securing the drinking water supply in the districts most subject to desertification or without water.
- Development of the traditional agricultural businesses (camel breeding, horse breeding).
- Development of the traditional hunting/fishing businesses and crafts (making yurts, garment, carpets, souvenirs, etc.) in the districts tending to desertification.
- Establishment of a Coordination Center for Implementation of Convention RIO-92 (to combat desertification, conservation of biodiversity and climate change) under the Ministry of Natural Resources and Environmental Protection and their synergism.
- Organization of environmental education, training and a higher information level amoung the population about desertification problems.
- Development of a system of especially protected zones.

These actions for the resolution of environmental problems in the region presume the involvement of communities and local populations in the processes of conservation and rational utilization of natural resources. Their implementation will secure a maximum attractiveness for the region for potential investors and the republican investments stipulated in the commitments on fulfillment of three conventions on the UNO ratified by Kazakhstan and contribute to internal political stability and strengthening of the national security.

References

Baitulin I, Karibayeva K (eds) (1999) National Strategy and Action Plan on Conservation and Sustainable Use of Biological Diversity, Ministry of natural resources and protection of environment, Kokshetau, pp 336
Baitulin I, in this volume

National Strategy and Action Plan to Combat Desertification in Kazakhstan

Isa O. Baitulin

Keywords. Aral disaster, socioeconomic development

The Republic of Kazakhstan is a huge Euroasian country. The country's area comprises 2.7 million ha, stretching for 2925 km from west to east from the Lower Volga and the Caspian to the Altai and China; 1600 km north to south, from the West Siberian Plain and the Southern Urals to the Tien–Shan ridges and the Kyzylkum Desert. Almost all landscape types of the globe are represented in the country's vast area, from subtropics and various types of sultry deserts to Alpine tundra and glaciers. The landscape diversity, as well as the great intracontinental seas and lakes, such as the Caspian, the Aral, Balkhash, Zaisan and Alakol determine the rich diversity of Kazakhstan biota. The flora of Kazakhstan numbers over 6000 vascular plant species; its territory is inhabited by 837 vertebrate animals, including 500 birds, 52 reptiles and 150 fishes. The diversity of invertebrates and cryptoganous plants amounts to tens of thousands of species.

However, the main typical feature of Kazakhstan is its intracontinental location in the centre of Eurasia and, determined by this fact, its harsh continental climate and the development of steppe and desert landscapes in the vast plains. Steppes take up about 45.5% of the country's total area, deserts 50%; their ecosystems, being vulnerable to human impact, easily lose stability and possess only low capacity tor self restoration.

Vast thinly populated areas were used for setting up military bases and test grounds, the mineral wealth for the development of extractive industry, the abundant pastures for livestock farming without due regard for meadowland capacity, and the fertile northern lands for ploughing up huge areas of virgin land and monoculture. Under the strict system of the former Soviet Union, Kazakhstan's economy was developed extensively, with the aim of maximum extortion of natural resources without regard for the environmental impact of such economic activities.

This resulted in a wide and deep ecosystem disturbance, desertification processes, and loss of biological diversity. The total area of desertified lands in the Republic of Kazakhstan makes up about 179.9 million ha or 60% of its territory, mostly in desert, semidesert and steppe zones. Almost all of the administrative regions suffer from desertification.

The environmental disaster in the Aral region is one of the gravest calamities of the present century. The Aral Sea was one of the unique large water bodies in the past. Being located in the centre of the desert, the sea exerted a favourable influence on the natural, climatic and environmental conditions of the surrounding territory and was a regulator of humidity over the vast Aral region, guarding against hot dry winds blowing from the Karakum Desert; it possessed a high biological productivity, and was important for fishing, hunting, transport and resort purposes.

Prior to 1960, the regime of the Aral Sea was relatively stable. The average level undulated within 53 m, the amount of flow into the sea being 56 km^3. The sea volume was 1060 km^3, the surface area 66 100 km^2, and the average salinity of the open sea 10 to 12 g l^{-1}.

The Aral Sea Basin with its total area of about 2.3 million km^2, covering the whole territory of Central Asia and the southern part of Kazakhstan, is one of the most important economic regions. This is the most ancient largest region of irrigated farming, with about 40% of irrigated lands in the former Soviet Union (7.4 million ha) and the production of 95% of raw cotton, similar for rice and 25 to 30% of vegetables and fruit.

The enormous land resources, plenty of sun and the rapid growth of the population in the region caused intensive development of the irrigated lands, whose area has increased since the early 1960s 1.5 times in Uzbekistan and Tajikistan, 1.7 times in Kazakhstan and 2.4 times in Turkmenistan. As a result, water resources of the Amudarya and Syrdarya Rivers were taken for irrigation. In the past 30 years the Amudarya and Syrdarya river flow has ceased. The sea received a small portion of the former river water and its level has dropped by 15 m, its area was reduced by 50%, its water volume by 67% and salinity increased by three times as much. The sea retreated from the coast and exposed over 30 000 km^2 of the sea floor from which in Kazakhstan alone some million tons of sand and salt aerosol are blown out annually. Most of it falls out from the currents near the source. From satellite imagery one can see that the trains of storm-induced efflux stretch for 200 to 400 km. Aerosols < 16 µm may be blown even beyond Kazakhstan. In fact, the sea has been divided into two water bodies – the Greater and the Lesser Aral Sea. All over the Aral region, spring river floods have ceased, tugai and reed vegetation is reduced, numerous maritime lakes with rich flora and fauna have dried up, sandy deserts have expanded, climate aridity has increased, air humidity has decreased by 10 to 18% and the frostfree period has been reduced by 30 to 35 days. As a result of salinization and swamping, in the Syrdarya River Basin alone, 10 to 15% of the irrigated lands are annually being withdrawn from agricultural use. The natural grass productivity of desert-pasturable massifs has reduced two to three times, agricultural crop capacity decreased, and huge pastures have become worthless.

The socio-econonic and environmental consequences of the Aral disaster are enormous. An extreme sanitary and epidemiological situation has arisen in the Aral region; in the past 15–20 years infectious diseases have increased sharply, especially typhoid (up to 29 times in different years), tuberculosis, viral hepatitus and cancer.

A grave radioecological situation has arisen in the Republic. Between 1949 and 1939 at the Semipalatinsk test ground 470 nuclear explosions were carried out, including 26 ground, 87 air and 354 underground tests. During the period of tests, radionucleids were released into the atmosphere with a total activity of about 45 million Curie. Additionally, between 1966 and 1987, 23 underground nuclear explosions occurred in different regions of Kazakhstan for geological exploratory purposes. Uranium mines make a considerable impact upon the radiation in the Republic. Over 40% of uranium raw materials were extracted here for the nuclear industry of the former Soviet Union. Especial impact is caused by the uranium mines in Southern Kazakhstan, where uranium is extracted by the method of underground leaching. Such a technology leads to the contamination of water -bearing layers with radionucleides. The extent of the contaminated belt covers an area of about 100 km in length and 10–15 km in width.

A significant problem for Kazakhstan is the localization and storage of radioactive wastes. In 1991, 109 anomalies were registered, 20 radioactive waste sites appeared as a result of uranium ore extraction and processing. Westwards of the Balkhash Lake, the military spilled fuel on the ground and contaminated near the lake coast over 500 ha by heavy petrol compounds.

Over 16 000 tons of satellite and missile scrap metal fell on the desert scarring the landscape.

A certain damage to the land reserve of the Republic is caused by technogeneous impact inevitable in construction activities and mineral extraction. The area of technogeneously disturbed lands covers about 1800 km^2.

Land productivity and soil fertility are also influenced by secondary salinization and erosion processes.

All this has resulted in an environmental crisis in many regions of Kazakhstan, in large-scale desertification – a process of ecological degradation of natural resources accompanied by loss of biological diversity, decrease in natural ecosystem productivity and health problems.

Thus, Kazakhstan faces the issue of preventing further environmental degradation and taking concrete measures to rehabilitate natural ecosystems and to create normal conditions for the human environment.

Having signed the Convention for Combating Desertification, the Republic of Kazakhstan committed itself to its steadfast implementation. In 1997, on the request of the Government, an expert team with wide public participation and with funding support from the UNEP, UMDP in the Republic, initiated the development of the National Strategy and Action plans to Combat Desertification.

The process of combating decertification will be carried out in accordance with the action plan, mostly through the elimination of the main causatory factors and the identification of measures to prevent and eradicate these phenomena. National plans will be supplemented by subregional and regional action plans, especially in cases when problems arise from the use of transboundary resources, such as lakes and rivers.

A novel starting point of the national strategy in implementing the Convention is the wide use of an approach based on activities planned at the local level, with active participation in decision-making by local people who, together with governmental officials, NGOs, will closely collaborate in the development of action plans. To promote the process, information is being disseminated on action plans to combat desertification through regional seminars, meetings and the mass media.

In the development of National Strategy and Action Plans scientists of the country, specialists of different departments, representatives from public organizations were all involved.

Combating desertification is mainly a problem of sustainable development, improving the well being of people by preventing land degradation, raising productivity, ensuring food, safety and political stability. One strategic objective of the governmental policy in Kazakhstan is to secure a favourable human habitat on the basis of optimum production development, wise use of natural resources under conditions of biological diversity and reproductive capacity conservation.

It was established that the main condition to achieve these goals is improving the system of environmental management, creating a system of efficient natural management based on environmental and economic problems balance. The governmental strategy for achieving sustainable socioeconomic development is based on the system capacity to account for ecological priorities when making economic and other decisions. Therefore, in order to implement effective environmental policy at national (the Republic), regional (oblasts) and local levels, a harmonious system of environmental legislation should be set up as well as modern regulative and economic bases of natural use.

In the course of realizing this ecological policy in the Republic, it is necessary to widely use economic tools of natural use based on preventing environmental damage, ensuring sustainable development.

Economic tools should stimulate wise natural use and present additional funding sources for environmental measures. Their main task is to make the exceeding of natural use norms economically unprofitable and to stimulate natural resources users with the help of economic incentives to undertake environmental action.

Of special importance as an economic tool are fines for excessive use of natural resourees and environmental pollution. Their amount depends on the degree of resources use as compared with the fixed norms and on the degree of pollution as compared with the limit.

To implement objectives for preventing ecosystem degradation, one should keep to the main strategic principles of natural use at national and local levels that bind all landowners and users to carry out:

- rational land organization, providing for environmentally and economically expedient distribution of lands and their internal arrangement accounting for pasture rotation, forest restoration and biological diversity conservation;
- land protection from water and wind erosion, mud flows, flooding, swamping, secondary salinization, drying up, hardening, pollution and choking up with industrial wastes, chemicals, fires and other degradation processes;
- restoration and increase of soil fertility, hayfield and pasture productivity as well as conservation of the fertile soil layer in face of works connected with

National Strategy and Action Plan to Combat Desertification in Kazakhstan 445

technogenous impacts upon lands, for using humus layer in reclamation of the lands.

The concern for conservation and increase of natural resources is a national task and it should be solved under the aegis of the relevant executive and legislative bodies of the Republic of Kazakhstan. Combating desertification will be fruitless without strict governmental regulation of economic activities. The activities should be regulated by special standards and requirements that reflect the criteria of ecological and economic stability. However, a clash of economic, social and ecological interests is inevitable, especially under conditions of limited funding.

A great ecological reform will have to be carried out when changing ownership forms in agriculture. Private ownership of land should be accompanied by creating special conditions for implementing measures to protect crops, surface and groundwaters to conserve biological and landscape diversity. In the desert zone with low environmental capacity for restoration, land privatization should be delayed, the lands must remain in the State Reserve under special conditions of protection.

Only by common efforts by all interested ministries and departments, in the spirit of partnership with overall support from the Government could this problem be solved. For this, the following measures should be taken at the national level:

- setting up of relevant bodies responsible for preparation, coordination and implementation of plans to combat desertification;
- analysis of the environmental state in ecologically disturbed areas, assesssment of causes, identification of priority actions, development of recommendations to mitigate desertification impact and to increase land productivity;
- organization of monitoring system of desertification processes;
- mobilization of financial and human resources to combat desertification;
- rising awareness of local people on degradation and desertification processes, on objectives and principles of the Convention and objectives of the National Action Plans to combat desertification.

The UN Convention for Combating Desertification gives much prominence to participation by NGOs in implementing its main provisions at a local level. Successful combat against desertification depends greatly on the response of local people to the critical environmental problems and to considered dissemination of actions to combat desertification. A purposeful action plan is needed for this.

In accordance with the Convention provisions, "Affected countries – parties to the convention, together with other parties and the international community – should cooperate to ensure the creation of a favourable international situation".

Of great importance for a solution to the desertification problem and migitation of its impact is information exchange and coordinated actions both in the affected country and at a subregional level. Here, the main activities should be linked to the

Asian Annex provisions to the UN Convention for Combating Desertification. Such countries of the region as China, Iran, Turkey and Israel which are already carrying out successful work for combating desertification, should be ready to share new technologies and expertise. It is useful to study Iran's experience on setting up and developing a network of training centres to combat desertification; its successful work has many times been noted by international organizations.

To prevent land degradation and to increase productivity, it is necessary to develop, first of all, concrete projects aimed at the elimination of desertification and at the mitigation of its social impacts in Kazakhstan. The conceptual basis of the developed national strategy consists in a harmonious combination of economic and enviromental problems, in the conservation and reproductivity of natural resources accounting for natural capacities, local, national and regional peculiarities. The strategy to combat desertification should become a part of the strategy for sustainable development of the sovereign Republic of Kazakhstan.

Concrete measures were developed for national strategy and action plans to combat desertification. They include:

- creating desertification map of Kazakhstan;
- monitoring of desertification within a common system of governmental ecological monitoring;
- development of desertification criteria to order lands by degree of degradation;
- land management providing for environmentally and economically expedient distribution and internal arrangement of lands on lands on landscape and environmental basic;
- protection of all lands from water and wind erosion, flooding, swamping, secondary salinization, dehumification, pollution by industrial and municipal wastes;
- restoration and increase of arable land fertility, hayfield and pasture productivity, reclamation of technogeneously disturbed lands;
- introducing crop rotation and soil-protecting technology for arable land cultivation;
- creating a system of especially protected natural areas;
- implementing forest-raising activities and forest planting in lands suitable for these purposes;
- planting of greenery in cities and towns, creation of forest shelter belts;
- fastening of creeping sands to protect pastures, populated areas and economic objects;
- introducing water-saving technologies for irrigation when cultivating agricultural crops or watering pastures;
- development of traditional agricultural branches (camel-, horsebreeding);
- conservation and sustainable use of natural resources and biodiversity;
- elaboration of legislations on standards and norms of land use;
- development of traditional trades (yurtas, clothes, carpet-making, souvenirs, etc.);

- drinking water supply in desertified and waterless regions;
- use of non-traditional energy sources (wind, sun, etc.);
- organization of ecological training, education and increased population awareness concerning desertification problems in Kazakhstan.

The Action Program to Combat Desertification cannot account for all the regional peculiarities and natural diversity in Kazakhstan. Therefore, for each oblast, concrete regional strategies should be developed on the basis of the program and with regard to local conditions; they must become constituent parts of a common plan of action.

All the economic plans for the development of the Republic of Kazakhstan, including the agroindustrial complex, mineral extraction, situating of enterprises and populated areas, laying communications etc. should have regard for environmental protection from desertification and for improvement of the environmental situation.

For each oblast and each district, lists should be prepared for protected plant and animal species, ecosystems and other wildlife. Background ecosystems, endemism centres, various plant and animal species, relict, rare and disappearing species and other natural objets should be protected.

Combating desertification in Kazakhstan and protecting natural resources is a national goal which can be achieved only with direct and active participation by administrative, legislative and executive bodies, public organizations and population.

Part VII:

Social and Economic Aspects

Economic-Demographic Strategies and Desertification: Interactions in Low-Income Countries

Beatrice Knerr

Keywords. Desertification, economic carrying capacity, migration, population growth

Abstract. This chapter analyzes the relationship between natural population growth, migration and desertification. Starting from results of the standard neoclassical economic theory which concentrates on forces and conditions that lead to an equilibrium after some exogenous disturbance has taken place, it challenges the hypothesis that demographic changes taking place in reaction to increasing discrepancies between carrying capacity and population density in arid regions tend to support the way to an equilibrium between both. The empirical analysis first considers comprehensive and general investigations about people's reactions to a deteriorating relationship between economic carrying capacity and population density. Based on this, it tests if the rules deduced from theories and general empirics are valid in rural regions affected by desertification. By doing this it will draw on experiences made in different low-income countries, relying largely on case studies from Asia, Africa and Latin America carried out by the author as well as by other researchers. The results demonstrate that predicting adjustment processes by applying conventional theoretical models may be insufficient or even misleading. For economically rational reasons the population may prefer to choose demographic strategies which lead to further deteriorating conditions and to cumulating downward processes characterized by a discrepancy between the short-term interest of the decision unit and the longer-term ecological interest. There is a high probability that they might contribute to a permanently unstable situation, implying a threat to the natural environment and to the people living in it.

"In a complementary way, ecologists and other natural scientists, are increasingly recognising that economic activity is "here to stay"; human activities are coming to dominate the global ecosystem and ecosystem analysis which does not explicitly include economic activities makes less and less sense. The stage seems to be set for a coming together of these two disciplines so that problems of resource use... in the global ecosystem can be discussed and assessed in a conceptual framework worthy of these problems." (Faber, Manstetten and Probst 1996, 24).

Introduction

Problems associated with desertification are a central issue of the future of mankind. At the beginning of the 21st Century arid regions[1] are among the most threatened living spaces, and worldwide they are increasing. Hence, humans as part of the ecosystem are faced with declining productivity of their natural environment. People who are not able to improve or at least maintain their living standard locally tend to move to where they suppose conditions are better. This development usually does not remain a problem of the people living in that region, but extends to their whole country, and might also spread to neighbouring as well as distant countries. As capacities to integrate additional in-migrants are already strained in many parts of the world, social conflicts tend to escalate within arid regions as well as far from them (Bächler 1995).

About one-sixth of the world's population is currently living in arid and semiarid regions, in countries with population growth rates of around 1.5% p.a. (World Bank 1999). Some arid countries like Jordan, Iraq, Morocco or Iran display growth rates of 2.2% and more, which are among the highest in the world (UNDP 1999). Within North Africa (Sahara, Sahel) where the major arid regions are located, for example, the population has doubled over the past three decades, while the strip of land with 100 to 150 ml of annual precipitation shifted by 6 km p.a. southwards between 1970 and 1990.

Under such conditions, people react. As these reactions have impact on the social, political, economic and ecological conditions, their implications must be predicted in order to shape preventive policy measures against undesired consequences, as far as this is possible. Within the spectrum of reactions, demography comes more than ever into the foreground. To collect, compare and draw conclusions from experiences made in this area is of vital and growing importance as increasing desertification is, in many regions all over the world, both a result of and a threat to human activities. It is developing into a central issue of the future of mankind, implying questions of food security, social peace and international conflicts.

Based on the results of the theoretical reflections, this chapter starts from the hypothesis that these demographic adaptations lead to a declining discrepancy between economic carrying capacity[2] and population density in arid regions (hypothesis 1). If we take the outside world as given, the carrying capacity of a certain region is a function of the available natural capital, the available physical

---

[1] According to the World Resources Institute's definition, arid regions are defined here as those with an average precipitation of less than 200 mm p.a. Semiarid regions are defined as those with a precipitation of 200 to 400 mm p.a. (WRI 1994).

[2] The economic concept of carrying capacity (cc) used in the present context refers to human beings. It explains carrying capacity as the ability of an economic space to maintain a certain population sustainably. From an ecological point of view the term „carrying capacity" denominates the maximum number of persons able to live in a certain region on the basis of the resources necessary for life while maintaining their living standards and their quality of life in the long term (Borchert and Mahnke 1973).

Economic-Demographic Strategies and Desertification 453

capital, the human resources, the human claims, the available technologies, the institutional arrangements, like land rights or habits of sharing among each other, followed by the population living in the special location, and the possibilities to exchange goods and factor services with the outside world and hence to make use of its comparative advantage in production (see Knerr 1998b).

The fact that cc and pd do not develop independently of each other is an essential aspect within the present analysis. Where pd grows, people might react by trying to increase cc, too. They might cope with the situation by introducing new technologies (Boserup 1981), by adapting the rates of fertility (Malthus) or by outmigration. These reactions may occur simultaneously, they may reinforce each other, or they may substitute for one another. Where pd/cc has increased to above 1, be it by declining cc or by increasing pd, natural capital would be used up if other adaptations are not chosen or are not successful. History has shown that migratory movements are the most common reaction. This chapter turns to the aspects of demographic movements which are lead by economic considerations, i.e. by the desire to improve one's material situation.

The empirical investigation carried out to test this hypothesis includes the family, the household farm unit, the region, the national and international level, where potentials, restrictions and institutional arrangements may coincide in a way which contributes improving the situation or turning it from bad to worse. By identifying conditions which promote adverse effects and reflecting on possibilities to modify them, the author intends to join in the effort to bring ecological and demographic developments in arid regions and in regions threatened by desertification on a way towards a socially accepted equilibrium.

The chapter is organized as follows. Following sections and test hypotheses about demographic reaction deduced from economic theories which emphasise forces that lead to an equilibrium after some disturbance has taken place. The second section concentrates on natural population growth and the third section on migration. This latter also includes some results concerning the combined effect of both. The empirical analysis considers comprehensive investigations as well as case studies about how people in arid regions react to a deteriorating relationship between carrying capacity and population density. It draws on experiences made in different arid regions of low income countries in Asia, Africa and Latin America carried out by other researchers as well as by the author. The chapter concludes with some policy recommendations to different levels of decision -making.

Adaptations by Natural Population Growth

According to the well-known theory of Malthus, natural rates of population growth would decline as the resource limit is reached, i.e. as soon as pd/cc > 1. The adaptation takes place via diseases, undernourishment and declining birth rates. As soon as pd/cc < 1, the population reincreases until a limit is reached, cc is transgressed. Hence, the system is always pushed to a cc/pd coefficient of around

1. This leads to hypothesis 2: birth rates react negatively to an increasing discrepancy between carrying capacity and population density.

This result has been challenged by the Old Age Security Hypothesis of Schultz (1974) and Willis (1980), who state that high rates of child mortality provoke high rates of fertility because more births are needed to have sufficient children who care for parents in old age. Based on this, this issue has been investigated by Nerlove for the case of environmental degradation under arid conditions. He shows that under mathematically strong but quite plausible assumptions, increasing degradation of natural resources leads families to optimize birth rates to having more children and not less.

Mainly two reasons are responsible for this development. Firstly, the increasing child mortality (at a given target number of surviving children) leads to higher birth rates. Secondly, under deteriorating conditions the desire to be supported in old age by one's children leads to desiring a larger number of children, as their expected future income, as well as one's own, will be lower, and therefore, more children are needed to provide security. Under such conditions, the limits of natural population growth tend to be set by the biological capacities of child -bearing and by the mortality of adults in reproductive age.

As each reproductive unit reacts by a strategy which increases its own security by bringing up more children, the community suffers by an increasing pd/cc coefficient, leading to a downward spiral of the system, driven by individually rational behaviour. Hence, the chances of attaining a stable equilibrium between human population and natural environment by adapting birth rates are rather limited. It can only be reached if fertility reacts negatively to the deterioration of the natural resources and, in addition, if this negative reaction is strong enough to compensate for the degradation of the natural environment. Nerlove's discouraging conclusion is that "unfortunately, there are ample reasons to suppose that in much of the Third World fertility is likely to react positively to increasing environmental degradation because parents perceive the benefits of having more children to be higher under environmentally more adverse circumstances than under more favourable ones" (Nerlove 1991).

A general explanation for these reactions, which are typically observed in poor countries, is supplied by Caldwell, who explains birth rates by intergenerational income transfers (Caldwell 1982). In traditional societies, he argues, there is a net wealth flow from children to parents, providing an incentive to bring up more children. In such a situation, as pd/cc still increases although it is already larger than 1, natural capital is used up. As a result, cc declines, which contributes to accelerating the downward process even more.

These results are confirmed by Murthy's investigations in Rajasthan, India's most arid region (Murthy 1998). Here, over the past decades, natural population growth has increased steadily, while droughts have accumulated since the early 1980s and problems of salinization have extended rapidly, implying decreasing productivity of agricultural resources. At the same time, the mortality of children as well of adolescent mothers is the highest in India. Problems are aggravated in rural regions; child mortality is almost double that in urban areas.

Economic-Demographic Strategies and Desertification   455

The fertility rate in Rajasthan is significantly above the Indian average. Girls tend to marry earlier, start to have children at a younger age and the reproductive period is extended. Almost 20% of the girls aged between 10 and 14 and 65% of those between 15 and 19 years of age are married. These are among the highest percentages in India. Birth rates among women older than 29 are dropping rapidly in the whole of India, but not in Rajasthan. The more arid the region, the younger the girls get married and the more children they will have. While in India the average age of women at their marriage is 16.7 years, it is 1 year less in Rajasthan and it is even 1 year less (14.7 years) in the district of Jalore, which is one of the most arid districts in Rajasthan.

The author provides further insights into the logic which drives the vicious circle of environmental degradation and increasing population pressure. In particular, he shows that when deciding about the number of births in the family, the fact that additional children provide additional labour force plays a decisive role. This last aspect gains importance when environmental conditions become worse. In arid regions, children are particularly required for tending animals, as the animals' survival becomes at the same time more important and more difficult when drought increases and desertification spreads.

In fact, looking at global statistics, it is striking that fertility rates in poorer arid countries are among the highest in the world and, in addition, in spite of decreasing availability of natural resources (UNDP 1999).[3]

Migratory Movements

According to the traditional neoclassical theory of migration[4], people move to where they expect to have a higher income, taking care of transaction costs (Schwartz 1973), risk and uncertainty (Greenwood 1969) and employment possibilities (Todaro 1976). In agricultural regions, the living standard of the population is largely determined by the natural resources. Spreading desertification then implies a higher pd/cc ratio and hence c.p., increasing outmigration. In addition, assuming that increasing pd/cc results in decreasing marginal productivity of labour, the hypothesis is deduced that migratory movements contribute to balancing the population densities between regions with a lower pd/cc coefficient and those which have a higher one (hypothesis 3).

In fact, extensive migratory movements out of arid regions are observed all over the world, and out-migration with the purpose of settling elsewhere seems to be the most simple reaction to increasing drought and/or population density. Yet the times when peaceful settlement migration on a large scale was possible for the population of regions hit by desertification are over, as in-migrants pose

---

[3] The author was impressed some years ago by the following statement made by a collegue from the Institute of Development Studies in Trivandrum, India: "Of course, Indians do family planning. But unfortunately they plan large families".

[4] There are a number of variations to this basic theory. For details see Knerr 1998a:31 ff.

economic, social and political pressure on the receiving regions. This applies, first of all, to migration across international borders, but increasingly also to internal migration (Knerr 1998a). Over the past decades, migratory movements have become more and more temporary and the outflow of labour is often accompanied by an inflow of capital brought by the migrants. Networking based on traditional relationships is an important social capital for pursuing migration strategies, as following paved ways reduces transaction costs.

The following case studies demonstrate the impact of outmigration, depending on the different socio-economic and cultural conditions.

Within arid regions, migration strategies to cope with the adverse environmental conditions have existed for centuries and millennia. In many regions, they have been essential for the survival of the population. It can be observed in many locations that they continue according to established patterns, although today, due to rapidly changing external conditions, they have assumed other dimensions.

Traditionally, migration plays a central role in the survival strategies of the people in Senegal. These strategies are predetermined by historically established patterns which typically differ between ethnics. This is demonstrated by Dia's detailed study of the migratory behaviour of the Kaskas, the Soninké, the Seres and the Haal Pular, who survive on migration/remittance strategies, each of them with their own way of supporting the reproductive unit in the region of origin (Dia 1992). A pattern common to all of them is that the migrants maintain their relationship to the social group (family, household, clan) at the place of origin. This implies joint household decisions with regard to migration and remittances to the family-household unit.

The Kaskas live in the Moyenne Vallée, where due to the climatic conditions, labour demand on the farms displays a pronounced peak over a short period. Insecure climatic conditions make irrigation an essential investment to increase and secure agricultural productivity. For the irrigated lands, external labour is particularly important due to the extremely narrow calendar of cultivation (sowing and harvest).

Migration among the Kaskas is extensive, with an average of 1.5 migrants per household. It has its roots in the colonial period, when forced migratory movements took place into the peanut basins as well as to the towns of Senegal and Mali, where labour was needed to build the railway from Dakar to Niger. The agricultural development strategies have included specialized forms of temporary migration which embraced mainly the younger age groups.

Migrants' income is the most important component of non-agricultural income among the Kaskas. On average, each household received 65 800 F CFA in 1988 (varying between 20 000 to 220 000 F CFA), corresponding to the salary for 188 to 268 working days. This average hides important differences between household groups with an impact which is decisive for the economic situation of the whole region. Three typical groups are discerned: (1) Households where about three -quarters of the men are migrants are able to support and subsidize their agricultural production by migrants' remittances to pay for inputs and for external labour over the season when requirements peak. (2) Large production units with few migrants and many men present. Here, in spite of large areas of land available per household (74 ha on the average), only a small part is cultivated (on the average 2.6 ha), due to lack of external income to hire labour force, buy inputs and finance irrigation parameters. Land productivity remains low, debts are common and the farm households suffer from an important food deficit[5]. (3) Small production units without migrants are the worse off. In spite of their small cultivated land areas, they are not able to satisfy their need for labour over the season of peaking labour requirements, for example for weeding. Although, in line with the Boserup thesis, they employ innovative technologies, like extensification, direct seeding etc., this is not enough to compensate for the existing lack of labour and inputs.[6]

As a consequence, families with migrants accumulate large land holdings.[7] The head of the production unit secures the farm management, and the family is able to pay the necessary input costs. The production units which have the most migrants are also those who enlarge their irrigated areas most, and mechanize their rice production more than others. The farm activities are not sustainable on their own yet, because the net profit of these investments is negative; it would not be possible to finance them out of the farm's production. Still, this strategy contributes to increasing the region's carrying capacity.

The Haal Pular have their origin in the Middle Senegal Valley. More than 90% of the men between 30 and 60 years of age have migrated at least once in their life; 58% of the migrants go to the towns of Senegal, 35% to those of Mauritania, and 6% move further away.[8] In the home villages, households on average consists of 1.4 men present, 2.2 women present and two absent men or women, not including the seasonal migrants. The close ties maintained within the clan support the organization of a strong seasonal migration pattern. Households of the clan The Seres have a rate of outmigration of 48.2%, largely in the direction of Dakar. These movements, which date back to the 1980s, have intensified significantly with increasing droughts. As a result, the Seres are distributed over Dakar, the Terres Neuves and their home region in Central Senegal, and between established in Dakar take over the responsibility for the young migrants, who constitute 30% of the Haal Pular in Dakar.

---

[5] The situation is different in Marabut families, which have large land holdings and few migrants. Here, much labour force is available because of the help provided by the Koran students, which justifies the presence of many men to organize the production process. Under these conditions, they usually are able to cultivate their land.

[6] The requipage of 0.36 ha, for example, requires 56 working hours, i.e. two whole working days for two men.

[7] This is possible because landuse rights are not tied to presence (which is the case in many other regions) but to belonging to a large family unit. The migrants are enlisted on the list of their parents' production unit.

[8] Over the colonial period, the Haal Pular of that region took over the new urban jobs in Dakar an Saint Louis. These jobs required a firm settlement and hence a resettlement of the production unit, without splitting the relationship to the rest of the clan.

458 B. Knerr

these three regions there is a lively coming and going which is supported by strong networks. Decisions about migration are usually taken by the Ndok yay, a subgroup consisting of the mother an her children, in coordination with the head of the farm-household unit.

The Soninké, who live at the Upper Senegal river, are specialized in long-distance migration. In the 1960s they took part in the labour force agreement concluded between Senegal and France[9]. In 1975, when 83% of the outmigrating Soninké went to France, the French government decided to stop in-migration from Africa. Some illegal movements to France went on, and in addition, new across-border paths established themselves, in particular to Central and Western Africa. At the Soninké the social group (resource unit) living, cultivating and consuming together, consists of on the average 16 persons. The oldest heads the group and decides about migratory movements, organizes the outmigration, and decides upon the use of remittances.

Similar strategies of maintaining the reproductive unit in the rural area by subsidizing agricultural activities by migrants' remittances are common in other arid regions of Sub–Saharan Africa, as, for example, in the Communal Areas of Zimbabwe, where the subsistence needs of the smallholder families can only be met because migrants' remittances are available to buy the required inputs (Hedden–Dunkhorst 1993).

These strategies are only successful, however, as long as the migrants are able to find employment which leaves them with a surplus to send or carry to their region of origin. An example where such a system has collapsed is Mauritania (see Fahem 1998). The country's arid nomadic areas are emptying, while the share of urbanization has increased from 8 to 47% between 1965 and 1988. This process has been promoted to a large part by the availability of water and food supplied by international aid organizations to urban centres and less by the availability of well-paid jobs. Here, too, outmigration from rural areas is highly selective in favour of the younger and productive male. Nevertheless, hardly any resources flow back there. In the rural areas women are significantly overrepresented, and in the urban areas underrepresented. One third of all households in the country are headed by women. In the face of increasing desertification on the one hand and continuing high population growth on the other, the largest part of Mauritania's population today is threatened by hunger and thirst. For a large part of them only international aid secures survival.

In addition, migration-cum-remittance strategies seem to support first of all those in the source regions who are in the possession of productive resources. It can be observed that in regions with many outmigrants holding profitable jobs somewhere else and keeping stable family ties to the home region, migration strategies do not provide a complete social net, and not all of those who have migrant family members are supported in case of need. As shown by Knerr and Schrieder for arid regions of Cameroon, migrants sent remittances first of all to those who hold some wealth, like land and animals, which the remittee might inherit, while the poor receive significantly less (Knerr and Schrieder 1999). The

---

[9] These agreements were also joined by the Soninké of Mali and Mauritania.

example of Cameroon demonstrates that migration strategies might be more a way to preserve productive capital, and hence, carrying capacity, than just to support the living standards of those left behind.

Remittances are less common in the South American context, with implications which are quite different from the African context. They are demonstrated by Müller (1993) for the Valle Grande in Bolivia, a smallholder region characterized by a net population loss and selective outmigration, with no significant remittances sent back; 76% of the population live in rural areas where non-agricultural income sources are largely lacking. With applied techniques of cultivation only 10% of the province area are suited to field crops. The past decades were characterized by repeated periods of drought. The major ecological problems which contribute to accelerating desertification are deforestation and soil erosion, which is mainly due to cattle holding in unregulated pasture economy. Between 1950 and 1992 the Valle Grande lost 20% of its population by outmigration. The high rate of outmigration is to a large part ascribed to the fact that Valle Grandinos dispose over economic alternatives to agriculture, which has always played a secondary role. Traditionally, they are engaged in trade and transport. Under these conditions, they have concentrated less on agricultural innovations to improve their farms. As a result, the region suffers from steep economic decline. Life expectancy is below and child mortality is above the Bolivian average.

Under these circumstances, outmigration had a negative impact on cc, as it leads to an erosion of the physical as well as the human resources.

The impact of outmigration on ecological carrying capacity tends to be negative. First, it did not reduce the number of cattle kept, but has only been accompanied by a stronger concentration of cattle holding. Secondly, deforestation accelerates due to lack of labour force: as the fields weed up extremely rapidly, and herbicides are very expensive as compared to labour, it is more profitable for the farmers to burn down forest areas in order to gain new fields instead of weeding. The net effect is increasing desertification.

The impact of outmigration on the social determinants of cc is most severe. It implies an erosion of the region's human capital with a self-enforcing downward effect. The major reason is that outmigrants are younger and more productive persons, while the weaker sections of the population tend to remain in the province. As the better qualified leave, the few demanding jobs, for example at the planning authority, cannot be filled adequately. With this equipment, it is increasingly difficult to push through the interests of the region against the Departamento. This pattern has persisted for decades, and it has entailed social problems, like overageing of the population, high dependency rates, alcoholism, high suicide rates, incest and mental illness, with negative consequences for economic development.

As unfavourable areas are increasingly emptying and isolated, the critical number of people for maintaining infrastructure and many public institutions does not exist there anymore, providing further incentives for outmigration.

Declining productivity has been particularly pronounced in agriculture. Land productivity has fallen far beyond the average of the Departamento and the country for almost all important crops for which Valle Grande up to some years ago had an almost monopolistic standing, e.g. maize. Technical innovations in agriculture are hardly observed.

As a result of this development, the economic potential of the region has dramatically declined and it is increasingly difficult for the remaining population to maintain itself. The shrinking purchasing power is also felt in the provincial capital, where businesses like smaller canning factories and many restaurants have closed down in large numbers.

Strategies of labour export from arid countries are also pursued on a large scale at the international level, often actively supported by their government. Arid countries are strongly involved in international labour migration and are among the world's major labour-exporting countries (see Knerr 1998b). Moreover, it is a striking fact that those arid countries which are not in a position to earn a significant amount of foreign exchange by oil exports of their labour force abroad are labour exporters[10]. In these countries, remittances are so high that they have a significant, and in many cases a dominant, influence on the economic development[11].

For the households involved, the monetary gain from international labour migration to rich countries is so large that the reactions of and consequences for the migrant's household differ significantly from those of intranational migration. Most often it is so profitable that workers' remittances to their household at home exceed all other sources of income. It is typical that in the regions of international outmigration, agricultural productivity declines because a large part of the younger male labour force is absent for a longer span of time, and remittances are seldom spent on productive farm investment. Typical investment categories are houses, furniture and vehicles. Pakistan might be taken as an example. The country has a strong tradition of temporary labour migration from rural to urban areas as well as to overseas. The migrant usually remains part of his household in the region of origin, and a large part of his income is remitted to the family. As Pakistanis are strongly involved in international migration, the question often arises, if the remittances are invested in a way which increases the cc of the home region. The hope that in the region of outmigration, remittances contribute to promote agriculture are usually disappointed. In Pakistan, savings out of remittances are spent for a large part for housing and for marriages of brothers and sisters (Batzlen 1999).

The conclusion that high rates of male labour migration, in particular if it goes to far-away countries, will contribute to lower birth rates in the region of origin might be misleading. For international migration from Pakistan, for example, a positive relationship between migration and the number of children has been demonstrated by Shahnaz and Khan: families with migrant husbands tend to have significantly more children, and also desire more children than those without migrants (Shahnaz and Khan 1997).

---

[10] Labour exporting countries are defined as those who receive more than 50% of their foreign exchange through migrants' remittances

[11] For details see Knerr 1998a

Conclusions and Policy Implications

On the grounds of evidence, hypothesis 1 has to be rejected. There are economically sound reasons for utility-maximizing individuals to react to desertification in a way which, under certain institutional conditions, contributes to still lower carrying capacities, so that the human adaptation process makes the situation worse than before. It cannot be expected that the population of arid regions adapts by demographic strategies which lead a new sustainable equilibrium.

Increasing birth rates as a reaction to worsening living conditions can cause or speed up a cumulative downward process. There is a high probability that birth rates in traditonal low-income regions react positively to increasing desertification, entailing even more pressure on the environment.

It cannot be expected that observed outmigration will contribute to a decline in the population density. As migration usually is positively selective, it may even entail a decline in the carrying capacity in the region of origin.

Nevertheless, in many arid regions, migration strategies are an effective means to support and subsidize the farms and the households associated with them, and hence contribute to counterbalance a declining cc/pd. As the survival of the farms depends on labour migration and remittances, agricultural production systems in arid regions more and more become a function of the prevailing migration strategies.

Migratory movements are a structural element of the economic behaviour of people in arid regions, and not a short-term reaction to short-term events. They are not a "mechanical" reaction to worsening environmental conditions, inducing people to move to places where their productivity is highest, but, when pressure arises, adaptations take place within these given structures. Traditional settings pre-determine the reactions to current changes and introduce inflexibilities. This has the advantage of reducing the transaction costs, but at the same time it is observed that much of that which had been useful for a community some years, decades or centuries ago, can be disastrous today, contributing to worsening instead of improving the situation.

Today, it is a matter of fact that in many regions which are hit by desertification the population can only be sufficiently supported by the remittances of outmigrants who make a living abroad.

By emphasizing the pathways of vicious circles, the author does not intend to deny that there also exist forces which lead to a new sustainable equilibrium. She only wants to point out the high probability that the adaptive processes which are initiated by increasing pd/cc might not improve the situation but make it worse.

Although much more research is required in this area to obtain a more complete picture, the available experience suggests some evident policy recommendations.

In recommending measures to influence birth rates, those outside a given culture have always to be careful. Suffice it to say that establishing systems of old age security, prohibition of child labour and promotion of schooling would help to reduce the incentive to have more children when environmental conditions become worse.

With regard to the implications of migration, any development policy, and in particular local development projects, have to take care of the fact that in regions with significant out-migration it might be expected that those remaining in the region are negatively selected with regard to their productivity. Where a large share of those present are women with small children and elder people, development projects should not be too ambitious with regard to raising the productivities. They might better have an emphasis on health care, schooling etc.

In general, trying to keep potential migrants at their place of origin or to bring migrants who have arrived in the towns back, might paradoxically increase it even more, as now there is no more migrant income to support the household in the source region. As it might not be able to subsist on the available resources, the result might just be that the whole household is forced to move.

References

Bächler G (1995) Umweltflüchtlinge. Das Konfliktpotential von morgen? Agenda, Münster

Batzlen Ch (1999) Migration and economic development. Remittances and investments in South Asia: a case study of Pakistan. PhD thesis, University of Hohenheim, Hohenheim

Boserup E (1981) Population and technological change: a study in long-term trends. Chicago Press, Chicago

Caldwell J (1982) Theory of fertility decline. Academic Press, London

Dia I (1992) Les migrations comme strategie des unités de production rurale. Une étude de cas du Senegal. Poverty and development. Sustainable development in semi-arid Sub –Saharan Africa, Ministry of Foreign Affairs, The Hague, pp 57–64

Faber M, Manstetten R, Propst J (1996) Ecological economics. Concepts and Methods. Elgar, Cheltenham

Fahem AK (1998) Population and desertification in Mauretania. In: Clarke J, Noin D (eds) Population and environment in arid regions (Man and the Biosphere Series 19). UNESCO, New York

Greenwood M (1969) An analysis of determinants of geographic labour mobility in the United States. Rev Econ Statist 51:189–194

Hedden–Dunkhorst B (1993) The contribution of sorghum and millet versus maize to food security in semi-arid Zimbabwe. PhD thesis, University of Stuttgart, Hohenheim

Knerr B (1998a) Labour migration from developing countries. Macroeconomic impacts and policy interventions. University of Kassel, Kassel

Knerr B (1998b) Impacts of labour migration on the sustainability of agricultural development in arid regions. In: Clarke J, Noin D (eds) Population and environment in arid regions (Man and the Biosphere Series 19). UNESCO, New York.

Knerr B, Schrieder G (1999) Migration and remittances as an income re-insurance mechanism for smallholder households in Sub–Saharan Africa: the case of Cameroon. Economic and Development and Cultural Change. University of Chicago, Chicago (in press)

Malthus T (1798) Essay on the principle of population. Cambridge University Press, Cambridge

Müller PM (1993) Tragfähigkeitsveränderung durch Bevölkerungsverlust. Beispiel Valle Grande/Bolivien. Geogr Rundsch 3:173–179

Murthy M (1998) Demographic development in Rajasthan. Paper presented at the Conf on Population and Environment in Arid Regions. Amman/Jordan, 24–27th October 1994

Nerlove M (1991) Population and the environment: a parable of firewood and other tales. Paper presented at the 5th Annu Meet Eur Soc Population Economics (ESPE), Pisa, 8 June 1991

Schultz TW (1974) Economics of the family: marriage, children and human capital. National Bureau of Economic Research, Chicago

Schwartz A (1973) Interpreting the effect of distance and migration. J Polit Econ 81:1153–1169

Shahnaz L, Khan AH (1997) Impact of male out-migration of female decision making: a case study of selected areas of Kharian. Paper presented at the Regional Workshop on Return Migration and Long-term Economic Development in South Asia, Islamabad, Pakistan

Todaro M (1976) Internal migration in developing countries. International Labour Office (ILO), Geneva

United Nations Development Programme (UNDP) (1999) www.undp/hdro/population.htm

Willis RY (1980) The old-age security hypothesis and population growth. In: Burch TK (ed) Demographic behavior: interdisciplinary perspectives on decision making. National Bureau of Economic Research, Boulder, Co,

World Bank (1999) World Development Report. Washington

World Resources Institute, World Resources Report. Var Issues, Oxford

Final Remarks

The Department of Ecology is grateful for the help of several people and institutions that enabled this successful international conference at Königswinter (near Bonn). Support of the BMBF and of the project management BEO made it possible to bring together such a broad scientific community. The help of the UNESCO, the help of the UNCCD-secretariat and the help of the Arbeitnehmer -Zentrum in Königswinter is greatly acknowledged. Special thanks are due to Dr. Joachim Kutscher, to Dr. Vefa Moustafaev, and to Dr. Antonio Pires for their contributions. Thanks are due to the staff of the Department of Ecology and to advanced students from the University of Bielefeld for their help.

This international seminar has covered various subjects, dealing with a variety of environmental problems and presentations were contributed by various representatives from different countries. Only their tight cooperation enabled us to bring together most of the contributions in this comprehensive book. The help of the Springer-Verlag and especially of Mrs. Gramm is gratefully acknowledged. The preparation of the diverse contributions as camera-ready manuscripts was done by Christiane Dalitz, Anja Scheffer, and Henning Todt. We owe them many thanks. Special thanks are due to Mrs. Anja Scheffer, without her this volume would have never been published.

We do hope that this volume is one important step forward to promote exchange between scientists from different fields and from various national and international organizations. One result of such collaboration was the recently formed "German Competence Network for Research to Combat Desertification" (Desert∗Net). This will serve as a source for new research efforts and will certainly serve as an information source for the COP 4, the Conference of the parties of the CCD, which will take place in Bonn at the end of 2000. It would also help to define future projects to combat desertification. It should concentrate not only on regional risks but also on broad areas which need help for the maintainance or development of an intact environment with agroecosystems for sustainable land-use in deserts. The ultimate goal is to provide a safer world for future generations.

S.-W. Breckle